INTERNATIONAL EN

n

es

ORGANISATION FOR ECONOMIC CO-OPERATION AND DEVELOPMENT

INTERNATIONAL ENERGY AGENCY

2, RUE ANDRÉ-PASCAL 75775 PARIS CEDEX 16, FRANCE

The International Energy Agency (IEA) is an autonomous body which was established in November 1974 within the framework of the Organisation for Economic Co-operation and Development (OECD) to implement an International Energy Program.

It carries out a comprehensive programme of energy co-operation among twenty-one* of the OECD's twenty-four Member countries. The basic aims of IEA are:

i) co-operation among IEA Participating Countries to reduce excessive dependence on oil through energy conservation, development of alternative energy sources and energy research and development;

ii) an information system on the international oil market as well as consultation with oil companies;

iii) co-operation with oil producing and other oil consuming countries with a view to developing a stable international energy trade as well as the rational management and use of world energy resources in the interest of all countries;

iv) a plan to prepare Participating Countries against the risk of a major disruption of oil supplies and to share available oil in the event of an emergency.

**IEA Member countries: Australia, Austria, Belgium, Canada, Denmark, Germany, Greece, Ireland, Italy, Japan, Luxembourg, Netherlands, New Zealand, Norway, Portugal, Spain, Sweden, Switzerland, Turkey, United Kingdom, United States.*

Pursuant to article 1 of the Convention signed in Paris on 14th December, 1960, and which came into force on 30th September, 1961, the Organisation for Economic Co-operation and Development (OECD) shall promote policies designed:

- to achieve the highest sustainable economic growth and employment and a rising standard of living in Member countries, while maintaining financial stability, and thus to contribute to the development of the world economy;
- to contribute to sound economic expansion in Member as well as non-member countries in the process of economic development; and
- to contribute to the expansion of world trade on a multilateral, non-discriminatory basis in accordance with international obligations.

The original Member countries of the OECD are Austria, Belgium, Canada, Denmark, France, the Federal Republic of Germany, Greece, Iceland, Ireland, Italy, Luxembourg, the Netherlands, Norway, Portugal, Spain, Sweden, Switzerland, Turkey, the United Kingdom and the United States. The following countries became Members subsequently through accession at the dates indicated hereafter: Japan (28th April, 1964), Finland (28th January, 1969), Australia (7th June, 1971) and New Zealand (29th May, 1973).

The Socialist Federal Republic of Yugoslavia takes part in some of the work of the OECD (agreement of 28th October, 1961).

FOREWORD

Energy conservation continues to be an important component of energy policy. The 1970s showed how vital our energy resources are in maintaining the economies and lifestyles for which we have worked so hard. It also became apparent that consumers, and thus society as a whole, were capable of using each barrel of oil or each kilowatt-hour of electricity more efficiently. It was not that energy consumers had been intentionally wasteful. They had formed the habit of using energy without constraint to meet growing demand generated by decades of economic development.

Since the early 1970s consumers, energy conservation service industries, energy supply companies and governments have taken a very active role in contributing to improved energy efficiency. This improvement was one of the main contributors to our current surplus of energy supply.

This present study represents the first comprehensive analysis by the IEA of the lessons learned from our past experience and the prospects for future efficiency gains. The study is timely because we currently face a changing energy situation which makes policy making difficult and yet which confirms our need for strong thoughtful action. Because markets will tighten again, this study should be a valuable contribution to the debate.

The Secretariat was helped in the preparation of this study by officials of Member governments and by an informal group of experts drawn from outside government*. I am most grateful to all of them for help without which the study could not have been completed. The study is however published on my responsibility as Executive Director of the IEA and does not necessarily reflect the views or positions of the IEA or its Member governments.

Helga Steeg
Executive Director

* A list of members of the informal group is at Annex H.

Table of Contents

List of Annexes

EXECUTIVE SUMMARY

Since the oil price increases of 1973-74 and 1979-80 energy conservation — the more efficient use of energy — has become an important component of energy policy. Until then, energy conservation had occurred but it was not an issue of central concern to government energy policies.

Even in the current easy energy markets it is important to maintain the momentum of energy conservation. Energy conservation is important for long-term economic well-being and security because:

- energy conservation will extend the availability of energy resources that are depletable;
- there is likely to be a return to tightening energy markets before the end of the century; energy conservation will delay and lessen its impact;
- energy conservation reduces the environmental consequences of energy production and use in a way which is consistent with energy policy objectives;
- investment in energy conservation at the margin provides a better return than investment in energy supply;
- investment in energy conservation can often be undertaken in small increments and is therefore flexible at a time when the energy outlook is uncertain.

The purpose of this study is to assess the role of energy conservation in the context of general energy policy after more than ten years of concerted efforts by consumers and governments. The study assesses the contribution that energy conservation has made to the current energy situation and the factors responsible for that contribution; the further economic potential for improvements in the efficiency of energy use; what, if anything, is preventing the achievement of that potential; and what measures are available to governments to promote efficiency.

Main Themes and Conclusions

Since 1973, governments, businesses and consumers have taken action. Examples abound of reductions in the energy needed to heat homes, drive cars or produce steel. For the IEA region as a whole, the amount of energy used to produce a unit of gross domestic product — referred to as energy intensity — fell by 20% between 1973 and 1985. If energy intensity had remained unaltered, energy demand would have been some 880 Mtoe higher. The biggest improvement in energy intensity was in the industrial sector mainly as a result of improvements in the efficiency with which energy was used. There were also significant improvements in the residential/commercial and transportation sector but little change in the efficiency of electricity generation. The increase in energy prices and long standing trends towards increased productivity were the driving forces in bringing about improvements but it was supplemented by government policies and programmes to promote energy conservation. Gradually declining real energy prices since 1982 have reduced, but not reversed, these trends towards lower energy intensity.

Three basic points for the future emerge from this study:

(i) There is considerable potential for further improvement on an economic basis in efficiency in energy use;

(ii) There are limitations in the energy conservation market, some inherent, some resulting from actors in the market and some resulting from government regulations, which prevent this potential from being fully realised;

(iii) Carefully planned government policies can be effective in reducing these limitations.

The key problem for government conservation policy is how to achieve more of the economic potential for energy conservation in market economies especially at times when short-term price signals do not reflect the long-term outlook. Its solution is difficult because energy conservation activities are very disaggregated. Inevitably decisions on investment and other energy conservation actions rests with industries and consumers. Government policies provide a framework to encourage such actions. Success requires a number of different approaches in order to reach and motivate a broad range of interests, for many of which conservation is a secondary activity. The major elements in such a policy are discussed below.

(i) **Potential for Conservation Improvements.**

While considerable energy efficiency improvements have been made since 1973, there remains a large potential for further efficiency improvements in all end-use sectors. If energy conservation measures which are now economically viable were fully implemented by the year 2000, energy efficiency would be more than 30% higher than current levels. The remaining potential for efficiency improvements is especially large in the buildings sector, but major opportunities exist in industry, transportation and transformation.

Based on current trends and government policies, up to three-fifths of this existing potential for conservation is likely to be achieved. The transportation sector and energy-intensive industries are likely to achieve more of this potential than other sectors, such as residential and commercial buildings and smaller industries. If the full potential for economically viable efficiency improvements were to be realised, energy demand in the year 2000 could be further reduced by at least 10% (more than 450 Mtoe per year from currently projected levels).

(ii) **Market Limitations.**

In theory, market clearing energy prices in a perfect market would produce an optimal economic allocation of resources in the energy sector, including an appropriate effort on energy conservation. In practice there are four main reasons why this does not happen:

(i) In many IEA countries proposals for investment in energy conservation are judged by investors against significantly stricter criteria than supply investment;

(ii) Market prices are inevitably affected by short-term influences and may not reflect the long-term outlook. Electricity and in some countries gas prices to consumers are not fully determined by the market. For wider policy reasons governments and other public authorities are prone to hold down energy prices, particularly those of electricity and gas;

(iii) Energy prices typically do not take fully into account the external costs and benefits, particularly the environmental and security costs, associated with energy production and use;

(iv) Many specific limitations impede the working of the energy market. They include lack of information and skills necessary to conserve energy, the invisibility of energy use, lack of confidence in new conservation services and products and separation of responsibilities for energy expenditures and conservation actions.

(iii) Conservation Policies and Measures.

(a) *Organisation of Energy Conservation Activities*

Many different individuals, businesses, and other organisations are involved in energy conservation activities. They are far more numerous and disaggregated than those involved in energy production. The organisations provide equipment, services, advice, motivation or incentives to consumers. There is scope for government policies to promote and co-ordinate their activities. In particular:

(i) Energy conservation service industries such as equipment manufacturers and installers, energy auditors, financing companies and plant builders have become more important as conservation requirements and opportunities become more complex. Governments can encourage the interests concerned to come together, and modest financial help can often play an important role in this process;

(ii) The energy supply industries in some countries play an important role in providing energy conservation services. For example, in the United States many gas and electric utilities have comprehensive conservation programmes, most of which are required through federal or state legislation. Much emphasis is being placed on a total integration of efficiency and supply options into the utility planning process;

(iii) In some cases, the industries have been ready to assume this role for their own commercial reasons, but in other cases, government encouragement or even legislation has been necessary, particularly for gas and electric utilities which are already under some form of regulation;

(iv) Non-governmental, non-profit groups such as special interest groups, service clubs or voluntary industry groups are valuable in providing services and sharing information. They understand the needs of specific localities and can link energy conservation to other regional needs such as the provision of employment. Small injection of public funds can much assist such local activities;

(v) Governments at all levels encourage energy conservation. The roles and responsibilities depend on the constitutional framework and are more complex in federal states. In governments themselves, there is a need for a strong central conservation policy group headed by a senior official who forms part of the top management of the department responsible for energy and for effective inter-departmental co-ordination of conservation activities. Strong political leadership and bureaucratic commitment is the key to the success of government conservation activities. All Member governments should re-examine energy conservation arrangements in their countries with a view to drawing upon experiences of others.

(b) *Energy Pricing and Taxation Policies*

There is general agreement among IEA countries that where world markets exist consumer prices should reflect the world market price; in other cases consumer prices should normally reflect the long-term cost of maintaining the supply of the fuel concerned; and proper weight should be given to energy policy objectives in tax policies. Energy prices should also internalise as far as possible certain externalities such as the environmental costs of energy production and use — the polluter pays principle.

Substantial progress has been made in relating the prices of oil and coal to consumers to the world market price. There are, however, still price controls, subsidies or other distortions on energy prices in a number of countries. There are arguments for and against these arrangements but, when the consumer price results in lower than world price,that will work against energy conservation.

The main problems lie with prices for electricity and gas. In theory the right basis for determining them is long-run marginal costs. There are, however, serious practical problems about applying long-run marginal cost pricing. The principle is used in a minority of IEA countries to provide an approximate basis for prices, particularly of electricity. In many IEA countries prices for electricity are based on some form of historical costs. For gas prices the most commonly used basis, particularly in Europe, is to link prices to those of competing fuels notably oil products.

Tax policies are determined by considerations going well beyond energy policy. The need to raise money is the most important. Only a few Member countries such as Denmark, Portugal and Sweden have tried to use energy taxation explicitly as an instrument to promote energy conservation. There are some instances of distortions in tax policy which work against energy conservation. For example, there are tax regimes which do not give equal treatment to supply and conservation options and also the tax regime for company cars in Sweden, the United Kingdom and to a lesser extent in other countries.

It is desirable that adequate importance be given to energy conservation in formulating prices and tax policies. In particular:

— remaining subsidies to or controls on oil prices should be eliminated or reduced as soon as possible;

— the conclusions adopted by the Governing Board on 27th March 1985[1] on electricity prices should be implemented and developed and their applicability to gas prices should be further considered where appropriate;

— adequate importance should be given to conservation in decisions on taxes;

— distortions in the tax regime which work against energy conservation should be eliminated or reduced.

---

1. The Governing Board urged Member Governments to strengthen policies, either directly or through discussions with other levels of government, the electricity industry and regulatory bodies, to ensure that electricity prices are set at a level which encourages efficient use, guides consumers to rational choices between electricity and other forms of energy and promotes optimal investment decisions by both producers and consumers.

(c) *Government Conservation Programmes*

Since the first major oil price increase in 1973, governments have put together a range of programmes to encourage consumers to undertake conservation actions. These include information programmes, financial incentives and regulations and standards.

Information programmes are the cornerstone of every energy conservation strategy. They motivate and create awareness, explain conservation opportunities, improve technical skills and publicise other government programmes. The major programmes include residential and industrial energy audits, appliance labelling, training and education and publicity campaigns. A summary of the use and effectiveness of information programmes is in Table A.

Financial incentives in the form of grants, tax incentives and soft loans have been valuable in improving financial attractiveness and access and introducing new technologies into the market. Initially, financial incentives were very popular among Member governments as a means to encourage consumers to take action. Since 1982, some governments have shifted emphasis away from large incentive programmes. Nevertheless, financial incentives have proven to be cost-effective if designed and implemented properly including minimising the number of free-riders (participants in a programme who would have undertaken the conservation action even in the absence of the programme). Grant programmes generally have positive benefit-cost ratios but they are often difficult to administer and have to be directed to specific audiences to be most effective. Tax incentives are easier to implement for both government and consumers. Industrial consumers particularly treat them as part of the array of tax incentives available. It is, however, more difficult to direct tax incentives than grants towards investors who would not otherwise have undertaken conservation investments. Soft loans can be valuable to consumers who need access to capital. However, they have been difficult to assess because of few national studies. A summary of the use and effectiveness of financial incentives is in Table B.

Regulation and standards are used to varying degrees in all IEA countries. They are useful because they provide long-term continuity during periods of price volatility. They can be most useful in the residential and transportation sectors where there is more standardization of equipment and where there are special segments such as the rental accommodation market or markets heavily influenced by style and advertising (e.g. automobiles). They ensure that minimum efficiency levels are met. Standards and regulations should be reviewed periodically to ensure that they are still up to date. A summary of the use and effectiveness of standards and regulations is in Table C.

Specific conclusions are:

- thorough analysis should be made of the remaining opportunities for energy efficiency improvements, the obstacles to their achievement and which decision makers will need to act;
- upon identifying areas of economic potential which are unlikely to be achieved by the market, governments should assess the full range of policy instruments in order to determine the most appropriate and cost-effective programme mix for each situation;
- in designing and implementing programmes, every effort should be made to ensure their effectiveness and maximise the incremental conservation actions that result;
- conservation programmes should be evaluated periodically to ensure they are meeting policy objectives and maximising effectiveness;
- information programmes should be the cornerstone of every conservation strategy: they can motivate and create awareness, explain conservation opportunities, improve technical skills, and publicise other government programmes;
- financial incentives should be used selectively to support the operation of the market by providing access to needed capital; motivating consumers to undertake economically viable conservation efforts; helping to introduce new technologies; and helping to develop a conservation service industry;
- regulations and standards can be valuable in some instances to maintain the long-term momentum and to overcome market limitations, particularly in special market segments (e.g. the residential sector which is the least price responsive end-use sector, rented buildings and markets heavily influenced by style and advertising (e.g. automobiles)). They ensure minimum levels of effort, are useful during periods of energy price fluctuations and should be reviewed periodically;
- energy efficiency objectives should be carefully integrated with industrial, social, fiscal and other policies that affect energy use.

(d) *Research, Development and Demonstration*

The main contribution to promoting energy efficiency over the rest of the century is likely to be made by the commercialisation and diffusion of existing technologies rather than research and development (R&D) into new technologies. Demonstration — the trial of newly developed technologies or applications under normal working conditions on a large enough scale to determine the feasibility of a full commercial application — is thus of particular importance. Further R&D is required to develop new technologies into the next century. Socio-economic research also has an important contribution to make to the formulation of energy policies.

The level and type of government activity in RD&D vary by industry and country. In the transportation sector, the automobile industry for the most part undertakes its own RD&D although governments are involved to some extent in basic fuel efficiency research and play a major role in aviation RD&D. In the industrial sector, companies often do their own R&D to improve their own efficiency. In the residential sector, the building industry performs very little R&D and faces major technology transfer problems. Extensive demonstration efforts are often necessary.

The formulation of conservation policies requires an understanding of the many non-technical factors which influence efficiency improvements and which can make policies more effective. The four main areas of research include consumer behaviour research, micro-economic research for improving evaluation of policy measures, macro-economic research to determine broad trends, impacts and influential variables, and techno-economic research to have a better understanding of how energy is used, to assess remaining potential and to develop better statistical indicators of efficiency improvements.

Specific conclusions are:

- those Member governments which do not support demonstration programmes or other appropriate forms of technology transfer should look again at their position in the light of the success of such programmes in other IEA countries and the European Community;
- the design, management and results of RD&D programmes should be carefully and regularly assessed; in the case of demonstration and technology transfer programmes, evaluations should examine

the effectiveness of the effort in overcoming market limitations to the adoption of new technologies;

- the lessons learned from assessments should be made available to other governments directly and through the appropriate international organisations;
- socio-economic research should be continued or expanded as appropriate to ensure better formulation and assessment of conservation policies and programmes;
- efforts should be made to improve the quantity and quality of data on energy consumption and to improve international comparisons of energy efficiency;
- there is a need for closer collaboration within government (since responsibility for RD&D is often in ministries that do not handle energy policy) and between government and industry to avoid unnecessary duplication and to optimise use of resources.

(e) *The Exemplary Role of Governments as Energy Consumers*

Governments use energy directly and face most of the same obstacles that confront other energy users. For example, many governments rent accommodation and face landlord-tenant concerns. In many ways, the task is more complicated than for many industries because of the autonomy of various government organisations and the lack of incentives for cutting costs. They have the opportunity to set an example of good conservation practices.

Governments also have a responsibility to manage their resources well. Like industry and commerce, many governments have developed energy management programmes. They have come to recognise that energy costs, constituting a significant proportion of operating costs, can be controlled and provide many financial benefits.

Government units have varying levels of autonomy in implementing programmes to achieve their energy efficiency goals. There remains, nevertheless, a need for the central co-ordination of government-wide programmes. Common functions of the co-ordinating body are to provide advice, guidance and assistance to the various participating organisations, to undertake training programmes, to report on total public sector energy consumption and to initiate measures to improve energy management. Often the point of co-ordination is within the ministry responsible for energy. Some governments go further by having comprehensive mandatory programmes which can require the appointment of departmental energy managers, specific energy use targets, temperature settings and speed limits for government vehicles.

Governments can and should set a positive example of good energy management. This is important since it helps motivate consumers and lends credibility to a government's conservation strategy. If a government does not take the lead, it cannot expect consumers to follow suit. Therefore, a government's own conservation measures should be both effective and well publicised. In particular:

- governments should show a commitment to energy management, including a vigorous investment programme;
- governments should centrally monitor and publicly report progress;
- responsibility for government energy programmes should be clear and the management should be held to account for results.

Table A
Summary of Information Programmes

Policies/Programmes	Primary Goal	Degree of Use	Market Limitations Addressed	Implementation Environment	General Conclusions
Publicity Campaigns	- awareness	- most Member countries	- lack of information - invisibility	- implemented usually during period of high price increases - many countries have continued them throughout	- valuable for awareness creation
Residential Energy Audits	- awareness - motivation	- Canada, United States and Sweden primarily	- lack of information - invisibility	- implemented during period of high price increases	- valuable to increase awareness on part of consumers and show cost-effective options - problem with cost-effectiveness of comprehensive audits
Industrial Energy Audits	- awareness - motivation	- Canada, Europe, Japan	- lack of information - invisibility	- initially during periods of high price increases	- valuable to create awareness - problem of degree of sophistication and technical rigour
Appliance Labelling	- awareness - motivation - provide unbiased information to aid purchase decision	- Canada, United States, Europe, Japan	- lack of information - invisibility	- initially during periods of high price increases	- biggest effect on manufacturing industry - has been cost-effective means to produce energy savings - has worked well as voluntary programme
Transportation Fuel Efficiency Information	- awareness - motivation - provide unbiased information to aid purchase decision	- most Member countries	- invisibility - lack of information	- initially during periods of high price increases	- awareness generally high - credibility problems with fuel economy ratings

Source: IEA Secretariat analysis.

Table B: **Summary of Financial Incentives Programmes**

Policies/Programmes	Primary Goal	Degree of Use	Market Limitations Addressed	Implementation Environment	General Conclusions
Industrial					
Grants	- stimulation of discrete conservation investment	- most countries	- financial attractiveness and access - confidence - lack of information	- largely initiated between two price increases in 1970s - some terminated when energy prices started declining	- expansion and acceleration of investment - introduced new technologies - improved financial attractiveness - good benefit-cost ratio, even given recent price declines - wide range of incremental investment - created awareness - administratively complex - targeting on incremental projects possible
Tax Incentives	"	- North America, Japan, some European countries	- financial attractiveness - confidence	"	- easy implementation - created awareness - application process fairly easy for companies - of little use for non-tax-payers
Loans	"	- Japan, Germany, Austria	- access to capital - confidence	"	- in practice, small interference in market - mainly easing access to capital (companies in poor financial situation) - incrementality difficult to assess

Source: IEA Secretariat analysis.

Table B: **Summary of Financial Incentives Programmes** *(Continued)*

Policies/Programmes	Primary Goal	Degree of Use	Market Limitations Addressed	Implementation Environment	General Conclusions
Residential/Commercial					
Grants	- stimulation of discrete conservation investment	- about half of Member countries	- financial attractiveness and access - lack of information - confidence - separation of expenditure and benefit	- largely initiated between two prices increases in 1970s - some terminated in early 1980s when energy prices started declining	- popular and visible - created awareness - provided information to consumers - improved financial attractiveness - helped develop conservation service industry - poor results in rental market - poorer benefit-cost ratio than industrial grant programmes - administratively complex
Tax Incentives	"	- Austria, Belgium, Denmark, Germany, Japan, Switzerland, United Kingdom, United States	"	- largely initiated between two price increases in 1970s	- lower government involvement - mainly used by higher income groups
Loans	"	- Denmark, Germany, Japan, Sweden, United States	"		
Energy Transformation Sector					
Grants	- stimulation of investment into CHP and for DH	- Denmark, Germany, Ireland, Italy, Netherlands, Sweden	- financial attractiveness		- subsidies effectively reduced investment risks - benefit-cost ratio similar to industrial programmes - rather high incrementality - often lack of utility co-operation
Tax Incentives	"	- Austria	- financial attractiveness		
Loans	"	- Austria, Netherlands, New Zealand			

Source: IEA Secretariat analysis.

Table C

Summary of Regulations and Standards

Policies/Programmes	Primary Goal	Degree of Use	Market Limitations Addressed	Implementation Environment	General Conclusions
Building Codes	- upgrade efficiency of new building stock	- all IEA countries	- invisibility of consumption - lack of information - separation of expenditure and benefit	- energy efficiency aspect of existing building codes added after major price increases - have been maintained even in periods of declining energy prices	- very effective in overcoming market limitations - low cost means of upgrading thermal quality of new building stock - provide long-term signals - easy to adapt to regional/local conditions
Appliance Efficiency Standards	- upgrade efficiency of new appliances	- Japan, United States	- invisibility of consumption - lack of information - separation of expenditure and benefit	- initially implemented when energy prices increasing	- insufficient information to draw conclusion - most countries more interested in appliance labelling programmes than efficiency standards
Fuel Economy Standards for New Passenger Cars	- upgrade efficiency of new passenger cars	- nine countries - only United States has mandatory programme	- lack of information	- initially when energy prices increasing for specific period - some kept after target period	- directed towards manufacturers and importers - work in parallel with transportation information programmes - attribution of effects is difficult yet countries have maintained momentum to improve efficiency even when energy prices declining - both mandatory and voluntary programmes have achieved targets

Source: IEA Secretariat analysis.

PART A

OVERVIEW

CHAPTER I

Introduction

Energy conservation dates back to the beginnings of energy use. Fires were built in enclosed spaces and then later replaced by ovens; man learned to build shelters for protection but also to provide warmth with less fuel. This movement towards improved efficiency in energy use, however, has often been overshadowed by increasing demands for energy services — to produce more products, to travel faster and further or to be more comfortable. Both trends continue — as they should. But over the past fifteen years there has been a new effort to examine broadly the potential economic and social benefits of energy conservation. This examination has led to the recognition of energy conservation, for the first time, as an essential national and international objective.

Energy conservation means using energy more efficiently, whether through behaviour, improved management or the introduction of new technology. It has sometimes been associated with efforts to curtail energy use at the cost of economic activity and living standards, but this study is concerned exclusively with energy conservation as a means of increasing economic benefits. The International Energy Agency (IEA) has never considered it to be desirable or appropriate, except in the event of a short-term disruption of energy supplies, to attempt to decrease final demand by either reducing economic growth or forcing changes in the types of products or services available in Member countries. By increasing efficiency, energy demand can be reduced without requiring structural changes or adversely affecting economic growth.

The high interest in energy conservation over the past fifteen years was stimulated by the oil price increases of 1973-74 and 1979-80. The current easy energy market, however, in no way lessens the importance of energy conservation. The reasons for this are that:

— most energy resources are depletable: increased energy conservation will extend their availability;

— with economic growth there is likely to be a return to tight energy markets before the end of the century: increased energy conservation will delay and lessen the impact of such tightening;

— investments in energy conservation often provide a better return than investments in energy supply: increased energy conservation will therefore improve the general efficiency of the economies of IEA countries;

— there is widespread public concern about the environmental consequences of energy production and use: increased energy conservation in general reduces those consequences in a way which is consistent with energy policy objectives;

— investment in energy conservation can often be undertaken in smaller increments than investment in energy supply: energy conservation offers needed flexibility when the energy outlook is uncertain.

These arguments do not imply that investment in energy conservation can replace investment in energy supply. Both are needed.

Since its inception in 1974, the IEA has given much attention to energy conservation. The Governing Board, meeting at Ministerial level on 6th October 1977, adopted Principles for Energy Policy which committed Member governments to allow domestic energy prices to reach a level which encourages energy conservation and to reinforce energy conservation on a high priority basis by providing the resources for and implementating conservation measures (see Annex A). These principles were developed in the Lines of Action for Energy Conservation and Fuel Switching adopted by the Governing Board at Ministerial level on 8th December 1980 (see Annex B). Ministers have since reiterated the importance of energy conservation on a number of occasions, most recently at the Governing Board meeting on 9th July 1985 (see Annex C). They then recognised that conservation remained a particularly important part of energy policy and that there was potential for further gains in all sectors of the economy which could best be realised through market forces and government policies complementing one another in a manner which depends on national circumstances. Ministers agreed that government policies remained important to continued progress in reducing energy intensity and that those policies should be selective, carefully planned, cost-effective and that their results should be periodically assessed. They also recognised that the more efficient use of energy on an economic basis was of primary importance for achieving the objectives of both energy and environmental policy.

This study is an attempt to assess the role of energy conservation in the context of general energy policy after more than a decade of concerted effort by governments and consumers. Its main purpose is to assess how resources can be used to best advantage by governments for conservation. The study has three main objectives:

— *To analyse the changes which have occurred in energy demand and efficiency since 1973 and possible developments over the rest of the century, including an assessment of the economic potential for further energy conservation.* The study attempts to answer several questions, including: what efficiency improvements have already been made and what opportunities still remain? What are the factors which have led to past efficiency improvements and what factors are likely to aid further gains?

— *To examine whether there is a role for governments in promoting energy conservation in market economies.* The study addresses the question of whether there are limitations and imperfections in energy markets which might prevent market forces alone from achieving the optimal economic investment in energy conservation.

— *To identify and assess the effectiveness of the policies adopted by Member governments of the IEA to promote energy conservation and to suggest how conservation measures could be strengthened, if the analysis shows that there is a role for government action.* The study attempts to identify the types of government efforts that are likely to be most effective in accelerating the achievement of efficiency improvements, including general energy and economic policies, as well as more focussed information, incentive, regulatory and research activities. It does not, however, attempt to prescribe the specific policies or programmes that would be most effective for individual countries.

The conclusions of the study are summarised in Chapter II. Part B begins with a discussion of developments in the energy consumption and energy intensity of IEA countries between 1973 and 1985 and the main reasons for changes over this period (Chapter III). Chapter IV assesses the potential for further energy conservation and discusses possible future trends in energy demand and efficiency. Chapter V examines market limitations for energy conservation and the external benefits and costs associated with conservation which cannot easily be reflected in prices. Part C examines the actions which governments can take to achieve greater energy conservation. Chapter VI examines the way in which energy conservation activities are organised in Member countries. Chapter VII discusses energy price and taxation policies which help give the correct market signals for the optimal allocation of resources between energy supply, energy conservation and other investments. Chapter VIII describes the main energy conservation programmes of Member governments and examines their effectiveness. Chapter IX discusses research, development and demonstration (RD&D) programmes and policies in the field of energy conservation and the contribution of socio-economic research to the formulation of conservation policy. Chapter X addresses the critical role of governments in promoting energy conservation in their own establishments.

CHAPTER II

Conclusions

Between 1973 and 1985, energy demand in Member countries of the IEA grew by only 5% while gross domestic product (GDP) grew by almost 32%. Consequently, energy intensity — the amount of energy used to produce a unit of GDP in IEA Member countries — fell by 20%. This fall has significantly improved the energy situation in IEA countries and the world. The drop in energy intensity was primarily due to more efficient use of energy, but changes in economic structure also were important. Higher energy prices have played an important role in improving energy efficiency in IEA economies, but government policies have also helped increase energy conservation. The gradual decline of real energy prices from 1982 slowed, but did not reverse, these trends through the end of 1985.

Three basic points emerge from this study:

(i) There is considerable potential for increased energy efficiency on an economic basis, although this potential varies among countries and sectors. It is conservatively estimated that with existing technologies energy efficiency could be gradually increased by more than 30%. If this were achieved, energy demand would be more than 25% (or at least 1 200 Mtoe per year) below the energy demand levels that would have resulted if today's energy efficiency levels remained unchanged. A major portion of this potential will be realised based on current trends and government policies, but if this potential were to be fully realised, energy demand beyond the year 2000 could be more than 10% (at least 450 Mtoe per year) below currently projected levels.

(ii) There are limitations in the energy conservation market which prevent this potential from being fully realised:

- — in many IEA countries more stringent criteria appear to be applied to decisions on investment in energy conservation than in energy supply;
- — there are serious obstacles to economic pricing of energy;
- — energy conservation has certain external benefits which cannot in practice be easily reflected in prices;
- — there are specific market limitations which prevent price signals from being reflected in the decisions of energy consumers.

(iii) Carefully planned and assessed government policies to promote energy conservation can be effective in reducing market barriers and providing conservation on an economic basis, although they may have unintended effects.

The key problem is how to achieve more of the economic potential for energy conservation in market economies especially at times when short-term price signals do not reflect the long-term outlook. The achievement is difficult because energy conservation activities are disaggregated. Decisions about energy investment and use are made by millions of energy consumers. Numerous types of businesses and institutions are involved in promoting energy conservation: companies providing energy conservation equipment and services, public utilities, private, non-profit groups, governments and other concerns for whom conservation activities are only a small part of their total activities. Many of the decisions that affect energy efficiency are made for non-energy reasons, such as home renovation, factory expansion and urban renewal. The environment in which these activities take place is complex and linked to wider economic, political and social factors. Conservation policies, therefore, need to be based on a combination of market forces and government action. This is the approach which has been endorsed by IEA Ministers and is, in practice, followed by all IEA governments.

Success in government energy conservation policy requires a number of different approaches in order to reach and motivate a broad range of interests, for many of which conservation is a secondary activity. Based on the study's analyses, the major elements of an effective government conservation policy include:

(a) effective involvement of the various organisations which work on energy conservation;

(b) energy price, taxation and other policies designed to give the right economic signals to consumers;

(c) programmes to reduce or counterbalance limitations which prevent the market from working effectively with respect to energy conservation;

(d) research, development and demonstration to develop and transfer more energy-efficient technologies; research in the social sciences to help provide a better understanding of factors influencing consumer behaviour and to improve the effectiveness of conservation programmes;

(e) a strong lead by governments in effecting conservation in their own organisations.

Effective energy conservation policies require a governmental organisation to oversee the conduct of the underlying policy analysis, to guide the development of specific government initiatives and ensure that objective programme evaluations are conducted. While there is no ideal method of organising government conservation efforts, it is essential that they be guided by a policy group headed by a senior official who forms part of the top management of the ministry responsible for energy. In addition, under all types of government conservation structures, there must be effective interdepartmental co-ordination of conservation activities.

(a) Organisations and services

Many groups and organisations are involved in energy conservation activities. They include the energy conservation service industries, the energy supply industries (particularly the utilities), non-profit making organisations and various levels of government. Their effectiveness in promoting conservation depends to a large extent on the groups themselves. There is, however, scope for government policies to promote and co-ordinate their activities. In particular:

— the development of energy conservation services including energy contract management companies will be increasingly important as conservation requirements and opportunities become more complex. Responsibility for this development rests mainly with the interests concerned but experience in a number of countries has shown that governments can encourage these interests to come together and that information programmes plus modest financial help can play an important part in the process;

— energy supply industries can play an important part in promoting conservation through direct contacts with consumers and their understanding of consumer needs. In some cases, the industries have been ready to assume this role for their own commercial reasons, but in other cases, government encouragement or even legislation has been necessary, particularly for gas and electric utilities which are already under some form of regulation;

— non-profit groups can be effective in promoting energy conservation, particularly through their understanding of the needs of specific communities, and their ability to link energy conservation to other local needs, such as employment. Some injection of public funds can much assist such activities;

— all Member governments should re-examine energy conservation arrangements in their countries with a view to drawing upon the experience of others.

(b) Energy price, taxation and other economic policies

Economic energy pricing is an essential prerequisite for effective energy conservation policies. As a general rule, energy prices should reflect the long-run marginal costs of supply, give the right signals to producers and consumers in order to facilitate an optimal allocation of resources, enable energy suppliers to finance economically sound capital investment and thus ensure the long-term security of energy supply. There are, however, severe theoretical and practical problems about determining and implementing economic energy prices. Policy on taxation of energy products is inevitably determined by considerations wider than those of energy policy alone. It would be unrealistic to expect Member countries to adopt uniform price and tax policies since the economic and social considerations vary among countries. It is, however, desirable that adequate importance be given to energy conservation in formulating price and tax policies. In particular:

— remaining subsidies to or controls on oil prices should be eliminated or reduced as soon as possible;

- the conclusions adopted by the Governing Board on 27th March 1985 on electricity prices should be implemented and developed and their applicability to gas prices should be further considered;
- adequate importance should be given to conservation in decisions on taxes on energy consumption;
- distortions in the tax regime which work against energy conservation should be eliminated or reduced.

(c) Government conservation programmes

A range of policy measures are used by governments to complement and strengthen market signals that stimulate conservation and to address many of the market limitations described in Chapter V. Different policy measures can be used to remove or reduce the market limitations in various ways. Their use depends on many factors including complexity and cost of implementation, traditional use of policy measures within administrations, legislative requirements and timing required for expected results. Often several policy measures or an array of programmes to suit different market segments is the most effective. For example, residential energy audits are most useful when combined with financial incentives and fuel efficiency standards accompanied by information guides, labels and brochures. Therefore, policy measures should not be considered in isolation but as part of an integrated approach.

From the analysis of policy measures:

- most policy measures have been implemented to either address the lack of information and technical skills or the lack of access to financing or the economic attractiveness of conservation activities;
- financial incentives have shown to have positive benefit-cost ratios with the best results in the industrial and transformation sectors;
- there is a wide range of incremental effects due to a variety of factors, including percentage of financial incentive, range of eligible items, implementation, and design of programme;
- more focussed programmes are more effective than general ones;

- — awareness and motivation are not static and need to be reinforced periodically or through feedback mechanisms to give the consumer a better understanding of his energy use;
- — standards and regulations have been most useful in the residential/commercial and transportation sectors where there is more standardization of equipment and where there are special market segments such as rental accommodation;
- — standards and regulations ensure minimum efficiency levels;
- — effectiveness of programmes depends on good implementation, requiring human and financial resources, co-ordination within administrations and with other levels of government;
- — in the industrial sector, a combination of programmes encourages businesses to develop good energy management. Some examples are energy audits, training, monitoring and targeting, technical materials and technology transfer programmes;
- — there is still too little known about the effectiveness of programmes and evaluations that have been undertaken have generally not been sufficiently comprehensive.

The specific conclusions are:

- — thorough analysis should be made of the remaining opportunities for energy efficiency improvements, the obstacles to their achievement and which decision makers will need to act;
- — upon identifying areas of economic potential which are unlikely to be achieved by the market, governments should assess the full range of policy instruments in order to determine the most appropriate and cost-effective programme mix for each situation;
- — in designing and implementing programmes, every effort should be made to ensure their effectiveness and maximise the incremental conservation actions that result;
- — conservation programmes should be evaluated periodically to ensure they are meeting policy objectives and maximising effectiveness;
- — information programmes should be the cornerstone of every conservation strategy: they can motivate and create awareness, explain conservation opportunities, improve technical skills, and publicise other government programmes;

- financial incentives should be used selectively to support the operation of the market by providing access to needed capital; motivating consumers to undertake conservation efforts; helping to introduce new technologies; and helping to develop a conservation service industry;
- regulations and standards can be valuable in some instances to maintain long-term momentum and to overcome market limitations, particularly in special market segments (e.g. the residential sector which is the least price responsive end-use sector, rented buildings and markets heavily influenced by style and advertising (e.g. automobiles)). They ensure minimum levels of effort, are useful during periods of energy price fluctuations and should be reviewed periodically;
- energy efficiency objectives should be carefully integrated with industrial, social, fiscal and other policies that affect energy use.

(d) **Research, development and demonstration**

Energy conservation RD&D is widely spread over governments and industry. Substantial progress has been made in the last ten years. More needs to be done, however, to demonstrate new technologies which are technically viable, to more effectively transfer these technologies to private manufacturers and end-users, to improve the assessment of government programmes and to develop social research relevant to the formulation of energy conservation policies and programmes. Specific conclusions to these ends are:

- those Member governments which do not support demonstration programmes or other appropriate forms of technology transfer should look again at their position in the light of the success of such programmes in other IEA countries and the European Community;
- the design, management and results of RD&D programmes should be carefully and regularly assessed; in the case of demonstration and technology transfer programmes, evaluations should examine the effectiveness of the effort in overcoming market limitations to the adoption of new technologies;
- the lessons learned from assessments should be made available to other governments directly and through the appropriate international organisations;

- socio-economic research should be continued or expanded as appropriate to ensure better formulation and assessment of conservation policies and programmes;
- efforts should be made to improve the quantity and quality of data on energy consumption and to improve international comparisons of energy efficiency;
- there is a need for closer collaboration within government (since RD&D are often in ministries that are not responsible for energy policy) and between government and industry to avoid unnecessary duplication and to optimise use of resources.

(e) **An exemplary role for governments**

Governments can and should set a positive example of good energy management. This is important since it helps motivate consumers and lends credibility to a government's conservation strategy. If a government does not take the lead, it cannot expect consumers to follow suit. Therefore, a government's own conservation measures should be both effective and well publicised. Specific conclusions are:

- governments should show a commitment to energy management, including a vigorous investment programme;
- they should centrally monitor and publicly report progress;
- responsibility for government energy programmes should be clear and the management should be held to account for results.

PART B

IMPROVING ENERGY EFFICIENCY TRENDS, OPPORTUNITIES AND OBSTACLES

The first step in the study of energy conservation policy is the examination of the actual opportunities for energy efficiency improvements combined with a review of past experiences and likely future trends towards the realisation of these opportunities. Such an analysis provides the basis for the development or revision of effective government conservation policies.

The following three chapters present an overview of such an examination for IEA Member countries. The chapters contain:

— a review of trends in energy demand and efficiency since 1973 by country and end-use sector (Chapter III and a more detailed examination in Annex D);

— an assessment of the remaining conservation potential and the likely future trends towards achieving this potential (Chapter IV, with data sources in Annex E);

— an analysis of the imperfections or limitations of market forces and government policies which are likely to prevent or slow the achievement of economically attractive efficiency improvements (Chapter V).

The analyses are intended to provide a basis for broad policy conclusions; they are not sufficient for the design of specific conservation policies or programmes for individual countries, which requires more detailed analysis on a national basis.

CHAPTER III

Developments in Energy Demand and Energy Efficiency 1973-1985

Between 1973 and 1985 total primary energy requirements (TPER) in IEA countries increased by 5% or 0.4% a year. At the same time gross domestic product (GDP) grew by 32%, an average of 2.4% a year. The result was a fall of 20% in energy intensity — the amount of energy used per unit of GDP (TPER/GDP). About one half of this change occurred from 1979 to 1982. Since 1982, energy intensity has continued to decline, but at a much slower rate.

The reduction in overall energy intensity was the result of structural changes in the economies of IEA Member countries and increased levels of energy efficiency. Each factor made important contributions to lower energy intensities, although by different means. During the past ten years, improved energy efficiency has had the largest overall effect. The relationship between them and their contribution to overall energy demand is depicted in Figure 1.

Some Key Concepts

It is important to distinguish between general trends in energy demand and those resulting from efficiency improvements. Trends in energy demand, together with domestic supply, determine net energy imports and strongly influence energy prices. However, only through increased

Figure 1

Energy Demand and Efficiency
Terms and Definitions

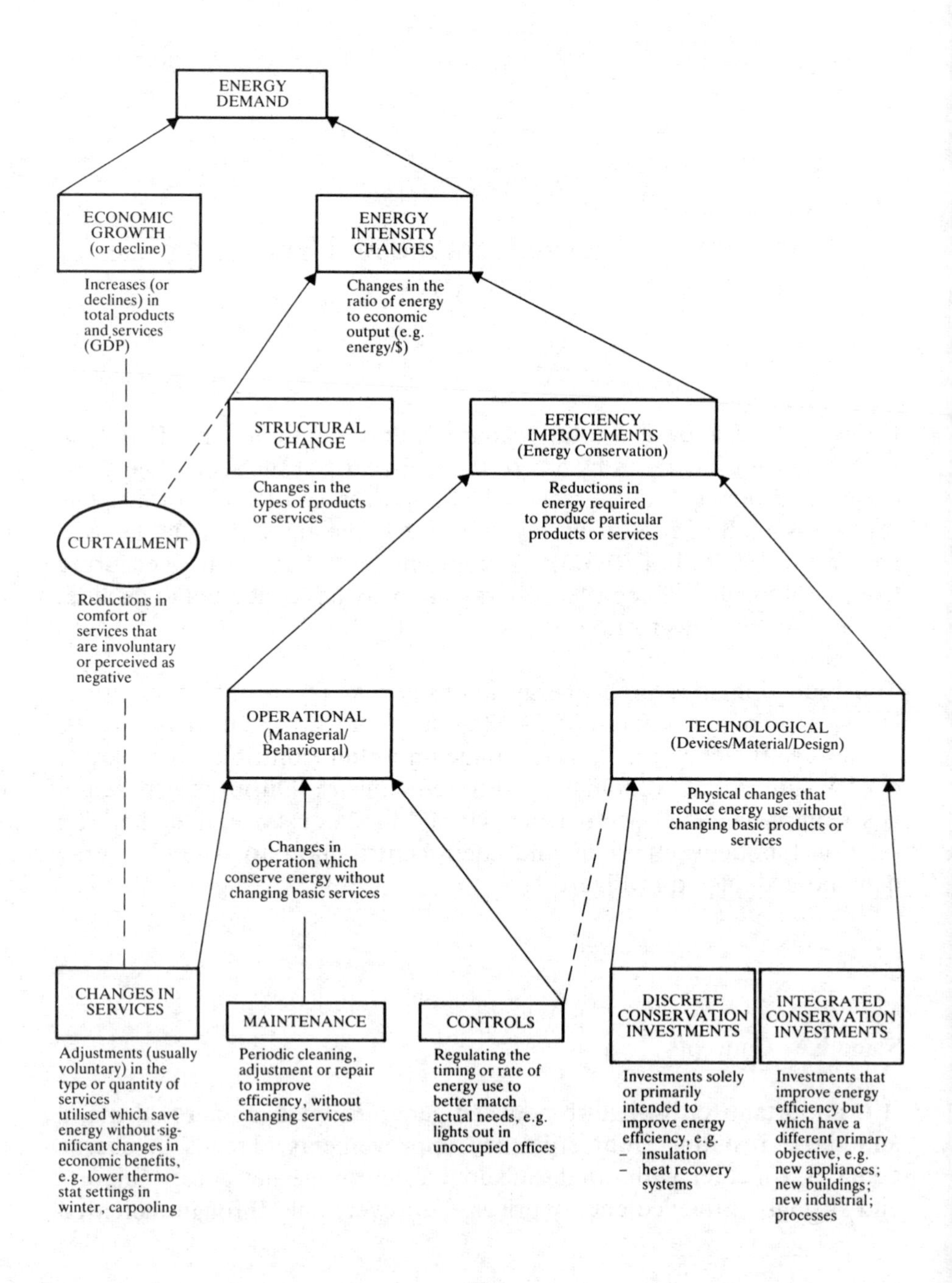

energy efficiency can demand be reduced, without lowering economic activity or changing services. That is why this chapter concentrates on energy efficiency improvements since 1973.

Energy management represents the comprehensive application of energy conservation and fuel substitution measures. Because it includes fuel choice, energy management has a somewhat broader meaning than "energy conservation", but the basic objective is the same — the reduction of energy costs.

Energy efficiency can be measured in many ways. For automobiles, it is generally measured as litres per 100 kilometres or miles per gallon, but more technically revealing measures of automobile efficiency usually exclude the effects of changes in passenger space and performance. For the industrial production of primary materials, it is often the energy required to produce a tonne of material. The best measure of energy efficiency varies depending on the end-use and even for a single end-use, there can be several different measures. This characteristic of energy efficiency makes it impossible to develop uniform statistical measures for whole economies or even sectors. As a result, the energy efficiency improvements that have occurred since 1973 must be examined case-by-case for each type of end-use.

A rough indicator of general progress towards improved efficiency is energy intensity — usually in the form of energy demand per unit of economic output (such as TPER/GDP). However, energy intensity reflects the combined effects of structural changes in economies and energy efficiency improvements. The economic structure of Member countries of the IEA differs from one country to another. Energy intensity varies depending on industrial structures, historical price developments, availability of indigenous resources, traditional use patterns, climate and geography. Countries which have harsher climates, more heavy industries or cheaper energy sources tend to have higher energy intensities, even though they may also be comparatively energy-efficient. A further difficulty with this measure is that it has the effect of indicating an increase (or a lesser decrease) in energy intensity in countries whose demand structure includes a greater use of electricity.

Even with these limitations, the rate of change in the energy intensity of IEA Member countries is the best readily available indicator of overall progress towards improved energy efficiency. This is especially true when changes in energy intensity are compared to the rate of change previously experienced by the country and to other countries with similar characteristics.

Trends in Energy Intensity and Efficiency

Changes in energy intensities were not consistent among IEA Member countries between 1973 and 1985. As Table 1 shows, in 1973 there were wide differences among Member countries and this undoubtedly has had an effect on relative rates of change. Canada, Luxembourg and the United States have been, and still are, the most energy-intensive Member countries; Denmark, Germany, Japan and Switzerland were the least energy-intensive Member countries in 1985, although each has shown various rates of progress since 1973. The countries whose annual rates of decrease in energy intensity were smallest (or even positive) have tended to be those that had more rapidly growing industrial sectors or had access to plentiful supplies of relatively low cost electricity[1]. Those countries with the largest decreases were already comparatively energy-intensive and were also experiencing significant structural changes, usually a stable or declining industrial sector. Both trends tended to bring IEA Member countries closer together in terms of average energy intensity by 1984, although major differences still exist.

Structural changes that have contributed to lower energy intensity levels have usually been the result of long-term trends of economic development. Higher energy prices have contributed to the acceleration of these trends by depressing demand for energy-intensive products and services and boosting demand for less energy-intensive forms of economic activity. Structural changes have occurred in many forms, from changes in product types within specific industries to shifts from one sector of economic activity (e.g. industry) to another (e.g. commercial).

---

1. The energy lost in the generation and distribution of electricity is reflected in the total primary energy requirements of individual countries. For statistical reasons, it is assumed that the energy generated from hydroelectricity is produced from primary energy at an efficiency rate of 38.5%. Largely because of this assumption, those countries which have found it economically attractive to increase greatly electricity use (usually because of the availability of low-cost hydroelectricity) have also increased their energy intensity (based on TPER/GDP).

Table 1: **Energy Intensity in IEA Countries**[1]

	1973	1979	1983	1985	Average Annual Change Rates 1973-79	1979-83	1983-85
Canada	.85	.86	.81	.80	.2	-1.6	-.7
United States	.79	.73	.64	.61	-1.3	-3.1	-2.9
North America	.80	.74	.66	.62	-1.1	-3.0	-2.7
Australia	.48	.51	.49	.45	.9	-1.2	-3.2
Japan	.42	.37	.30	.29	-2.2	-4.9	-1.3
New Zealand[2]	.40	.46	.46	.50	2.3	.2	3.8
Pacific	.43	.39	.32	.32	-1.7	-4.2	-1.4
Austria	.38	.36	.32	.33	-1.0	-2.6	1.3
Belgium	.47	.42	.34	.36	-1.6	-5.6	2.7
Denmark	.33	.31	.24	.27	- .7	-6.5	5.1
Germany	.38	.36	.31	.31	-1.3	-3.4	.3
Greece[2]	.38	.41	.42	.44	1.5	.4	2.3
Ireland	.52	.48	.43	.44	-1.3	-2.8	1.3
Italy	.41	.38	.34	.33	-1.2	-2.7	-1.1
Luxembourg	1.15	.88	.63	.65	-4.3	-8.0	1.6
Netherlands	.44	.42	.35	.36	-1.0	-4.5	1.8
Norway	.47	.44	.42	.40	-1.2	-1.3	-2.2
Portugal[2]	.41	.48	.48	.49	2.9	.0	.3
Spain	.31	.35	.33	.32	2.1	-1.5	-1.0
Sweden	.43	.42	.38	.40	- .4	-2.5	3.2
Switzerland	.24	.25	.25	.25	.8	.3	- .7
Turkey	.63	.56	.57	.56	-1.8	.2	- .2
United Kingdom	.44	.40	.35	.35	-1.6	-3.5	.0
Europe	.40	.38	.34	.34	- .9	-3.0	.4
IEA Total	.57	.53	.47	.45	-1.1	-3.3	-1.4

1. Energy intensity is defined as TPER/GDP or toe per $1 000 of GDP at constant 1980 prices and constant exchange rates in order to reduce the distortions introduced by exchange rate fluctuations. These indicators must be qualified for countries with high proportions of comparatively inexpensive electricity, such as hydro. It is economically attractive to increase electricity demand in such countries. Rapidly increasing electricity demand has a comparatively greater impact on TPER because of actual and assumed energy losses in electricity generation. For the electricity generated from fossil fuels, the actual thermal efficiencies of conversion are used. But for hydro and nuclear generated electricity assumed values are used which approximate the equivalent amount of thermal fuel which would be required to generate the same electricity. For example, the primary energy input to the production of hydroelectricity is calculated for most IEA Member countries using a theoretical generation efficiency of 38.5% and therefore the primary energy intensity of countries with large, inexpensive hydroelectricity generation tends to be overstated.
2. In some countries such as Greece, New Zealand and Portugal, energy intensity has increased since 1973 due to an increase in energy-intensive industries.

Source: Energy Balances of OECD Countries.

There have also been changes in personal lifestyles and other social characteristics resulting from higher incomes and shifting product prices, among many other factors.

The energy efficiency improvements that are reflected in the changes in overall energy intensity have resulted from several related factors, including: rising energy prices, general increases in economic productivity and government policies to promote the more efficient use of energy. The actual efficiency improvements that have occurred fall into two broad categories:

— technological, which includes investments in equipment and materials which improve energy efficiency;

— operational, which encompasses a wide range of behavioural and managerial changes and the way in which energy is used.

There are literally thousands of different types of technological and operational conservation measures which have been applied in energy end-use sectors. The basic categories of conservation actions are also depicted in Figure 1.

Response to Movements in Energy Prices

Increases in real energy prices, especially from 1978 to 1982, have been the most important single factor behind the substantial improvements in energy efficiency over the past ten years, the corresponding reductions in energy intensity and the slowing of energy demand growth. Figure 2 depicts the changes in the relative prices of labour, capital and energy in OECD countries since the 1960s[1]. It indicates that energy prices rose by nearly 80% in just four years after 1978, while the cost of capital rose only 36% and that of labour by even less.

Rising energy prices have not only accelerated the adoption of energy-efficient technologies, but they have also provided added stimulus to structural shifts towards less energy-intensive technologies and slowed economic growth. In the short term, steeply rising energy prices may have had a large impact on energy demand by depressing economic growth. But over the whole of the last ten years, their greatest

1. Total OECD data are used when IEA statistics are unavailable.

impact probably has been on the basic efficiency of energy use and, to a lesser extent, the rate of structural change in the economies of IEA Member countries.

Figure 2

Trends of Production Factor Prices
(1968 = 100)

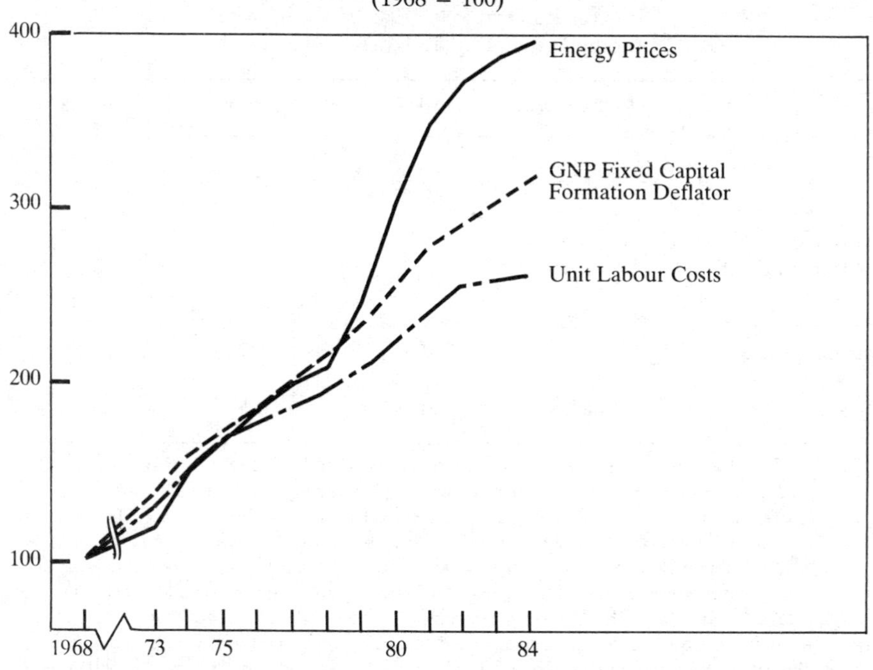

Source: "Historical Statistics 1960-1984", OECD, Paris 1986.

The responsiveness of energy demand and efficiency to rising energy prices has varied by end-use sector, region and type of price movement. Table 2 provides estimates of average own price elasticities by end-use sector for OECD and its three major regions. These average elasticities were based on a Secretariat analysis of the long-term response of energy demand during periods of rising fuel prices over the past thirty years. The analysis attempted to exclude the effects of income growth and major structural changes that were not primarily or directly dependent on end-use energy prices. The table indicates that energy demand appears to respond strongly to rising retail energy prices, especially over the long term (up to ten years), in every sector and region. On average, a 10% increase in end-use energy prices results, over time, in about a 5% decrease in demand (an own price elasticity of 0.5) and, as other analyses have shown, most of this reduction is achieved through improved energy efficiency. A similar analysis of the responsiveness of energy demand to falling energy prices indicates that the elasticities are generally significantly smaller; that is, a 10% decrease in end-use prices usually results in a long-term increase in demand of significantly less than 5%.

Table 2
Average Own Price Elasticities[1] of OECD Energy Demand by End-Use Sector and Region

	North America	Pacific	Europe	All OECD
Industry[2]	0.34	0.50	0.49	0.42
Resid./Commercial[3] (including agriculture and other)	0.60	0.70	0.55	0.59
Transportation[4]	0.63	0.30	0.70	0.61
All Sectors	0.53	0.47	0.57	0.52

1. Based on the long-term own price elasticities of specific fuels and electricity in response to upward movements in end-use prices observed periodically over the past thirty years. The periods examined varied by fuel type and region, although most energy prices did rise during the mid-1970s and from 1979 to 1982. The long-term effects of price increases have been observed to continue to increase incrementally for up to ten years, although the size of the annual incremental effects diminishes. The specific fuel and electricity elasticities were averaged using 1980 OECD energy consumption data.
2. Industrial fuel own price elasticities were calculated using a method, which relied in part on industrial production indices, designed to exclude the effects of major structural changes and other energy intensity changes that are not directly induced by energy prices.
3. Residential/commercial own-price elasticities were also calculated by methods designed to limit the effects of non-price-induced structural changes. These methods included the use of data on the changing penetration of central heating, among others.
4. Transportation fuel own price elasticities were calculated using a method designed to reflect primarily the induced changes in the ratio of energy use to distance travelled. However, because rising prices also have some long-term effects on total distances travelled, the elasticity values indicated for transportation probably understate the true values.

Source: IEA Secretariat.

Table 2 also indicates that industrial energy demand is less price-elastic than other end-use sectors. This may be because industry's long-term trend towards improved energy efficiency is more the result of broad economic forces, than solely energy price movements. Other analyses indicate that industry is generally quicker to respond to rising energy prices than other end-use sectors, but some of these responses involve product changes, or the elimination of production in high energy cost regions which may not be reflected in energy price elasticities. Also, while the economically attractive potential for energy savings per unit of output is often smaller in industry, in terms of percentage reductions, than in other sectors, there are basic changes in industrial processes which have been gradually introduced over very long periods. These types of very long-term technological changes can have important effects on energy use, but may also not be fully reflected in price elasticities. Since 1973, industry has experienced the largest increase in end-use energy prices and the greatest percentage drop in energy intensity. Although regional differences are generally small, it does appear that price elasticities are lower for industrial fuels in North America and transportation fuels in the Pacific and that they are higher for residential/commercial fuels in the Pacific. These elasticities may reflect regional differences in economic structure, climate and the influence of non-energy factors.

Analysis of End-Use Sectors

A. *Industry*

Industry is the most energy consuming end-use sector, accounting for 37% of total final energy consumption (TFC). Industrial energy consumption declined by 1.3% per year from 1973 to 1984, while industrial output increased at an average annual rate of about 2%. This resulted in an overall reduction of industrial energy intensity (the ratio of industrial energy use to value added) of 30%. From 1979 to 1983, the rate of annual decrease in industrial energy intensity was about three times higher than in 1973-79.

Energy intensity has been decreasing in the industries of most IEA Member countries for more than thirty years. For example, for manufacturing industries of OECD Member countries, the annual rate of decrease in energy intensity from 1960 to 1973 was two-thirds of the rate between 1973 and 1979.

Energy efficiency improvements appear to have been the major cause of decreased industrial energy intensity from 1973 to 1984, although structural changes within the industrial sector also played a very important role. Table 3 summarises the results of a recent study of the factors contributing to the decline in energy intensities in Europe. It indicates the major role of energy efficiency improvements. Several other studies of both United States and Japanese industries came to similar conclusions. In all cases, energy efficiency improvements

Table 3

Determinants of Energy Demand in Manufacturing Industry in Four Major European Countries[1]

(in petajoules)

	Total Final Consumption in 1979	Annual Change in TFC in 79/83	Change in TFC in 1983/84	TFC in 1984[2]
Fuels	6 590	- 375	+ 53	5 120
by factor:				
- Efficiency improvements		- 253	- 163	
- Structural change		- 60	+ 79	
- Changes in activity level		- 54	+ 146	
- Interfuel Substitution[3]		- 8	- 9	
Electricity	1 554	- 27	+ 64	1 510
by factor:				
- Efficiency improvements		- 11	+ 2.1	
- Structural change		- 3.4	+ 18.9	
- Changes in Activity level		- 12.8	+ 42.2	
- Interfuel substitution[3]		+ 0.1	+ 0.6	

1. Germany, Italy, United Kingdom, France.
2. Differences to actual TFC mainly due to stock changes.
3. Including intra-industrial structural change.

Source: ISI-Karlsruhe, Germany (based on EUROSTAT).

accounted for more than half of the decline in energy intensity for the periods examined. The periods assessed varied from 1972-81 to 1979-84.

The energy efficiency improvements made in industry encompassed the full range of conservation actions, but integrated conservation investments (those which were an integral part of new industrial equipment and processes) probably made the single largest contribution. Operational improvements, discrete conservation investments and government measures also had major impacts.

A method widely used by industry to achieve efficiency gains during this period was the institution of energy management programmes. A central feature of such programmes has usually been the designation of a senior manager with a broad range of energy-related responsibilities, including the operation and maintenance of energy-using systems, the identification and evaluation of major efficiency investments and decisions on fuel choice. There has been a general trend towards the establishment of such energy management programmes in large, energy-intensive industries over the past ten years. This trend has been accelerated in several countries through both voluntary programmes promoted by the government, such as the current British efforts, and mandatory requirements, such as in Japan where industries whose energy consumption is over a fixed threshold must appoint energy managers.

The best indicator of industrial efficiency improvements is the ratio of energy consumption per unit of product output, such as gigajoules per tonne of steel. Figure 3 displays trends in the efficiency of five Japanese industries from 1979 to 1985. While Japanese industry has been particularly successful in making energy efficiency improvements, similar progress has been made by major, energy-intensive industries in most IEA Member countries. Energy-intensive industries usually have been especially quick to adopt more energy-efficient processes and equipment.

B. *Residential/Commercial Sector*[1]

The residential/commercial sector is the second largest user of energy and the fastest growing. For the IEA as a whole, per capita energy use in

1. Includes public and agricultural use.

this sector increased by 2.5% between 1973 and 1979, but declined by about 12% between 1979 and 1984. The trend was again reversed in 1985 when per capita energy use rose by 2.1%.

Figure 3

Energy Efficiencies of Selected Industrial Products in Japan
(Index 1973 = 100)[1]

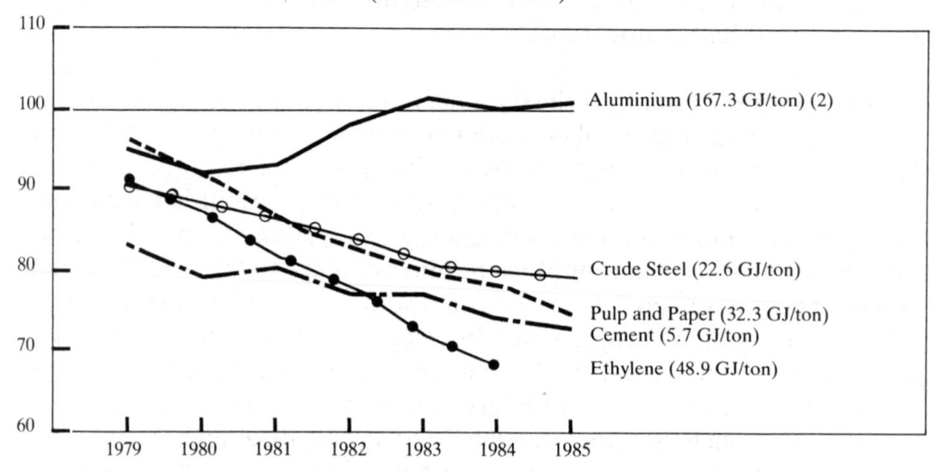

1. Specific energy consumptions in 1973 in brackets. All figures are by Japanese Fiscal Year, which is 1st March to 30th April.

2. Electricity consumption including conversion losses. The stable efficiency of the aluminium industry during this period reflects progress made in earlier years, absence of major technological advances and a period of decreasing production in Japan. In 1984, the Japanese aluminium industry was still among the most efficient in the world.

Source: Ministry of International Trade and Industry, Tokyo, Japan.

Residential Subsector. Energy use per capita in the residential subsector decreased continuously from 1973 to 1983 and increased slightly from 1983 to 1985. However, these trends masked major structural and efficiency changes. One study[1] has calculated that structural changes in the residential sector would have increased energy use per dwelling between 1972 and 1982 by about 10-15% in North America, Denmark and Sweden, 15-20% in Norway, the United Kingdom and Germany and more than 20% in Japan and Italy. However, in almost all Member countries these increases have been more than offset by large

---

1. Schipper and Ketoff, "Explaining Residential Energy Use by International Bottom-up Comparisons", *Annual Review of Energy,* Volume 10, 1985, p.382.

improvements in energy efficiency. As an example, Figure 4 shows efficiency improvements in specific useful energy for space heat between 1970 and 1982.

Commercial, Agricultural and Public Subsectors. The commercial subsector, which encompasses a broad range of economic activities, is the fastest growing subsector in most IEA Member countries. Unfortunately, a lack of data on energy use and the physical characteristics of this sector has hampered analysis of energy efficiency trends. Nevertheless, a recent study[1] has concluded that substantial efficiency improvements were made in this subsector from 1970 to 1982. One indicator, energy use per employee, showed:

	Percentage Change 1970-82
Denmark[1]	-19
Germany	-12
Norway	-13
Sweden	-15
United Kingdom	-12
United States	-24

1. For Denmark 1972-82 was used.

The energy-efficiency measures which have contributed to these reductions include a wide range of conservation actions in both new and existing buildings. Efficiency gains in new commercial buildings have been especially large and the efficiency of certain end-use technologies, such as lighting, has also increased substantially.

C. *Transportation Sector*

Oil accounts for close to 99% of energy consumption in this sector, and road transport is responsible for 80% of this oil use. Most measures of transportation energy efficiency have improved since 1973, but overall energy use trends have not been consistent. Transportation energy

1. Schipper, Meyers, Ketoff, "Energy Use in the Service Sector: An International Perspective", *Energy Policy,* June 1986.

Figure 4

Space Heat. Useful Energy[1] per Degree-day per Square Metre
(1970-1982)

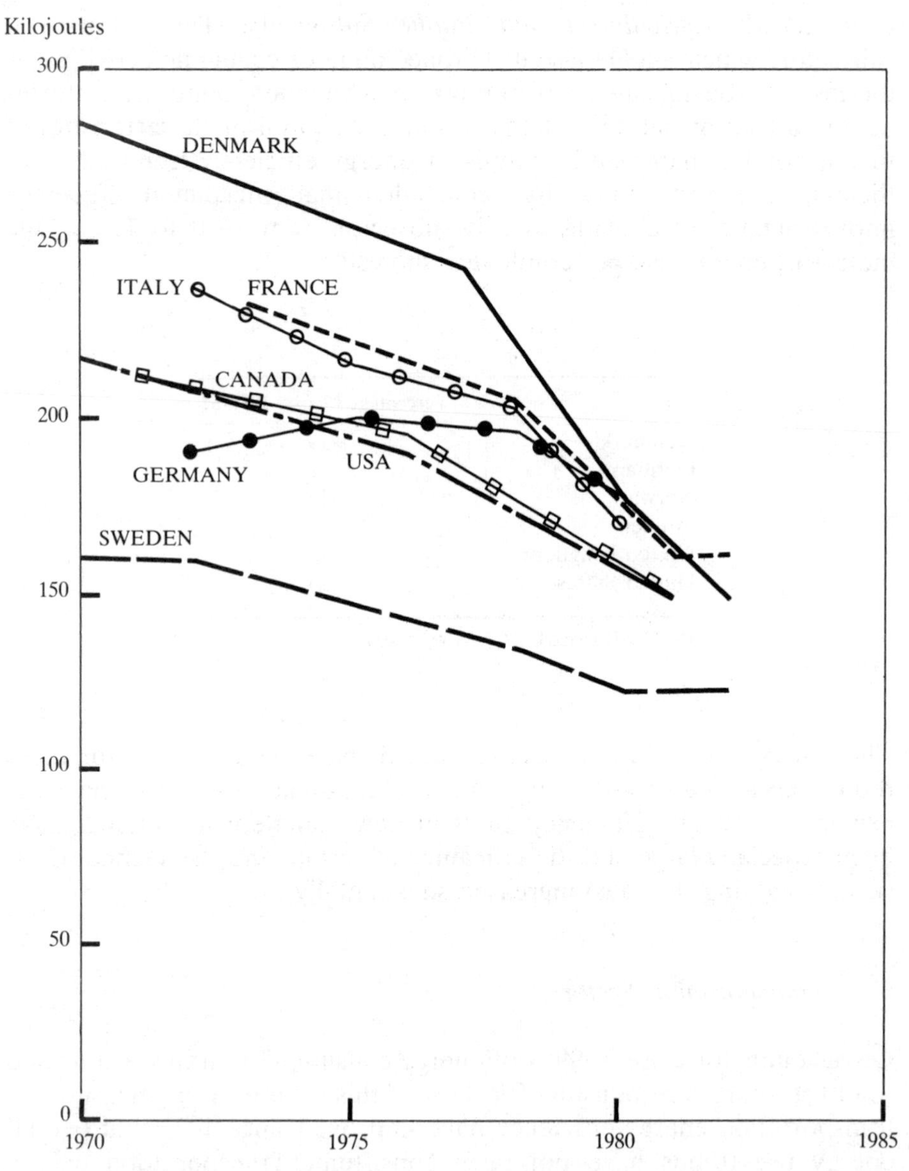

1. Useful energy is energy consumption for space heating multiplied by an average furnace efficiency for oil/gas burners of 66%, for solids of 55%.

Source: Schipper, Ketoff and Kahane, "Explaining Residential Energy Use by International Bottom-Up Comparisons", *Annual Review of Energy* 1985. 10:341-405.

demand rose between 1973 and 1979, declined from 1979 to 1982, and has since again risen slowly. While fuel efficiency in passenger cars has improved, better economic conditions have acted as a stimulus to consumption, particularly in commercial transport. Passenger cars accounted for almost 67% of consumption in the road transport subsector in 1973, and 61% in 1983, while consumption by commercial vehicles increased from 33% to 39%. Based on analysis by the IEA Secretariat, of the 19.6% reduction in consumption per passenger and commercial vehicle observed between 1973 and 1983, 65% was due to efficiency improvements and 35% to reduced average distance travelled per vehicle.

Vehicle fuel efficiency in the passenger car subsector has improved since 1973. Voluntary and mandatory standards governing new car fuel economy in nine countries have played a role as has the comparatively rapid turnover of vehicle fleets (about every ten years). Table 4 describes the improvements made in new car fuel efficiency since 1973. There have been improvements in all weight categories, and there are now smaller differences among countries within each category. While technical efficiency improvements have been incorporated in the new vehicles in all countries, these improvements have often been offset by trends towards larger and heavier cars, especially in some European countries. For commercial vehicles, technological progress in vehicle engine design rather than fuel prices is the key. Fuel efficiency improvements have been slower because units tend to be big and have longer lifetimes. Goods vehicles have shown more improvement than buses, and also the relative use of diesel versus gasoline has been responsible for different levels of fuel economy across IEA countries. Efficiency improvements in the commercial transport subsector, however, have been more than offset by increased vehicle kilometres driven — a reflection of general economic growth.

The maritime, aviation and railway subsectors all depend on petroleum products for energy. Together they accounted for about 17% of oil consumption in the transport sector in 1973 and 1984. Higher fuel prices and other factors resulted in significant efficiency improvements in each of these subsectors during the past decade.

D. *Energy Transformation Sector*

The transformation sector encompasses the conversion of primary fuel into more useful energy forms, as well as the distribution of energy to

Table 4

Actual New Passenger Car Fuel Efficiency

(Gasoline consumption in litres per 100 kilometres)

								Projections/Targets		
	1973	**1978**	**1979**	**1980**	**1982**	**1983**	**1984**	**1985**	**1990**	**2000**
Australia	n.a.	11.8	11.2	10.1	9.8	9.5	9.5		8.5[1]	
Canada	n.a.	11.5	11.5	10.3	8.5	8.5	8.4	8.2	7.4	6.8[2]
Denmark	9.0[3]	n.a.	n.a.	8.6	n.a.	7.3	7.0	7.1		
Germany	10.3	9.6	9.4	9.0	8.3	8.0	7.7	7.5		
Italy[4]	8.4	8.3	8.3	8.1	8.3	8.0	n.a.	7.8	7.4	
Japan	10.4	8.8	8.6	8.3	7.7	7.8	7.8	7.8		
Netherlands	n.a.	9.2	8.9	8.8	8.5	n.a.		9.1	8.6	7.8
Sweden	n.a.	9.3	9.2	9.0	8.6	8.6	8.5	8.5		
United Kingdom	11.0	9.1	9.0	8.7	8.1	7.9	8.8	7.6		
United States	16.6	11.8	11.6	10.0	8.9	9.0	8.9	8.7	8.8	8.2

1. 1987 target.
2. 1995.
3. 1975.
4. The figures for Italy represent average efficiency of the total car fleet.

Source: IEA Country Submissions.

end-users. In IEA energy balances, the energy demand of the transformation sector is the difference between the input of primary energy (TPER) and secondary energy forms produced (TFC). It therefore represents exclusively the losses in the conversion processes and in the extraction and distribution of energy. This sector is now larger and growing faster than any of the end-use sectors. The transformation sector is dominated by electricity (close to 80%), with most of the rest accounted for by energy use and losses in oil refineries.

Figure 5

Average Efficiency of Steam-Electric Plants in the United States

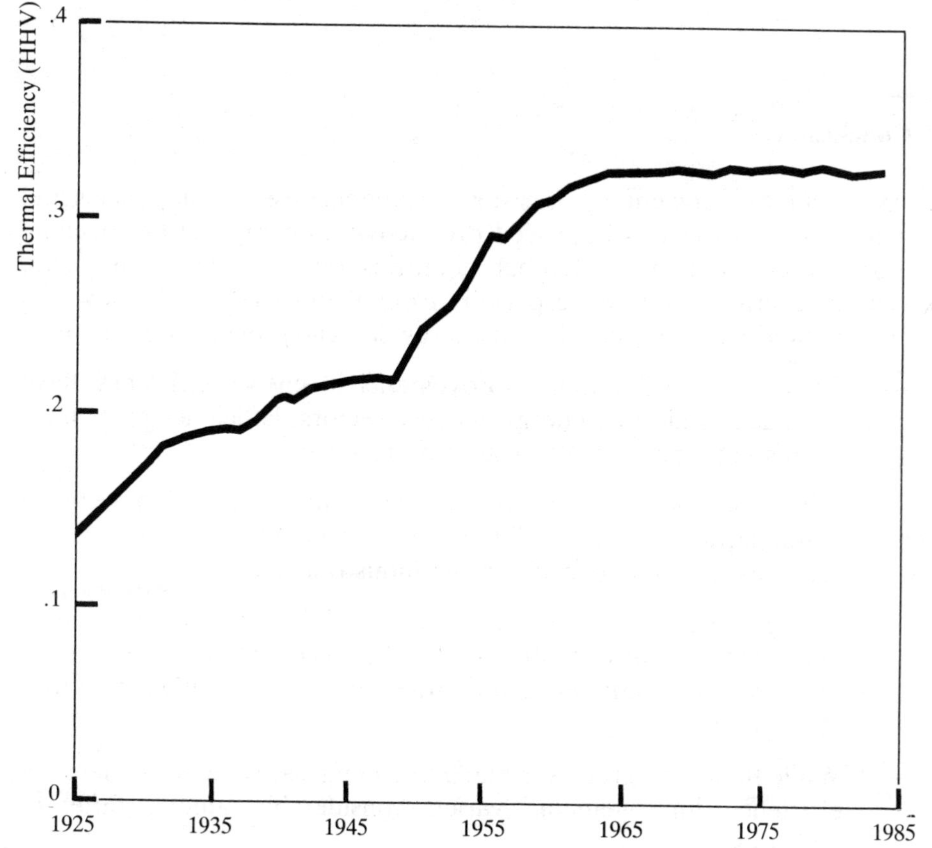

Source: IEEE Technology and Society Magazine, United States, 3/86.

The rapidly increasing demand for electricity has resulted in an increase in the energy losses that make up the transformation sector. Although the early development of electric power resulted in major increases in energy efficiency, these regular gains virtually disappeared during the 1960s, as revealed in Figure 5. Since that time comparatively little progress has been made and, in a few cases, energy efficiency has actually declined as slightly less efficient coal-fired power plants have increased their share of total electricity generation.

On the other hand, in the petroleum refining sector, there have been major efficiency improvements since the mid-1970s. These resulted in a 17% decline in the percentage of refinery input consumed internally from 1973 to 1984, despite higher input requirements to maintain gasoline octane levels while phasing out lead content.

Conclusions

As a result of rising energy prices, government conservation policies and general trends towards improved productivity, energy use has become more efficient in IEA Member countries over the past ten years. Improved efficiency rates, especially from 1979 to 1983, have lowered energy intensity ratios and slowed energy demand growth. In particular:

— Since 1973, substantial improvements in energy efficiency have been achieved in all energy end-use sectors. Efficiency gains have been small only in the transformation sector.

— While the energy efficiency improvements since 1973 have been large, particularly after 1979, they represent only an acceleration of previous trends, especially in the industrial sector.

— Rising energy prices, especially after 1978, were the single most important factor contributing to the acceleration of efficiency improvements, but numerous other factors also influenced conservation actions.

— While major progress was made in all end-use sectors, the level of efficiency improvements varied considerably among different end-uses.

— In industry, the largest efficiency improvements were made by energy-intensive industries, such as primary metals, which usually had the financial resources, technical capabilities and motivation to conserve.

- In buildings, major progress has been made in new building design and certain categories of equipment. These efficiency gains were not made consistently throughout the sector and were often outweighed by increasing space conditioning and home appliances.
- In transportation, the technical energy efficiency of new cars has been improved significantly in almost all countries, but trends towards larger vehicles and increased travel have offset these gains in some regions.
- In the transformation sector, little progress has been made towards improving the efficiency of electricity generation and distribution, but large gains have been made in the efficiency of the petroleum refining industry.
- Recent energy demand data for 1984 and 1985 suggest that stable or declining world oil prices and higher levels of economic growth have reduced, but not reversed prior trends towards declining energy intensity. These most recent trends deserve further analysis.

CHAPTER IV

Potential Efficiency Improvements and Future Trends

The previous chapter has shown that substantial efficiency improvements, combined with structural changes and slow economic growth, resulted in much lower rates of growth in energy demand over the past ten years than during the previous several decades. Chapter III also identified many of the causes of these changes and some of the more recent trends towards improved technical efficiency.

The purpose of this chapter is to consider probable future progress towards improved energy efficiency that is likely to occur in response to changing energy prices, increased productivity, other market forces and current government policies. The chapter also assesses whether this progress will be sufficient to achieve the economic potential for energy savings that exists. This section begins with an assessment of the current economic potential for energy efficiency improvements. It is followed by an assessment of likely future trends in efficiency improvements. The chapter concludes with an assessment of further opportunities for efficiency improvements, in addition to those that are likely to be achieved as a result of market forces and current government policies.

A. **Current Potential for Efficiency Improvements**

Assessing the remaining potential for efficiency improvements is important because it can help determine to what extent future progress is likely or possible. But even if it can be shown that a large potential for savings still exists, it is necessary to analyse carefully where the potential exists and why in order to develop effective government policies to achieve it.

The first essential step in this analysis is to establish energy price assumptions and investment criteria to be used in determining whether or not particular actions are considered desirable. Ideally, the energy price assumptions used in calculating the economic attractiveness of conservation investments should reflect the full costs of energy supplies during the life of the investment. This is usually referred to as the long-run marginal cost of energy, including appropriate external costs, such as the security and environmental effects of energy use. Similarly, to ensure a balanced allocation of capital resources, the criteria used in assessing conservation investments should resemble those used for energy supply investments or others which are comparable. In general, estimates of conservation potential should include those actions which are less expensive than energy supplies, while still providing the desired service. Since energy prices, external costs, conservation expenses and desired services vary, the actual values used in such analyses should be based on data from individual Member countries.

For conservation actions that involve changes in energy services, such as thermostat adjustments, the only true measure of cost-effectiveness is the value placed on the change by those affected. Because it is virtually impossible to determine independently such values, most assessments of the potential for further efficiency improvements exclude those actions which result in significant changes in services to energy users.

In examining the potential for energy efficiency improvements, it is also useful to review the broad range of actions that can be taken to conserve energy and some of the factors affecting their implementation.

Changes in Services: These types of actions, such as lower indoor temperatures in winter, are often responses to higher energy costs or to the introduction of individual metering for utility costs. Sometimes they result in permanent lifestyle adjustments, such as a shift to public transport. However, changes that are perceived as undesirable tend to be replaced by other types of efficiency improvements or simply phased out over a period of years.

Maintenance: Improved maintenance of energy-using systems can increase efficiency without affecting services. Because maintenance can be improved quickly and relatively inexpensively, it is often one of the first types of measures taken in response to higher energy costs. On the other hand, unless such practices are regularly pursued the efficiency gains can also be quickly lost.

Controls (and Monitoring): Manual and automatic controls help match the timing and rate of use of energy systems to consumers' needs. In this way, energy use can be significantly reduced for a relatively low cost and without affecting the desired service. Once an automatic control system is in place and working properly, the resulting efficiency gains are likely to be long-lasting. Such control systems are often supported by meters or other types of monitoring equipment. While such feedback systems do not directly affect energy use, they do encourage improved maintenance, control and investment efforts.

Discrete Conservation Investments: These are investments that are primarily directed at improving end-use efficiency, such as installing insulation in existing houses or heat recovery equipment on industrial processes. They can range from small, very cost-effective investments, to very large outlays which have comparatively low rates of return. Once installed, the resulting efficiency gains often last for many years. Decisions on such investments are often affected by short-term energy prices or other economic changes. If lower prices threaten the economic viability of the investment, it is likely to be rejected. On the other hand, such investments are usually not time-sensitive; that is, they can be reconsidered if prices stabilize or rise again.

Integrated Conservation Investments: These types of investments are made primarily for reasons other than conservation, such as replacing an old appliance or building a new home. The category includes the manufacture and purchase of new, energy-efficient automobiles and investments in new technologies, such as industrial processes, that are primarily designed to reduce total production costs or to achieve other, non-energy objectives. It also includes many technologies, such as thermal windows, that are sometimes implemented as discrete conservation investments, but are usually more cost-effective when implemented as part of a larger investment activity, such as major building repair or renovation. Efficiency improvements from this type of investment are less likely to be affected by short-term energy price changes than are discrete investments, described above.

Examples of each type of action can be found in all energy end-use sectors. Their relationship to overall efficiency improvements is depicted in Figure 1.

A full assessment of the remaining potential for energy efficiency improvements in Member countries of the IEA would be a truly enormous task — far beyond the scope of this study. It would require current and detailed data on the number and efficiency of energy-using devices, such as appliances, buildings, vehicles, industrial processes and other energy conversion systems. It would also require information on local energy prices, climate and the characteristics of existing stocks of energy systems in order to determine their suitability for retrofit. Such a detailed analysis has seldom been performed for individual sectors or localities, and even move rarely on a national or international basis. To extend such an assessment into the future would be even more difficult, requiring projections of numerous economic and energy conditions that are inherently uncertain.

A large number of more limited studies of the potential for further energy efficiency improvements have been made. These studies fall into two broad categories:

— Technology studies which analyse the efficiency of particular categories of existing end-use technologies;

— Sectoral studies which attempt to assess the broad potential for conservation in a major end-use sector or sometimes a country or region. These studies often rely upon technology-specific analyses.

The results of a range of technology and sectoral studies are summarised in Tables 5 and 6. They indicate the broad potential for economically attractive energy savings in most IEA countries, but they are not sufficiently comprehensive or precise to represent fully the many differences among Member countries. Because most of these studies were completed before 1986, they generally do not reflect the effects of the 1986 decline in world oil prices and they do not take into account the efficiency improvements achieved since their completion. However, for the following reasons these limitations are not expected to affect substantially the basic conclusions drawn from these studies:

— the high rate of return of most conservation actions means that even significant energy price decreases would not change their cost-effectiveness;

- the actual reductions in retail energy prices have been comparatively small (about 15% on average) for most end-use sectors and energy forms;
- long-term energy price expectations have not changed substantially;
- new technologies have emerged which were not fully considered in earlier potential estimates;
- the rate of efficiency changes over the past several years has not been sufficiently great to diminish significantly the potential for savings.

The data included in the tables, while neither comprehensive nor precise, do provide some general insights. First, the remaining potential for efficiency improvements seems to be especially large in the buildings sector. Second, the potential for savings remains large for all energy forms and sectors, and most energy uses or subsectors, regardless of region. However, the efficiency of some new industrial and transportation equipment approximates the efficiency of the best available technology, which suggests that the emergence of new technologies in these sectors may be key. Finally, while the remaining potential is large in all regions, it appears to be comparatively greater in those countries or regions that have traditionally been more energy-intensive, such as North America.

While these studies often differ in scope and assumptions about economic viability, they provide support for a conservative estimate that energy efficiency improvements of 30% could be realised over the next ten to twenty years if all economically viable conservation investments were made in IEA Member countries. The estimate does not assume the development of new technologies, nor does it include the further improvements that could be achieved through the complete replacement of major physical stocks (such as existing buildings), which will still be economically useful for much more than twenty years. Also the estimate is not an indicator of potential reductions in current energy demand levels. Assuming continued economic growth, there could still be an increase in TPER during this period, even if most of the existing potential for efficiency improvements were to be achieved. This is especially true for those countries with high rates of economic growth and for those energy forms experiencing larger increases in demand, such as electricity. However, a 30% increase in energy efficiency would,

Table 5

Energy-Efficient Technologies and the Economic Potential for Conservation[1]

Energy End Use/ Technology	(a) % of IEA TPER	(b) Existing Stock Average Efficiency (Units)	(c) New Stock Average Efficiency (% Savings)	Best Available Technology (d) Efficiency	(e) % Savings	(f) Average Useful Life of Technology	(g) Notes
RESIDENTIAL	20-25%						
- U.S. (All electric)		1 501 (Watts per capita)		328	- 78%	Over 30 years	
- Sweden (All electric)		1 242 (Watts per capita)		266	- 78%		
Heating and Cooling - Building shell thermal efficiency	8-12%					Over 30 years	
- U.S. (winter)		160 (KJ per m² per degree day)	100 (- 37%)	50	- 70%		
- Sweden (winter)		135 (KJ per m² per degree day)	65 (- 52%)	35	- 74%		
Heating - Oil/Gas - System Efficiency	8-12%					10-20 years	
- U.S.		65-70% (% of TPER converted to useful heat)	75-80% (- 13%)	84-94%	-23-26%		
Cooling - Central a/c -	1-2%					10-20 years	
- U.S.		7 (Energy Efficiency Rating)	9 (- 22%)	14	- 50%		
Refrigerators/freezers	2%					10-15 years	
- U.S.		1 500 (kWh/year)	1 300 (-13%)	750	- 50%		

1. Documented in Annex E.

Table 5

Energy-Efficient Technologies and the Economic Potential for Conservation[1] *(Continued)*

Energy End Use/ Technology	(a) % of IEA TPER	(b) Existing Stock Average Efficiency (Units)	(c) New Stock Average Efficiency (% Savings)	Best Available Technology (d) Efficiency	(e) % Savings	(f) Average Useful Life of Technology	(g) Notes
- Germany		About 400 (kWh/year)	(- 20%)		At least - 20%		
- Japan		35 (kWh/month)	28 (-20%)		At least - 20%		
Water Heating	3-5%					15 years	
- U.S.		4 000 (kWh/yr)	3 600	1 700	- 57%		
COMMERCIAL	15-20%						
Heating and Cooling	10-12%					30 + years	
- U.S.		1.31 (GJ per m²/yr)	0.73 (- 44%)	0.32	- 75%		
- Sweden		1.04	0.76 (- 27%)	0.25-0.46	-55-75%		
Large Office Buildings	5%						
- U.S.		270 (KBtu/ft²year)	200 (- 26%)	100	- 63%	30 + years	
Lighting	3-5%					1-10 years	
- U.S.							
Ballast/Tubes		64 (lumens/Watt)	73 (- 12%)	86	(- 26%)		
Controls					(-20-30%)		
Total					-40-50%		
TRANSPORTATION	20-25%						
Automobiles	10-13%					10 years	40 mpg may be technically possible at cost of less than $1/ gallon saved

1. Documented in Annex E.

Table 5

Energy-Efficient Technologies and the Economic Potential for Conservation[1] *(Continued)*

Energy End Use/ Technology	(a) % of IEA TPER	(b) Existing Stock Average Efficiency (Units)	(c) New Stock Average Efficiency (% Savings)	Best Available Technology (d) Efficiency	(e) % Savings	(f) Average Useful Life of Technology	(g) Notes
- U.S.		19.0 (miles per gallon)	26.1 (- 34%)	31.5	- 46%		
- Japan		[11] (km/1)	13 (- 15%)				
Other road transport	7-10%						
Air Transport	2-3%						
- All		[25] (passenger miles/gallon)	30 + (-20%)	40 +	- 40%	15-30 years	
Rail/Marine/Other	2-3%						
INDUSTRY	35-40%						
Chemicals	6-8%						
- U.K. (Inorganic)					- 13%		
Iron and Steel	5%						
- U.S./Japan/U.K./Neth.		22-24 (GJ/tonne)	17-18 (-20-25%)	N.A.	At least -20-25%	10-30 years	
Non-ferrous metals	3%						
- OECD (Aluminium)		15-17 (mWh/tonne)	13.5 (-10-20%)	N.A.	At least -10-20%	20-30 years	Technology being developed could reduce consumption 30-40%
Paper	3%						
- U.K. (Paper & Board Making)					- 30%		
Stone, Clay and Glass	2%						
- U.S./France/Switz./U.K. (Bricks/Pottery)		2.5 (MJ/kg)	1.5-2.0 (-20-40%)	N.A.	At least -20-40%	10-30 years	0.5 MJ/kg theoretical minimum

1. Documented in Annex E.

Table 5

Energy-Efficient Technologies and the Economic Potential for Conservation[1] *(Continued)*

Energy End Use/ Technology	(a) % of IEA TPER	(b) Existing Stock Average Efficiency (Units)	(c) New Stock Average Efficiency (% Savings)	Best Available Technology (d) Efficiency	(e) % Savings	(f) Average Useful Life of Technology	(g) Notes
- France/U.K./Switz./Germany (Cement)		3.6-3.8 (MJ/kg)	3.3 (-8-13%)	N.A.	At least -8-13%	10-30 years	1.5 & 3.0 MJ/kg theoretical minimums for dry and wet processes
Food	1%						
Space Heating, Cooling, Water, Heating, Lighting	2-3%						
ALL SECTORS							
Electric Motors	20%	75-90% (% converted to motive power)	80-92% (-2-7%)	85-93%	(-3-12%)	10-20 years	More efficient motors
					(-10-20%)		Variable speed controls
					-15-30%		Total Potential Electric Motor Savings
Central and On-Site Electricity Generation	35%						
- U.S. (Gas Turbines)	N.A.	30% (% converted to electricity)	35% (- 15%)	39-41%	- 25%		Best combined cycle and steam injected gas turbines Projected 1990 best: 44-48% efficient; total savings: 32-37%

1. Documented in Annex E.

Table 6[1]

Sectoral Studies of Conservation Potential

Country	(a) End-Use Sector	(b) Year of Study	(c) Estimated Economic Potential for Demand Reductions %	(d) Remaining Potential After Projected Effects of Market Forces %	(e) Year in which Economic Potential could be Achieved	(g) Notes
United States	Residential/Commercial	1981	- 50%		2000	
United States/Texas	Residential/Commercial	1986	- 50%		2000	Electricity only
United States	Residential/Commercial	1985	- 27%	- 17%	2000	
United States	Transportation	1985		- 30%	2010	
United States	Industry	1984	-35-40%		2000	
		1985		- 18%	2010	
United States	All Sectors	1983		- 22%	2000	
Canada	Residential	1986	- 33%	- 21%	2000	
	Commercial	1986	- 22%	- 24%	2000	
	Industrial	1986	- 15%	- 9%	2000	
	Transport	1986	- 20%	- 18%	2000	
	All	1986	- 22%	- 17%	2000	
Japan	All	1983	- 15%		1990	
		1984	- more than 20%		1995	
United Kingdom	Industry	1984	-21-29%		2000	United Kingdom study estimates that 21-25% efficiency gain is likely to occur. 29% estimate based on "technical potential"
Netherlands	Industry	1985	- 21%		2000	
	Residential	1985	- 21%			
	Commercial	1985	- 38%			

1. Documented in Annex E.

Table 6[1]

Sectoral Studies of Conservation Potential *(Continued)*

Country	(a) End-Use Sector	(b) Year of Study	(c) Estimated Economic Potential for Demand Reductions %	(d) Remaining Potential After Projected Effects of Market Forces %	(e) Year in which Economic Potential could be Achieved	(g) Notes
Austria	Industry	1981	- 10%			Heat pumps only
	Residential	1984	- 25%			
Norway	Industry	1984		- 12%		
	Residential/Commercial	1984		- 10%	2010	
Sweden	Residential	1985	- 50%		N.A.	
	Commercial	1985	- 40%		N.A.	
	Industry	1983		-20-50%	1990	
European Community	Residential/Commercial	1984	- 30%		1995	
	Small-/Medium-sized Industry		-10-20%		N.A.	
	Industry	1986	- 25%		2000	
Western Europe (EUR-9)	All Sectors	1983		- 19%	2000	
	Industry	1985	- 30%		2000	

1. Documented in Annex E.

if achieved, reduce future energy requirements by about 25% from the levels that would have been reached if no further efficiency improvements were made. This could be the equivalent of more than 1 200 Mtoe per year in 2000.

The potential for conservation is like the potential for additional energy supplies. With stable energy prices and current end-use technology, energy efficiency improvements are limited. But as energy prices fluctuate and new technologies are developed, the potential for cost-effective efficiency improvements changes, just as the economically recoverable reserves of energy change with prices and technology.

There are many reasons why there remains a large potential for economically attractive efficiency improvements. First, a major portion of the potential for energy conservation can only be realised through integrated investments that are dependent on the gradual replacement of the existing, less efficient stocks of energy-using equipment, vehicles, processes and buildings. It is often not technically possible or economically practical to modify existing equipment to achieve the full potential for energy savings, but major energy efficiency improvements have become integral parts of many new products.

Second, new technologies always require time to be introduced and even afterwards, they usually need many years to penetrate fully any market.

Third, there are large segments of key end-use sectors that have not yet taken significant conservation actions because of distorted energy prices or market limitations, such as lack of information or absence of direct user responsibility for energy costs (see Chapter V).

These characteristics of the market for conservation technologies help explain why improvements in the energy efficiency of whole economies are gradual and not spasmodic. They also help explain why trends towards improved energy efficiency are likely to continue in the future — even without higher energy prices or expanded government efforts. The following section examines likely future trends in energy efficiency and the extent to which they will achieve the existing potential for conservation.

B. **Future Trends in Energy Demand and Efficiency**

Many factors will determine the levels of energy demand and efficiency in IEA Member countries over the next twenty years. The two most important are likely to be economic growth and energy intensity. Future levels of energy intensity will, in turn, be determined by the rate of structural change in the economies of IEA Member countries and the rate of energy efficiency improvements. The most important factors influencing future energy efficiency levels will be:

— *Energy costs and availability* — reflected in energy prices and the ease of access to supplies;

— *Conservation technology costs and availability* — the other half of the equation which is assessed by investors in energy conservation. Recent trends have increased the availability and reduced the costs of conservation technologies;

— *Conservation services* — the availability of the technical and financial services most energy users require in order to take effective conservation action. They range from the availability of basic information on appliance efficiency to the level of knowledge about conservation technologies held by consulting engineers to the availability of contract energy management services for building owners;

— *Turnover of capital stocks* — the rate at which existing energy using systems are replaced by new, usually more efficient systems, determined primarily by the physical characteristics of the individual systems, but influenced by the rate of economic growth. Some examples include conventional incandescent light bulbs (often less than one year), automobiles (ten years), home appliances (ten to twenty years) and commercial buildings (more than thirty years); and

— *Economic growth* — which will affect the resources available for investment in more energy-efficient systems.

Energy and related government economic policies will also have very important effects on future energy demand, usually by influencing one of the factors cited above (see Chapters VII and VIII).

Based on an assessment of recent trends and the likely future direction of the above factors, a discussion of some probable future trends in energy demand and efficiency follows. After the sector summaries, there is an assessment of the opportunities for conservation which are not likely to be achieved based on the trends currently anticipated.

I. Industry

Over the next twenty years, industrial energy demand can be expected to grow, but in most IEA countries it will probably grow at a much slower rate than total GDP and slower than other end-use sectors. This is the result of both lower overall growth in industrial production (and faster growth in the service sector), and reduced energy intensity levels. Future reductions in industrial energy intensity are likely to be caused by continued shifts towards less energy-intensive industries (i.e. industries with lower energy to value added ratios) and basic improvements in industrial process technologies. The more widespread use of existing, more energy-efficient industrial processes and equipment will probably represent the largest share of industrial efficiency improvements over the next ten years. Discrete conservation investments, which are more sensitive to short-term energy price trends, probably will have a less significant impact than they did during the past ten years, but improved operational practices already introduced are likely to be sustained and even strengthened through the wider application of computer-based control technologies.

Improvements in energy efficiency are likely to continue to occur in large, energy-intensive industries which are also experiencing moderate to rapid growth. But such industries are not very common in IEA Member countries. While these industries have the motivation and the technical and financial resources to improve energy efficiency, industries which are smaller, less energy-intensive or expanding less rapidly are likely to make efficiency improvements more slowly. One indication of this can be drawn from a survey of European industry managers which was commissioned by the IEA Secretariat. The survey indicated that industries were likely to pay significantly less attention to efficiency improvements if energy represented less than 10% of the firm's manufacturing costs.

Individual countries or regions which have well developed industrial sectors may experience significant reductions in energy intensity as a result of structural shifts towards new, less energy-intensive industries. However, these countries may also find it difficult to stimulate conservation investments in those existing industries which are stable or declining. Countries with effective mechanisms for bolstering industrial energy management programmes and otherwise encouraging the accelerated adoption of new, more energy-efficient technologies may be able to overcome some of this natural resistance. Regions that experience high growth in more traditional, energy-intensive industries are likely to experience large efficiency gains, but are less likely to witness significant overall reductions in industrial energy intensity.

Short-term decreases in world oil prices, which have a greater effect on end-use prices for industry than for other end-use sectors, will reduce the rate of improvement in industrial energy efficiency. Lower oil prices are likely to reduce significantly the expected rate of return of many discrete conservation investments, but they are not likely to have a similar effect on established operational practices or on the adoption of efficiency improvements which are an integral part of new industrial technologies and processes. Companies with established energy management programmes are not likely to abandon them because of comparatively low short-term prices. These assessments have been partially confirmed by a recent IEA survey of the response of selected European industries to the sharp drop of world oil prices in 1986. The lower product prices that result from reduced energy costs are also likely to bolster demand for energy-intensive products, but again this effect is unlikely to change basic trends towards less energy-intensive industries.

II. Residential/Commercial

As a result of growing populations, increasing incomes and a rapidly expanding service sector, energy demand in residential/commercial buildings is likely to continue to rise but, as in the other end-use sectors, these increases will probably be moderated by improving energy efficiency.

a. *Residential Buildings*

The rate of growth in energy demand in the residential subsector will vary significantly among Member countries depending on three different factors:

— the rate of growth in the number of households;

- the rate of increase in a few key energy using characteristics, including conditioned space per capita, central heating and major appliances;
- the rate of increase in overall efficiency, but especially the rate of increase in the average efficiency of new energy-using equipment installed in both new and existing buildings.

Increases in personal income will have an important but limited effect on total residential energy demand unless the rate of household growth is high or the present penetration of major energy-using residential equipment is low. In countries where neither characteristic is present, increased personal expenditures are likely to go towards substantially less energy-intensive domestic services (such as home electronic equipment) or towards non-residential energy use, such as transportation. Although the market penetration limits of certain residential energy features (such as space, central heating, and clothes washers and dryers) are very difficult to predict, analyses of major appliance penetration rates suggest that likely increases in such features over the next ten to twenty years will be smaller for most IEA Member countries than was the case during the last twenty years.

Because of more energy-efficient residential heating equipment, major appliances and new building designs, the efficiency of the residential subsector will continue to improve in the future. The more energy-efficient designs already adopted by the housing construction industry and equipment manufacturers are likely to be maintained regardless of short-term changes in market signals. On the other hand, consumer behaviour, which can have a major effect on residential energy use, and the rate of discrete conservation investments, is likely to be significantly affected by energy prices.

A number of new, more energy-efficient technologies have been or will soon be introduced in the residential market, such as pulse combustion furnaces and compact fluorescent light bulbs. The success of these technologies will have important effects on long-run residential energy efficiency and demand.

The establishment of more effective public or private services able to assist residential building owners in selecting and implementing cost-effective conservation measures could be especially influential. The development of a more effective service industry is, however, likely to be hindered by the fragmentation, lack of sophistication, and limited resources of this sector. Energy services or contract management companies have begun to serve effectively portions of the multi-family housing sector, but it is unlikely this model could be extended to small residential buildings. For smaller buildings, a first step may be the introduction of more effective energy monitoring or feedback mechanisms which would enable and encourage consumers to exercise more control over their energy use. A very conventional, but still very effective form of feedback is individual meters for utility costs. Experience in Europe and the United States has indicated that the introduction of such individual metering usually reduces energy demand by 15-25%.

Current and future government actions will have major impacts in this subsector. Many governments of Member countries have traditionally played an important role in the residential subsector through the establishment of building construction standards and the ownership or support of lower income housing. In addition, government conservation efforts, such as information programmes and financial incentives, have been particularly numerous in the residential subsector.

b. *Commercial Subsector*

The commercial subsector, which includes all non-residential and non-industrial buildings, agriculture, the public sector and other miscellaneous energy uses, encompasses a broad range of structures, energy uses, and owner/occupant characteristics. Therefore, generalisations about future trends in energy demand and efficiency are difficult. There is a segment of the commercial subsector which is likely to respond to market forces in much the same way as major industries. This segment includes those large corporations and building management companies that have available skilled technical resources and the financing and motivation necessary to pursue vigorously improved energy efficiency. But this segment represents one-third or less of the total floor space and energy use of the subsector.

Another major segment of the subsector is composed of government-owned or institutional buildings. Some governments and institutions have had access to the required technical resources, but they often do not have easy access to the necessary financing (because of annual budget restrictions and a tendency not to borrow money for these types of efficiency investments). In many cases, the agencies or individuals undertaking conservation measures do not secure financial benefit from it, thus reducing the motivation towards improved efficiency.

A third segment is made up of medium- to small-sized businesses, for which energy usually makes up only a small percentage of their total costs and which often do not own the buildings where they are located. These businesses usually have neither the technical resources nor the motivation to make significant conservation improvements. This segment probably accounts for one-fourth to one-half of energy use in the commercial subsector.

As new buildings are constructed and existing energy-using systems replaced, efficiency improvements that have already been introduced will have a larger and larger impact. These trends have been reinforced in many IEA Member countries by the establishment of energy conservation construction standards for new commercial buildings. Such standards will have a particularly large effect because of the high growth rate of the commercial sector. The extent to which energy efficiency measures are adopted during major renovations of existing commercial buildings, which often occur once or more during the building's lifetime, will also have a major effect on the long-term trends in this subsector. But such renovations are often not covered by existing energy conservation building standards. For this subsector especially, government actions will be critical because government-owned or supported buildings represent a large fraction of the total sector. Most IEA Member governments have instituted and maintained conservation programmes for these buildings. For these reasons, efficiency is likely to continue to improve in this sector, even during an extended period of stable or declining energy prices. This trend might even be accelerated if energy service or contract energy management companies penetrated a large portion of the market.

III. **Transportation**

Transportation energy demand will increase over the next decade or two, partially reflecting major increases in the distance travelled by road vehicles and aircraft. The overall increases in demand, however, will be reduced considerably by improved energy efficiency. Long-term trends in the energy efficiency of transportation are expected to be primarily determined by trends in petroleum prices and the emergence of new technologies, but major efficiency improvements already incorporated into many new automobiles, planes and other transportation equipment will not have their full effect for at least another ten years.

As noted in Chapter III, the transportation sector almost entirely depends on petroleum products and is heavily dominated by road vehicles, especially automobiles. In response to steeply rising oil prices in the late 1970s, major efficiency improvements were introduced in new automobiles, such as increased engine and transmission efficiency, reduced vehicle weight and improved vehicle aerodynamics. These efficiency improvements are likely to be further increased by the widespread adoption of technologies that have already begun to be introduced in some new car models. These new car efficiency improvements will continue to improve the overall fleet fuel economy of many IEA countries until the mid to late 1990s. However, in some countries the recent improvements in new car efficiency have been largely offset by shifts to larger (and heavier) vehicles. In these countries, average fleet fuel economy should remain relatively stable. Improvements in new car fuel economy beyond the mid-1990s will be determined by a mix of different factors, including energy price trends at the time, the success of current research efforts and trends in automobile sizes.

The improvements in the fuel economy of new trucks and other service vehicles are not likely to be as large, but past improvements, nevertheless have been sufficient to have some of the same effects on average fleet fuel economy over the next ten to fifteen years.

Other transportation energy uses represent less than 20% of the sector's total energy demand and, for the most part, have not experienced the major efficiency improvements that have occurred in new automobiles. One significant exception is commercial aircraft, where new planes generally consume 20% less per passenger mile than the average for existing fleets.

The rate of growth in overall transportation energy demand and the rate of change in average automobile fuel economy are expected to continue to vary greatly among IEA countries. In most European countries, vehicle sizes, the number of vehicles per capita and the annual vehicle miles travelled have all been considerably lower than in the United States and Canada. With economic growth, these measures have tended to increase more rapidly in Europe than elsewhere, and this has been reflected in increasing energy demand. In addition, higher European gasoline prices and very competitive automobile markets have resulted in the earlier adoption of fuel efficiency technologies than in North America — so future increases in European automobile fuel efficiency may be smaller.

IV. **Transformation**

Unlike the other end-use sectors, major efficiency improvements have not occurred in the transformation sector over the past ten to fifteen years. This sector is also the one which is growing most rapidly and which is likely to continue to grow the quickest in the future.

The basic energy efficiency of the transformation sector is not likely to change substantially over the next ten to twenty years, although there are several technologies which will offer opportunities for significant improvements in certain areas. The most important existing technologies that could improve the efficiency of electricity generation are combined cycle and steam injected gas turbines, and combined heat and power (CHP). These technologies could reduce the total primary energy used by 20-35% in specific applications. In the case of gas turbines, they require the use of natural gas or oil and, in the case of CHP, the siting of facilities near industrial, commercial or residential users of heat energy. These characteristics will most certainly limit their use to only a portion of total generating capacity, but these more efficient generating technologies could still make a significantly greater contribution than they do today. Combined cycle or steam injected gas turbine technologies might be effectively used in areas where gas supplies are plentiful or where siting or environmental constraints make coal or nuclear power impossible. CHP could play a larger role in serving the electricity and heat requirements of industries and large commercial or residential complexes.

Even with wider use of more efficient conversion technologies, total energy demand by the transformation sector is expected to grow as a result of increasing electricity demand. The rate of increase in electricity demand over the next ten to twenty years is likely to be considerably less than it was during much of the past thirty years. This is true not only because of increasing efficiency in the use of electricity, but also because of changes in some of the underlying factors that have led to increased electricity use in the past (such as slow growth in the electric intensity of industry and buildings, and reductions in the relative price advantage of electricity, compared to oil and gas).

D. Summary of Efficiency Trends and Remaining Opportunities

The previous sections have described the current potential for energy efficiency improvements and the extent to which this potential is likely to be achieved in the future. There also has been an identification of end-uses where substantial potential for efficiency improvements exists, but based on current and anticipated trends seems unlikely to be fully achieved. Table 6, column (d) summarises some estimates of the economic potential for conservation that will remain, even after accounting for the efficiency improvements which are likely to result from changing energy prices, general productivity improvements and other market forces. Many of the reasons why these economically attractive efficiency improvements are not likely to be achieved are discussed in Chapter V. Ways to increase the achievement of these remaining opportunities are the focus of Part C.

In order to provide an overview of the likely remaining opportunities for efficiency improvements, it is useful to review current and expected trends in three broad areas:

— the extent to which existing, more energy-efficient technologies have already been incorporated into the design or manufacture of new energy-using equipment, vehicles, buildings and processes;

— the rate at which existing buildings and energy-using systems are being modified to utilise energy more efficiently;

— the extent to which energy-efficient operating practices, especially improved maintenance and control of energy-using systems, have already been adopted.

New Product Efficiency. Table 5 is not only a useful indicator of the current potential for energy efficiency improvements, but also shows the extent to which the most energy-efficient existing technologies are being utilised in new products, including industrial processes, electric motors, home appliances and buildings. Comparisons of columns (c) and (d) indicate that in many cases market forces and existing government efforts do not appear to be resulting in the adoption of the most efficient technologies. But in a few areas, such as new buildings and home appliances, the gap appears to be especially large. On the other hand, the gap does not appear to be very large for some technologies, such as new automobiles. While Table 5 is not a comprehensive review of all technologies or energy end uses, it does tend to support the general conclusion that the buildings sector is probably less responsive to market forces than the other sectors and that government efforts to date do not appear to have fully compensated for this. It also indicates that energy-intensive industries and the transportation sector in general appear to be generally more responsive to market forces and/or government efforts. However, gaps in the data available and the major differences among Member countries mean that country specific studies are required to provide truly useful information on the most important remaining opportunities for efficiency improvements.

Retrofit of Existing Facilities. Data on the rate of conservation investments in existing energy-using systems, buildings and other facilities are very incomplete. It is also difficult to identify the economic limits to the potential for efficiency improvements through retrofit alone. Table 5 provides a broad measure by indicating the difference between the best available technologies (column (d)) and the average efficiency of the existing stock (column (b)). Again, the gap is especially large in the buildings sector. But these differences overstate the savings that could be achieved through retrofit. The investments required to make existing facilities as efficient as new facilities are often not cost-effective. In some countries, data or less quantitative information are available on the rate of efficiency improvement in existing buildings. Existing rates of annual investment appear to be well below 5% of the total economically justified investment in energy efficiency. This means that based on current trends and technologies, it would take over twenty years to achieve most of the current potential. The rate of investment also appears to vary considerably from one portion of the building sector to another and, of course, from country to country. For example, those segments of the residential/commercial sector that have a high percentage of tenant occupied space, such as multi-family buildings, have tended to make conservation investments more slowly. So again, country-specific analyses are required. Retrofit of existing facilities is also important in the industrial sector. But for industry, the data on the rate of investment are even poorer. There have been some indications that major energy-intensive industries have made very substantial investments in existing facilities. However, the remaining opportunities for economically attractive investments also appear significant. More data would be necessary to estimate the rate at which the remaining investment opportunities are being achieved.

Improved Operating Practices. Efforts to improve the operating efficiency of energy-using systems through better maintenance and controls have been under way for many years. These types of actions are usually the most economically attractive and easiest to implement. This was reflected in the strong upsurge of such activities as one of the first responses to rising energy prices. Despite strong economic incentives, however, there are still many industries, commercial businesses, institutions and individuals that have yet to adopt such improved operating practices. This fact in itself is an indication that market forces and existing government efforts have not been fully effective.

It is not possible to estimate precisely the magnitude of the efficiency improvements which are not likely to be achieved as a result of market forces and existing government policies. But based on the estimates contained in Table 6, column (d) and other available information, it is reasonable to conclude that there would be economic potential to reduce further the projected energy demand in 2000 by more than 10% (at least 450 Mtoe per year) if existing opportunities for efficiency improvements were fully exploited. Some of the ways in which government might strengthen existing market forces or programmes in order to achieve these savings are addressed in Part C.

E. Conclusions

Despite substantial improvements in the general energy efficiency of the economies of IEA Member countries over the past ten years, there remains a large potential for economically justified efficiency improvements in virtually every region and sector, as well as in the use of every form of energy. Some energy efficiency improvements will continue to occur in virtually every end-use sector, even without higher energy prices. Efficiency improvements, which have already been made an integral part of many new appliances, equipment, vehicles, buildings, and industrial processes, will have increasing effects for ten to twenty years, even if energy prices are stable. Behavioural conservation actions and discrete conservation investments are expected to continue, but at a slower pace because many of the easiest conservation actions have already been implemented and because future energy price increases are likely to be more moderate than those experienced during the past ten years.

The analysis suggests that the potential for economic energy conservation measures is not likely to be fully realised or even approximated in the foreseeable future. The extent of the shortfall will vary between sectors and regions:

- the residential/commercial sector has the largest current potential for energy efficiency improvements and is also the sector likely to achieve the least of its potential based on anticipated trends;
- based on available technologies and current energy prices, some industries and types of transportation equipment may approximate the limit of cost-effective efficiency improvements in new stock within the next ten years (although efficiency improvements would continue as existing stocks are replaced);
- the potential for energy conservation, while large in all IEA regions, still appears to be greatest in those regions that have been traditionally the most energy-intensive, such as North America.

CHAPTER V

Market Limitations and Externalities

The analysis in Chapter III has shown that rising energy prices were the main impetus to the improved energy efficiency of the economies of the IEA countries between 1973 and 1985. In theory, market clearing energy prices in a perfect market would produce an optimal economic allocation of resources in the energy sector, including an appropriate effort on energy conservation. In practice there are a number of reasons why this does not happen. First, in many IEA countries different criteria appear to be applied for decisions on investment in energy conservation than those used in energy supply. Second, there are obstacles to economic pricing in the energy sector. Third, energy conservation has certain external benefits which in practice cannot be easily reflected in prices. Fourth, there are specific market limitations which prevent price signals from being reflected in the decisions of energy consumers.

I. Imbalances between Investment in Energy Conservation and Energy Supply

Investment in energy supply is much higher than in conservation. This is not surprising. The expenditure of relatively small amounts of capital on conservation can often bring big savings in energy costs. Many improvements in the efficiency of energy use are obtained from more general investments in new equipment, mainly for reasons other than energy conservation. Supply investments, on the other hand, are often part of large projects with long construction periods and long lives. Very large investments in energy supply, particularly electricity generation, will be needed if IEA countries are to meet demand and maintain energy security.

There is, however, also evidence that investment in energy savings today may be more cost-effective than investment in new energy production. In particular, Canadian work undertaken in 1984 attempted to define the real resource costs of energy conservation and energy supply alternatives. The results of this work are shown in Table 7.

Table 7

Comparative Costs of some Energy Conservation and Energy Supply Options

(1983 Canadian dollars, 7% Real Discount Rate)

Cost of Energy Saved by Various Energy Conservation Investments (Primary Energy)	$/BOE[1]	Value of Energy from Various Sources — Existing Supply	$/BOE[1]
New Housing:			
Super Energy-Efficient	$15-30	Electricity	$35-65
		Natural Gas	$15-35
Existing Housing:		(Short to Long term	
Reduce Average Consumption:		Export Prices)	
By 30%	$13		
By 40%	$28	International Crude Oil	$22
		at Montreal	
High Efficiency Gas Furnace			
(Compared with Moderate		**New Supply**	
Efficiency Gas Furnace)	$10-13		
		Oil Sands Plant	$35-60
Steam Pipe Insulation	$8-10	Offshore Oil	$30
		Nuclear Electricity	$60
High Efficiency Commercial		Offshore Natural Gas	$35
Lighting	$15-30		

1. BOE = Barrel of oil equivalent (energy equivalent of one barrel of oil): costs and values assumes a zero security of supply premium for oil.

Source: Energy, Mines and Resources, *Economics of Energy Conservation in Canada,* Ottawa, Winter 1984.

It is important to secure optimal allocation of resources so that in general the same criteria are applied to decisions on investments to increase energy efficiency or supply. There are grounds for thinking that, in practice, the system of decision-taking is different and that as a result conservation investment is judged against significantly stricter criteria than those for supply. This arises from the fact that investment in energy supply is central to the activities concerned, while investment in conservation is often seen as more peripheral. Thus when resources are tight, enterprises concentrate on their main activity.

Major supply projects are assessed with sophisticated techniques, such as discounted cash flow analysis. Available evidence suggests that most private companies in IEA countries require, before agreeing to such investment, nominal internal rates of return of 10 to 20% after tax, with 15% as a central value. Depending on inflation rates, this is equivalent to real rates of return of about 5 to 15%. Much large supply investment in the public sector requires real internal rates of return of between 5 and 10%. Proposals for investment in a peripheral activity like conservation, on the other hand are often judged against simpler and seemingly more stringent criteria. The payback periods for current conservation investments appear to run mainly between one and three years. In the early 1980s even shorter payback periods were often required (one to two years), but it is possible they may have lengthened with the improved economic climate in some regions. Payback periods of one to five years are equivalent to a nominal internal rate of return of close to 100% to about 15% — well above those for supply investments. Their use means that potential investors are undervaluing the stream of returns which continue after the end of the payback period until the end of the investment's life which in some cases is comparable to energy supply investment. This imbalance in favour of supply investment is reinforced by the fact that energy suppliers, particularly those which are government-owned, can often raise funds for investment more easily and at lower cost than many of those wishing to invest in energy demand.

The criteria applied by private home owners to conservation investment appear to be less stringent than those used by companies, probably involving payback periods between two and ten years. In some countries, such as the United States where the average house is sold every seven years, required payback periods are closer to the lower limit; thus many home owners are unwilling to invest with payback periods of five to seven years. The problem of raising finance, however, is probably more severe in the residential compared to the private industrial sector. Moreover, because of lack of sufficient knowledge of how to evaluate the economics of different kinds of energy investments, investment in the residential sector is often less effective than originally intended.

II. **The Obstacles to Economic Pricing of Energy**

At their meeting on 9th July 1985, the IEA Governing Board at Ministerial level agreed on the need for energy pricing policies which permit consumer prices to reflect world market prices. Where world markets do not exist, Ministers agreed that prices should reflect long-term costs of energy supply and which interfere as little as possible with the operation of market forces, particularly through direct controls or subsidies. However, there are difficulties in applying these principles.

First, prices are inevitably influenced by short-term market considerations and thus may not reflect the long-term outlook which is important for investment decisions. This may have more serious consequences for conservation than energy supply. The bulk of supply investment is undertaken by large energy organisations many of which have sophisticated forecasting techniques to map long-term trends in energy prices. On the other hand, decisions on conservation investment are taken by organisations which are generally small and less knowledgeable about energy trends, and by individual consumers.

Second, the energy sector has a strong tendency towards monopoly although this tendency has been mitigated by competition between different sources of energy. During the last decade energy markets have become much more competitive. This is particularly true of oil and coal where the current surplus capacity in recent years has led to tight competition and, in the case of oil, to important structural changes. However, for certain uses there is no alternative to electricity. In addition, while the organisation and regulation of the electricity industry in IEA Member countries are diverse, they generally have one thing in common — public utilities have a monopoly of the market. They control supply but must accept a measure of public regulation and an obligation of delivery to the consumer. Gas is in a middle position. Delivery to consumers is often in the hands of monopoly utilities subject to public regulation. The effects of this monopoly are reduced by the fact that gas competes with other forms of energy — oil, coal, district heat and electricity — for all end uses; also several gas suppliers are vying for the same market — the competition is intense in the United States but more limited elsewhere. Developments in a number of IEA countries show a tendency to introduce more competition in energy markets. But monopoly characteristics are likely to remain, particularly in electricity and gas distribution. The implications of this situation for energy conservation will depend on the way in which prices are set. Artificially high prices could promote conservation but could provoke a misallocation of resources. Cross subsidisation between groups of consumers on the other hand could work against conservation.

Third, there are severe practical problems about determining what economic energy prices are and how to implement them, particularly for electricity and gas, although to a lesser extent. These problems are discussed in Chapter VII.

Fourth, government and other public authorities are prone to intervene in the level of energy prices. This intervention has led to price distortions in many countries, including general attempts to hold down energy prices in the interests of other policies, such as the promotion of industrial development, the reduction of inflation or to help the poor. These attempts were particularly marked following the oil price increase of 1973-74. Despite improvements in recent years many of these policies are still in place. This subject is also discussed further in Chapter VII.

III. Externalities and Indirect Benefits and Costs

The market-place allocates resources on the basis of costs and benefits reflected in prices. It may not always take into account wider social costs and benefits to which economic activity gives rise. A number of such externalities are particularly relevant to energy conservation.

(a) *Energy and the Environment*

The more efficient use of energy on an economic basis is of primary importance for achieving the objectives of both energy and environmental policy. Means to promote the efficient use of energy are — with very few exceptions — free of environmental disadvantages and the environmental problems associated with energy production and consumption are therefore reduced if less energy is used.

This view was confirmed by the meeting of the IEA Governing Board at Ministerial level on 9th July 1985. The Ministers agreed to strengthen, as appropriate, their policies to promote the efficient use of energy by economic energy pricing, removing barriers to the effective operation of price signals through the market and adopting specific measures and programmes.

There are exceptions to the general rule that measures to conserve energy have no adverse effects on the environment. For example, residential conservation can be improved with more insulation. However, indoor air quality may worsen due to both the effects on indoor air quality of reduced air circulation in a "tight" building and emissions from the building construction itself (e.g. formaldehyde, glass fibres). However, this is mainly a design problem which can be overcome by improved ventilation. To an increasing extent, heat exchange in combination with ventilation has become standard in new constructions. Potential conflict also exists between controlling emissions and obtaining optimum energy efficiency in motor vehicles; however, with advancing technology, gains can be made on both fronts. Emissions of hydrocarbons and nitrogen oxides can be reduced without incurring penalties on fuel economy or engine performance. Basically, the same argument also holds for large boilers in industry and power plants and concerns continue to arise about burning waste to produce heat and/or electricity, particularly near urban areas.

An undertaking which pollutes the environment will not normally have to assume directly the costs of the clearing up that pollution causes to buildings, health or nature. Many undertakings now take into account wider social responsibilities but in every IEA country protection of the environment is necessarily dealt with by regulations or planning consent systems. These regulations often incorporate the "polluter pays principle" under which the cost of measures to limit environmental damage, for example by emission controls, falls on the polluter. That means that those costs will be internalised in prices and reflected in the decisions of energy consumers and producers. However, controls do not eliminate environmental damage. There will remain external costs which can be reduced through energy conservation measures which would not normally be accounted for in energy price systems.

(b) *Energy Security*

Heavy reliance on imported fuels from insecure or unstable sources results in external costs to national economies that are also not fully reflected in market prices. Such costs may be reflected directly in the costs of maintaining fuel stocks or indirectly in national expenditures related to defence, but they are most often only clearly visible when those imported supplies are interrupted or manipulated. The pursuit of energy conservation can reduce the burden which other energy policies (such as oil exploration, oil stock piling and the development of alternative energy sources) would otherwise assume to secure stable energy supplies.

(c) *Social Policy*

The inability of poor people to pay for energy to heat their homes adequately is a serious problem in a number of IEA countries. Programmes to insulate homes can help to reduce this problem and, in some circumstances, can also reduce government expenditure. For example in the United Kingdom, government expenditure on fuel subsidies to the poor have risen from £11 million in 1973 to £443 million in 1985 — a tenfold increase in real terms — and a further £2 billion is estimated to be spent on energy by households on supplementary benefit. On the other hand, government expenditure on capital account for the poorest households (20% to 25% of the total) was only about £15 million in 1985 for basic insulation measures. An increase in the latter expenditure, as well as helping its recipients, might significantly reduce government expenditure on revenue account. The United States Weatherization Programme is the only grant programme in the IEA which is specifically designed for low-income groups.

(d) *Employment*

To the extent that certain energy conservation activities and investments involve significant net employment creation, they can generate external social benefits not fully reflected in purely economic calculations. A number of studies in Member countries have examined the effects of energy conservation programmes on employment, and several have concluded that net gains could be expected. The findings of some of these studies are:

— In the Netherlands, official macro-economic analysis based on experience until 1983 showed that by continuing conservation subsidies alive through 1985-90, employment could be boosted (3 000 jobs per year net);

— In the United Kingdom, the employment potential of an energy conservation programme directed principally at reducing space heating has been assessed over a ten-year period for two levels of investment: £10 billion for the base case and £24.5 billion for the maximum case[1]. The annual average employment generated by the two programmes is estimated to be 50 000 for the base case (building up to 68 000) and 124 000 (building up to 155 000) for the maximum case. Of the total number of jobs created, two-thirds would be employed in installing or manufacturing energy conservation equipment and materials. The jobs created would be largely unskilled and semi-skilled, with a high proportion of the jobs occurring in areas of high unemployment. The negative employment impacts upon the energy supply industries would be small — 1 300 in the base case and 2 750 in the maximum case at the end of the period. It is estimated that the cost to the government per job created would be around £10 000, which compares favourably with the job creation cost resulting from other government policies (around £12 300 from the regional incentives policy);

— A study for the Commission of the European Communities (EC) has estimated[2] that a strong energy demand policy in the ten EC Member countries[3] may lead to an overall average yearly net employment increase of about 520 000 person years by the year 2000. Here, negative effects caused by the substitution of conventional energy technologies and by decreased energy production and distribution have already been taken into account. By including the effects of the alternative use of funds which would be made available through reduced energy costs, net effects on the economy were calculated. This estimate is based on an analysis of the effects on employment in four European countries of increased market penetration of four major energy conservation technologies as well as biogas plants and solar collectors. The results of this analysis were then extrapolated to cover the entire additional energy-savings potential which is technically feasible in the ten EC countries.

1. Association for the Conservation of Energy, *Jobs and Energy Conservation*, London, February 1983.
2. Commission of the European Communities, *Employment Effects of Energy Conservation Investments in EC Countries*, Brussels, November 1984.
3. This study was undertaken before Portugal and Spain joined the European Community.

Similar results have also been found in the United States[1] and Canada[2] in 1978. Several studies examined, to some extent, the effects of conservation on employment, after accounting for the lost jobs in supply industries. One of the reasons why there was a positive net effect is that the conservation investments considered were more cost-effective than increased energy supplies — thus they had positive effects on overall economic activity. Another reason is that conservation employment tends to be national or locally based, whereas a portion of the energy supplied is often imported. Actual employment effects will encompass a wide range of industries and vary depending on local circumstances. Further study is needed also on the employment effects of energy supply strategies. In comparing the net employment of demand strategies on a local and regional level, factors such as labour intensiveness of supply and demand investments, workforce skills and degree of subsidisation would have to be investigated. Such analysis can only be made on the basis of specific programmes in specific situations.

IV. **Specific Market Limitations**

Lack of information and technical skills

Adequate information about what should be done to improve energy efficiency and how to do it is essential and yet often lacking. This is true both for consumers and for those providing energy-using equipment and conservation services and equipment.

Each of the end-use sectors has particular needs for information and skills. Utilities and large energy-intensive firms are aware of their primary energy inputs and costs, and they have the information and skills necessary to minimise them. On the other hand, small- and medium-sized and even large companies in non-energy-intensive industries often lack sufficient information and probably cannot provide sufficient skills themselves to accomplish potential energy savings.

---

1. Douglas L. Dacy et al., *Employment Impacts of Achieving Energy Conservation Goals Pertaining to Automobile MPG Efficiency, Home Retrofitting and Industrial Energy Usage for the Period 1978-85,* Institute for Defense Analysis, May 1978.
2. David B. Brooks, *Economic Impacts of Low Energy Growth,* Economic Council of Canada, Discussion Paper 126, Ottawa, Canada, 1978.

The residential/commercial sector has greatest diversity due to many different building types, locations and uses; different heating/cooling, lighting, and cooking systems and requirements; and different types of users. Many building users or inhabitants know little about conservation techniques and most lack the skills necessary for anything but the simplest equipment installation. Information is largely lacking, confusing or inappropriate. Help comes from a variety of sources — neighbours, merchants, utilities, consumer associations and governments — but the advice can be conflicting[1]. When that happens, consumers will commonly opt to do little or nothing. Moreover, consumers frequently do not know what measures are likely to be most cost-effective. A United Kingdom study on investment in 1982-83[2] showed that obtainable rates of return in residential conservation fell between 5% and more than 100% in 1982-83, but that the least economic investments, such as double-glazing, were implemented on a much broader scale than highly cost-effective projects because of active market pressure for the higher-cost options. Service groups such as manufacturers, builders, insulation installers, furnace repairmen, architects, engineers, and equipment operators can lack sufficient knowledge of available energy-efficient equipment and energy management techniques including total systems approaches.

Invisibility of energy consumption and conservation

Most energy consuming systems and services do not measure energy consumption. Thus there is little feedback to encourage consumers to conserve. Individual appliances used by residential consumers — including heating systems — seldom monitor energy use and the central meter for the dwelling gives poor feedback. The bill for energy use is in many cases received weeks or even several months after the event. The results from the industrial energy bus programme (see Chapter VIII) have also shown that industrial consumers also have a very poor understanding of energy flows within their enterprises. The automobile is probably the most obvious exception since the consumer must regularly refill the tank and is thus more aware of consumption.

---

1. See Stern and Aronson, *Energy Use: The Human Dimension,* National Research Council, 1984, Chapter 3.
2. Evidence provided before Energy Committee of the House of Commons, 30th March 1983.

A related problem is that the results from a conservation measure may not always be obvious. Undertaking a conservation action such as retrofitting a house does not necessarily produce noticeable results in either energy savings or in energy costs since there are many variables such as climate or behavioural changes (e.g. increasing thermostat settings) or changing energy prices that can affect energy consumption or costs. Thus the consumer may not see significant changes in the energy bill.

Confidence

Since 1973 there has been a barrage of new conservation products and services trying to break into the market. Some, predictably, have not yielded the expected results in terms of costs, reliability and energy savings, and a credibility problem has developed. An example is the use of urea formaldehyde insulation in housing which was later banned in some countries because of health and performance problems. With a rapidly expanding market, many new service companies have also come into existence. While most of the industry has matured in a normal fashion (in fact, part of it was well established before the 1970s), there have been examples of fraud, misleading advertising and substandard performance which have affected consumers. Industry has tried to regulate itself through trade associations. The United States' National Association of Energy Service Companies (NAESCo), for one, has been very active in enhancing the credibility of the new third party financing industry.

Separation of expenditure and benefit

A person who uses energy often does not pay directly for his consumption. In other cases, the person who uses and pays for the energy does not own — and therefore is unwilling or unable to invest in — the building, vehicle or energy-using equipment involved. The best known examples of these problems are in rental accommodation where those who pay for a conservation measure often do not receive the full benefit. Many rental agreements stipulate that the property investment is the responsibility of the landlord whereas the benefits derived from conservation investment go to the tenant who pays the energy bill. In theory the landlord can recover his investment through higher rent and an increase in the capital value of his property. In practice this is often difficult particularly in those countries where there are rent controls. In other cases the landlord pays some or all of the energy bills and the total cost is part of the total rent. There is no incentive for the tenant to conserve energy, although the landlord can reduce operating costs by undertaking conservation improvements.

In most Member countries this separation of expenditure and benefit also arises in the public sector. In the United Kingdom, for example, a public hospital saving energy through investment may find its funding for running costs reduced. The same holds true in Spain. This was also true for the health sector and schools in the Netherlands but this has recently been reversed and replaced by a financing system which allows for energy conservation investment for hospitals or according to the type of school.

Other, perhaps less important, aspects of the separation of expenditure and benefit include the billing of costs for space heating and/or tap water according to non-consumption related parameters (e.g. apartment size). To overcome this some governments have required individual metering depending on actual energy consumption (Austria, Germany). Another example is the issue of the company car. In some countries, particularly Sweden and the United Kingdom, there is an increased use of company cars for personal use. This practice discourages energy-efficient driving habits and regular maintenance, and almost invariably lead to the purchase of vehicles larger than the individual would normally own, and to increased total mileage.

Access to Capital

There is in general no shortage of capital in IEA countries. However, in some cases consumers may have difficulties in raising funds for investment in energy conservation especially when investment has to be made in discrete blocks. For example:

— in the residential sector there are financing problems with low-income groups and the elderly, as well as some multi-unit building owners; in some countries, building societies, and other sources of finance for the purchase and improvement of houses may be reluctant to advance money for energy conservation investments;

— in the industrial and commercial sectors small- and even medium-sized concerns may lack capital reserves and depending on their general financial situation may find it difficult to raise outside finance on terms which they can afford.

Obstacles to Technology Development

Market limitations which tend to slow the development and introduction of more energy-efficient technologies include:

- *Industrial fragmentation.* Many industries which are primarily concerned with the manufacture of energy-using technologies are so fragmented that no individual company can afford to support a significant research, development and demonstration effort. This is especially true in the building design and construction industries. Fragmentation also often means that individual companies find it especially difficult to capture the economic rewards of introducing more efficient products. This further reduces the incentive for innovation.
- *Long-term or high risk technologies.* With respect to a few conservation technologies, the type of research required involves commitments of resources for many years with a comparatively high risk of failure. Private companies of almost any kind are often unwilling to support such research. A recent Canadian study showed that energy conservation suppliers spent an average of 6% of sales on R&D, but the majority spent less than 2% of sales. Private support is further limited when the technology being advanced is so general that it would be difficult for the developer to capture the economic rewards of the research, even if it were successful.

Conclusions

Appropriate energy prices set the tone and provide the broad signals for conservation measures to be undertaken. There are, however, a number of factors which prevent the market for energy conservation from working with full efficiency. They include:

- different systems and criteria for taking investment decisions which would result in a bias in favour of marginal investment in energy supply rather than conservation;
- major difficulties for the energy sector to set prices at the right economic level;
- the existence of external costs in energy use and external benefits in energy conservation which are not and cannot be fully reflected in prices; they include environmental, security and social costs and benefits;
- specific market limitations such as the lack of the information or skills necessary to conserve energy, the invisibility of energy use, costs and conservation, a lack of experience with or confidence in new conservation products, the separation of responsibilities for energy expenditures and conservation actions, and a lack of access to capital by many energy users.

As a whole, these market limitations represent a formidable obstacle to future progress towards improved energy efficiency. Part C addresses how best to overcome these barriers to energy conservation.

PART C

CONSERVATION POLICIES AND THEIR EFFECTIVENESS

The preceding chapters have shown that there is substantial scope for further energy conservation on an economic basis, but that there are obstacles to achieving this potential which will not be overcome fully under present policies. The key problem is to develop policies to overcome those obstacles which are suitable for market economies.

Achievement of the potential for energy conservation depends on the combined efforts of consumers, conservation service industries, energy supply companies and all levels of government to bring about:

— better energy management;

— increased investment in proven conservation technologies;

— development and demonstration of new technologies, particularly those which are close to the commercial stage.

National governments have the task of establishing a policy framework which will encourage all concerned to strengthen their efforts to achieve energy conservation. Government policies need to be based on thorough analysis of where the main potential for energy conservation is, to identify the obstacles to achieving this potential and to examine how these obstacles can be overcome. Their implementation requires the effective involvement of organisations outside central government which can work in energy conservation. The policies themselves fall into four main areas:

- energy price and taxation policies to provide the correct economic signals necessary for the optimal allocation of resources;
- policies and programmes to remove or counterbalance other market limitations;
- research, development and demonstration to spread existing relatively unproven technologies and to develop new technologies;
- an example by government through the promotion of energy conservation in the establishments under their direct control.

The remaining chapters examine the organisation of energy conservation activities and then each of the four policy areas. An effective government conservation strategy will need to be broad and include a series of elements drawn from all of them.

CHAPTER VI

Organisation of Energy Conservation Activities

Many different individuals, businesses and other organisations are involved in energy conservation activities. They are far more numerous and disaggregated than those involved in energy production. The organisations provide equipment, services, advice, motivation or incentives to consumers. Without them, the consumer undoubtedly would not undertake conservation actions as confidently or as carefully. These organisations — whether private or public — must be effectively involved in any government effort to improve energy efficiency. For this reason, this chapter describes some of the characteristics and background of these organisations, and how they have already been involved by governments.

I. **Energy Conservation Service Industries**

Many types of professionals and companies are involved in providing energy conservation services, often as part of wider operations. They include insulation manufacturers and installers, energy auditors, consultants, architects, boiler retrofitters, heat pump manufacturers, financing companies and plant builders. Some of these companies have existed for a long time although they may have had to learn new techniques. Others have only developed since the mid-1970s and are now maturing. Because they have direct contact with consumers and can have a positive effect in getting consumers to take energy conservation actions, many governments have supported or encouraged their activities.

Many existing trade associations in these industries promote energy conservation. In addition, new associations specifically related to energy conservation have been formed in recent years. For example, in the United Kingdom the Association for the Conservation of Energy was formed in 1981 to represent 15 conservation service companies and the Energy System Trade Association was formed to establish and maintain high professional standards and to provide a focal point for energy users seeking information on energy efficiency goods and services. In the United States the National Association of Energy Service Companies (NAESCo) was formed in 1983 to represent "both corporate and public sector organisations concerned with developing, building, financing and managing third party financial energy projects". In early 1986 a European Federation of Energy Management Associations, representing 14 national energy management organisations in ten EC countries, was formed. The federation will encourage exchange of information and experience in Europe.

A recent innovation has been the introduction of energy contract management companies which manage energy use in enterprises by providing technical, managerial and financial resources required for a retrofit project. They receive a return on their investment from the achieved savings. These companies have become prominent in the United States in large part because of the favourable tax climate. The United States has recently enacted changes to federal procurement policies allowing government agencies to enter into multi-year contracts with energy service companies. There has been less progress in other countries, but similar companies are now beginning to develop in Canada and in Europe, where the EC is seeking to encourage their development. Their success will, however, be much influenced by tax practices, which vary among Member countries.

II. **Energy Supply Industries**

The energy supply industries in some Member countries play an important role in providing energy conservation services through their direct dealings with consumers and their marketing and technical skills. Conservation activities are sometimes undertaken to improve the commercial outlook particularly with respect to the need and cost of new capacity. Sometimes conservation activities occur under government pressure or legislation. Most of the time, energy supply industries provide conservation services in response to both factors.

The gas and electric utilities in the United States have by far the most comprehensive conservation programmes in the IEA. As discussed in a recent IEA study, these programmes have developed largely owing to "special difficulties and financial burdens faced ... in providing new capacity, in particular where the costs of new capacity are much higher than historical costs" [1]. The main federally mandated programme was the Residential Conservation Service (RCS) which required utilities to provide a variety of conservation services including energy audits to residential consumers. The federal law was repealed in August 1986 although many states still require such programmes. Large utilities have now been given a federal mandate to provide an audit service for small-sized commercial businesses and apartment buildings. Each of the states also has a regulatory body for utilities which also puts forward its own requirements.

A great emphasis now is being placed on an integration of efficiency and supply options into the utility planning process. For many this is considered the least-cost strategy. About 20% of all utilities in the United States now have some form of least-cost strategy and many more are considering it. Generally these strategies are being introduced through requirements of state public utility commissions. The federally-owned Bonneville Power Administration (BPA) was one of the first utilities to treat conservation as a resource similar to traditional forms of supply. BPA, for example, spends approximately $100-120 million a year on conservation activities. Many state regulatory bodies also are requiring utilities to develop distinct conservation strategies. For example, utilities in Texas are compelled to examine the role of conservation in meeting future demand through cost-effective rather than information programmes [2]. Some utilities are undertaking conservation programmes because of load management concerns.

Many other energy supply companies in Member countries have active conservation programmes. A few of the more illustrative examples are:

— the Italian Government uses its state-owned supply companies — the National Commission for Atomic and Alternative Energy Sources (ENEA), the Electricity Board (ENEL) and the Oil and Gas State Holding Company (ENI) — to implement many of the national conservation programmes. ENEA is responsible for

1. IEA, *Electricity in IEA Countries — Issues and Outlook,* Paris, 1985, p.52.
2. See Association for the Conservation of Energy, *Lessons from America: The Regulation of Gas and Electric Utilities in the USA,* London, February 1986.

training, information and advisory services for the government. These state-owned organisations have virtually taken over the government role of policy making;

— The Norwegian Government believes that the electric utilities should play a central role in encouraging energy conservation. For example, the Oslo Electricity Works has a target of 15% saving in energy consumption by the year 2000 relative to 1980. This is being achieved by a combination of information programmes, grants and loans;

— the United Kingdom Government has encouraged energy supply industries to provide conservation services. The government has solicited the support of all energy supply companies, among others, for a co-ordinated awareness campaign. For many years British Gas has had a conservation programme which includes training, information, awards and RD&D. The electricity supply industry has been active by providing information, encouraging manufacturers to develop more energy-efficient appliances and encouraging customers to use cheaper off-peak electricity. Oil companies such as Shell and British Petroleum have formed energy contract management subsidiaries.

The role of the supply companies can best be decided in the light of specific national or local circumstances. Their involvement in energy conservation activities following those developed in the United States, particularly the requirement to consider measures to reduce demand as an alternative to increasing supply, however, could be worthwhile in other Member countries.

III. Non-governmental, Non-profit Activities

Many non-governmental groups are trying to encourage improvements in energy efficiency with varying degrees of support from government. First, there are special interest organisations concerned with energy conservation that both provide services and act as pressure groups on government. A good example is the Alliance to Save Energy in the United States, which is a private, non-profit coalition of business representatives dedicated to increasing the efficiency of energy use through research, demonstration projects, public education and policy advocacy. There are also groups which have broader interests than energy: consumer groups, service clubs, building societies, non-profit companies and local initiative groups, and environmental and ecological groups. These groups offer a variety of services including information, labour to weatherize the homes of the poor and elderly, help to people applying for available government incentives, testing of appliances and even third party financing for multi-family dwellings. For example, the Citizens Conservation Corporation in Boston, Massachusetts, provides a third party financing service using funding from a variety of sources including petroleum overcharge funds, the Massachusetts Housing Finance Authority and the Federal Department of Housing and Urban Development. In the first three years, their projects covered 2 500 apartments with energy savings estimated at $3 million. In Germany, the Foundation for Comparative Testing has carried out tests on household appliances, automobile accessories and heating systems since 1978. In the United Kingdom a network of over 250 local groups has been brought together in Neighbourhood Energy Action. In addition, there are voluntary industry organisations which have been formed to share information about conservation opportunities and monitor progress within industry subsectors. The Canadian Industrial Energy Conservation Task Forces were established through a voluntary arrangement between government and industry in 1975-76 for such purposes. They set targets, monitor progress and share technical information. The Task Forces also provide an annual report to the government.

Many of these groups receive financial and technical support from governments. For example, the State of Victoria in Australia provides funds to local groups for energy-related projects such as public education, public workshops or conferences, preparation and production of submissions to government on energy issues, or planning of community projects. German federal and Länder governments provide funding for advisory services and comparative testing (e.g. for appliances and heating systems) by the Association of Consumers and the German Housewives' Association in approximately 170 towns and cities. Neighbourhood action groups in Britain receive support from the Department of Health and Social Security to help insulate the homes of the poor; they also are funded by the Manpower Services Commission to provide training for the unemployed; and the Department of Energy helps finance administrative expenses. In this way the objectives of manpower, social and energy policy are all advanced.

IV. **Government Organisations**

Governments at all levels are involved in bringing about energy conservation. Their roles and responsibilities depend on the constitutional framework. The sharing of responsibilities is generally the most complex in federal systems. In some cases the federal government has very little authority in the area of energy conservation (for example, Switzerland). In other countries the roles of central, state or local government may overlap.

Central government organisation for energy conservation has to strike a balance among conflicting requirements. A strong central organisation can ensure that conservation activities are vigorously pursued and properly co-ordinated across the whole range of government, not just as part of energy policy but also as part of other policies, such as housing. However, there is a danger that over-centralisation of responsibility may reduce the commitment of those responsible for implementing conservation measures in sectorally-oriented agencies.

Energy conservation was a relatively new field for most governments in 1973. Energy ministries were supply-oriented, and co-ordination of conservation activities across the whole range of government was lacking. Since 1973 and particularly since 1979, it has become increasingly accepted that the strategic co-ordination of government-wide conservation activities is important. For instance, after the 1982 Rayner Scrutiny in the United Kingdom[1], the British Government established the Energy Efficiency Office within the Department of Energy to increase visibility and strategic co-ordination, but did not vest the Office with the responsibility for implementing all programmes. Some countries have set up organisations for co-ordination which are independent of the normal departmental structure and which have more flexibility in developing and implementing a conservation strategy: France's Agence Française pour la Maîtrise de l'Energie (AFME) is the best known example. Austria has an independent Energy Efficiency Agency controlled by a committee headed by the Federal Chancellor and financed largely by public funds. This agency has a more limited role than the AFME and is mainly responsible for information, motivation and education in the area of energy conservation. Other bodies have less policy influence. The Netherlands has three separate agencies. SVEN

1. E.G. Finer, *Rayner Scrutiny: How The Government Handles Energy Conservation,* London: Department of Energy, September 1982.

(the Dutch Energy Conservation Information Agency) is the central agency for information on energy conservation which provides information mainly for businesses or business organisations. Non-government groups are represented on the management committee of SVEN. There is also the Dutch Energy Development Company (NEOM) for demonstration and introduction of new techniques and the Management Office for Energy Research (PEO). Denmark has the Energy Conservation Committee which undertakes information campaigns and conducts an annual survey on consumer attitudes. Japan's Energy Conservation Centre is a quasi-autonomous institution outside government, established and maintained by the major energy consuming companies. It supplements government activities through general research, information, advice and other activities.

Inter-departmental co-ordination is not easy. Many governments have experienced difficulty undertaking or re-orienting initiatives when two or more ministries are involved. The Netherlands has had a special inter-departmental steering committee for energy conservation since 1971. The question whether to centralise is an ongoing concern. Canada found that it had to centralise full responsibility for its Canadian Home Insulation Programme (CHIP) under the Department of Energy, Mines and Resources because the housing agency initially responsible did not sufficiently emphasize energy priorities. Consideration is now being given to de-centralising other federal programmes. In the United Kingdom, the Rayner Scrutiny had recommended that the Home Insulation Scheme (HIS) be given to the Department of Energy for similar reasons, but it was decided that the department responsible for housing should keep the programme as HIS was considered an integral part of the home improvement policy as a whole.

Conservation measures cannot be implemented at the national level alone. Implementation must be carried through to regions and localities. The most appropriate level of government to implement programmes depends on national circumstances but the closer implementation lies to consumers, the more effective it usually is. Many energy departments have established regional organisations. These offices can implement programmes directly with consumers (Canada and the United Kingdom), co-ordinate national programmes which are managed at the state level (United States) or provide advisory services (Switzerland). Local or state governments often administer national programmes.

There are particular problems in federal states. In Switzerland, for example, most of the responsibility for developing and implementing policies rests with the Cantons. The recent effort to combat environmental damage by stronger energy conservation policies has required effective co-ordination and co-operation between the Federal Government and the Cantons. In many federal states, the division of responsibility is not well articulated and can lead to duplication of services. With the change in the federal government in Canada in 1984, federal-provincial negotiations have led to a sharing of responsibilities between the two levels. In the United States, many states have also created autonomous programmes; public utility commissions which are under state authority have in many cases been instrumental in requiring utilities to develop conservation strategies.

Conservation programmes and strategies ultimately depend on political will. The personal commitment of Ministers can overcome many obstacles, as has been shown recently in the United Kingdom where the Secretary of State for Energy is personally leading the current conservation drive. Political leadership needs to be supported by a strong bureaucratic organisation, high quality, committed staff and an official as high-ranking as those concerned with energy supply to ensure that conservation receives as much weight as supply considerations in the formulation of energy policy.

Parliaments can play an important part in giving a political impetus to energy conservation. In the United Kingdom persistent criticism by the House of Commons Select Committee on Energy was a major factor in bringing about the establishment of the Energy Efficiency Office. The Select Committee has subsequently maintained its constructive criticism of various aspects of the Government's conservation programmes. In the United States, Congress has consistently provided greater funding for conservation activities than has been requested by the Executive Branch.

Conclusions

Success in energy conservation depends on bringing together and motivating a large number of groups and companies which are often not primarily concerned with conservation. The way in which this can best be done will vary among countries according to constitutional structures, government organisations and traditional energy conservation activities. But in all this chapter suggests:

(a) The development of energy conservation services including energy contract management companies will be increasingly important as conservation requirements and opportunities become more complex. Responsibility for this development rests mainly with the interests concerned but experience in a number of countries has shown that governments can encourage these interests to come together and that information programmes plus modest financial help can play an important part in the process.

(b) Energy supply industries can play an important part in promoting energy conservation through direct contacts with consumers and understanding of consumer needs. In some cases the industries have been ready to assume this role for their own commercial reasons but in others government encouragement, even legislation has been necessary, particularly for gas and electric utilities which are already under some form of regulation.

(c) Non-profit groups can be effective in promoting energy conservation, particularly through their understanding of the needs of specific communities and their ability to link energy conservation to other local needs, such as employment. Small injections of public funds can much assist such activities.

(d) In governments, there is a need for a strong central conservation policy group headed by a senior official who forms part of the top management of the Energy Department or related group; there should also be effective inter-departmental co-ordination of conservation activities. Strong political leadership and bureaucractic commitment is, however, the key to the success of government conservation activities.

Much progress has been made since 1973 in improving the organisation of conservation activities, but it has been uneven and in no case is it wholly satisfactory. It is important that all Member governments should re-examine energy conservation arrangements in their countries with a view to applying the organisational lessons learnt by others.

CHAPTER VII

Energy Pricing and Taxation Policies

Chapter III has shown the changes in energy prices which have occurred since 1973, and the response of energy demand to these changes. Chapter V has shown that there are a number of obstacles to economic energy pricing and to the reflection of external costs and benefits in energy prices. However, IEA governments agree that sound energy pricing and taxation policies are the basis for efficient promotion of energy conservation and further reduction of oil import dependence. This chapter considers what should be the content of such policies, focussing on energy demand and conservation rather than on energy supply.

A. Elements of Effective Energy Pricing and Taxation Policies

Only limited progress towards establishing a common international basis for developing appropriate energy pricing and taxation policies has been achieved in the IEA. In 1981 IEA Ministers discussed proposals for the economic pricing of energy and agreed on the need to carry forward earlier IEA decisions regarding energy pricing policies. These discussions demonstrated the importance of developing a consistent approach to consumer energy prices, and reflected widespread views that as an objective this should centre on the following elements:

(i) where world markets exist, not only for oil, consumer prices should reflect the world market price;

(ii) where world markets do not exist, consumer prices should normally reflect the cost of maintaining supply in the long term;

(iii) subsidies of consumer prices and other interventions which discourage conservation, high levels of domestic production and substitution away from oil should be avoided, and a thriving energy trade should be developed;

(iv) electricity tariffs should not prevent utilities from raising the revenues necessary to provide capacity to meet future requirements;

(v) in considering tax policies, proper weight should be given to energy policy objectives;

(vi) energy prices should be characterised by transparency so that consumers and producers can make economically efficient decisions.

National policies in this area are strongly influenced by the nature of specific fuel markets and by different energy situations. In addition, energy pricing and taxation policies are generally influenced by much broader considerations than those of energy policy alone. Broader concerns such as the short-run impact of energy prices on inflation, employment, growth prospects, international competitiveness and equity considerations must be taken into account. Energy prices must also take account of the need to internalise certain externalities, such as the general costs of energy production and use to society. This has become increasingly important with the rising concerns about environmental protection.

B. Oil Prices

Substantial progress has been made in relating the price of oil to consumers to the world price level. In many countries oil price controls and subsidies have been removed, and this has had a major impact on improving the energy position of these countries. Examples include the deregulation of oil prices in the United States and Sweden in the early 1980s, the raising of Australian crude oil prices to world levels in the late 1970s, and the deregulation of Canadian crude oil prices in 1985.

However, a number of instances still exist where consumer oil prices are distorted by government intervention, through price controls or subsidies. For instance:

— in Greece, the current level of oil product prices does not reflect fully the cost of imported crude oil;

— in Italy, despite considerable progress in liberalising price controls, there is still substantial government intervention in energy pricing;

— in Portugal, the government still controls energy prices but most of the subsidies on oil products were removed at the beginning of 1986.

There are wider policy reasons for these arrangements but from the point of view of energy conservation it is desirable that they should be reduced or eliminated.

C. Coal Prices

There are few if any instances of government policies which deliberately hold the price of coal below the world level. Where departures from the world level exist, they arise rather from the need in certain countries to retain high cost domestic coal production for social, regional and security reasons. Coal prices in Germany and the United Kingdom, and to a lesser extent Japan, are influenced by heavy subsidies and limitations on coal imports. For example, in Germany and the United Kingdom electric utilities purchase indigenous coal at prices higher than the world price partially to cover higher costs of production. There are arguments for and against these arrangements but, in general, they do not work against conserving energy.

D. Electricity and Gas Prices

(a) *Level of Tariffs*

Chapter V has pointed out that there are particular difficulties about determining and implementing economic prices for electricity and, to a lesser extent, gas. There is only limited agreement among IEA countries about the proper basis for determining prices in these cases. In many IEA countries prices for electricity are based on some form of historical costs. For gas prices the most commonly used basis, particularly in Europe, is to link prices to those of competing fuels notably oil products.

In theory, the right basis for determining prices is long-run marginal costs. In a minority of IEA countries, these are used to provide an approximate basis for prices, although the results may be adjusted, for example, to maintain consumer satisfaction or to meet revenue or funding requirements. There are, however, serious practical problems about determining long-run marginal costs. In other IEA countries, the fact that prices based on long-run marginal costs may lead to substantial profits or losses on a historic accounting basis is seen as a decisive objection to this approach. Prices are based on some form of accounting costs with adjustments for reasons similar to those applied to long-run marginal cost pricing. Prices so derived will not necessarily give the right long-term signals to consumers. Particular problems for the development of consistent national energy pricing policies exist in some countries with a federal organisation, especially Australia, Canada and the United States, where legislative and administrative responsibility regarding energy prices is divided between national and state levels.

Some governments have experimented with policies of fixed price relativities, generally to encourage substitution from oil to more abundant (often domestic) energy resources. Examples include Australian liquefied petroleum gas (LPG) policy, Denmark's natural gas pricing policy, New Zealand's compressed natural gas (CNG) policy and Swedish energy taxation. However, it is rare that prices based on arbitrary relativities can be maintained in the face of unexpected and often contrary movements in cost.

These difficulties in determining the appropriate level of electricity and gas prices have been complicated by governments and tariff regulators who have taken other factors into account such as anti-inflation policy, or the perceived needs of social equity.

(b) *Tariff Structures*

The structure and the level of tariffs for electricity and gas are important for energy conservation. Tariff structures are usually designed in two parts to reflect the capital costs (the "capacity charge") and the fuel costs (the "energy charge") of making supply available to a consumer. Tariffs in two parts like this are a useful way of passing separate price signals for the "lumpy" element of costs (investment in new capacity to meet expected peak needs) and for the more continuous energy element. From the point of view of energy supply investments, such tariffs can offer a rational approach to cost recovery. For the largest gas and electricity consumers it is possible, using complex metering, to charge separately for the capacity and the energy elements in the tariff. In practice, more approximate tariffs are designed for medium-sized and smaller consumers in the domestic, commercial and industrial sectors using simpler meters, based on two-part or multi-part tariffs. These tariffs are usually based on the time-profiles (load curve) of consumption for average consumers in each class. Typically, the average price of electricity or gas falls as consumption rises since the capacity used is spread among a greater number of units consumed. This aspect may act as a disincentive to energy conservation. Attempts have been made to modify electricity tariffs so that additional tranches of energy are priced higher than the base tranche according to social requirements such as certain amounts of heating or lighting. A conflict can arise then between the design of tariffs to reflect the costs of supply versus the promotion of energy conservation. The progressive tariff of Japan was introduced to reflect the progressivity of supply costs and they can contribute, as a result, to promoting energy conservation.

(c) *Transparency*

Another matter of concern, particularly in the area of electricity and gas tariffs, is the transparency of energy prices to consumers. Behind this broad concept is the aim of securing that price variations among classes of customers are based on the differences in service costs and that the utilities justify differences in a generally understandable way. In the case of electricity and especially gas, the cost of providing distribution and storage to supply residential consumers' seasonal heating and time of day cooking demands is much higher than the cost of supplying a large industrial user with stable annual or at least daily demand.

Therefore, it can be to the advantage of the residential customer, for example, if utilities can maintain a certain load of industrial customers, because this will reduce the cost of serving the residential customers' needs. For gas, this will remain the case even if prices to large (industrial or electric utility) customers have to be very much lower than those to residential customers in order to compete effectively against heavy fuel oil, and tariffs need to be flexible and capable of diverging enough to accommodate changes in competing fuel prices.

As a kind of check, utilities are often required to publish tariffs. With a few exceptions, complete electricity tariff schedules are published in many countries including Italy, the United Kingdom and the United States. In the United Kingdom, the Electricity Act of 1947 requires that tariffs be framed so as "to show the methods by which and the principles on which the charges are to be made". On the other hand, electricity tariffs for the largest industrial customers in Germany are not published. It is desirable that more emphasis should be given in Member countries to securing the transparency of tariffs. Although any allocation of costs between classes of customers is subject to interpretation, transparency is a prerequisite for an appropriate public control of the differentiations of tariffs between groups of customers. Transparency also makes it easier to avoid, to a substantial degree, cross-subsidisation between various groups of consumers.

(d) *Policy for Electricity and Gas Prices*

It would be unrealistic to expect all Member countries of the IEA to adopt a single tariff-setting principle. The general aim should be to find pragmatic methods of fixing electricity and gas tariffs which will give the right signals according to resource allocation and enable utilities to finance economically sound capital investment. To this end prices should be fixed in a way which takes into account past and present costs, plus those expected in the future. Possible methods include relating prices to long-run marginal costs, in particular where these are increasing in real terms, using current cost accounting or allowing a substantial part of the costs of construction in progress to enter the rate base.

The Governing Board, on the basis of a study by the IEA Secretariat, adopted Conclusions in relation to electricity on 27th March 1985 [1]. These Conclusions are also generally applicable to gas prices. The Governing Board urged Member governments, either directly or through discussion with other levels of government, the electricity industry and regulatory bodies to:

---

1. IEA, *"Electricity in IEA Countries - Issues and Outlook"*, OECD, Paris 1985.

— see that changes in fuel and operating costs are reflected promptly in electricity tariffs;

— avoid cross-subsidisation between consumers and the use of electricity prices in a way inconsistent with energy policy in order to promote social, industrial and other policies;

— develop tariff structures which will promote optimal use of investment in the electricity industry and rational patterns of electricity demand.

The IEA has not formally adopted conclusions specifically with respect to gas pricing. However, the general principles set out at the beginning of this chapter are applicable, and the conclusions adopted for electricity have considerable relevance to gas. The separation of energy and capacity charges, and variations in tariffs to different types of consumer to reflect the cost of service as well as energy supplied, are generally recognised and applied. But the wide range of institutional circumstances relating to gas tariff policy is likely to hinder the further development of consistent pricing principles among different countries.

E. Taxation of Energy Use

Through taxation, governments can affect consumer prices and, hence, energy demand and efficiency. However, tax policies generally reflect much broader concerns than energy objectives, in particular, fiscal and different macroeconomic, social, regional and industrial considerations. Generally the need for governments to raise revenue in a convenient and effective way is the primary consideration in virtually all Member countries. As a result there is a wide variation among Member countries and among energy commodities in the proportion of final energy prices represented by taxation. Taxes are heavily concentrated on petroleum products, particularly gasoline, where they are easy to collect and where there is an easily explained justification — the need to construct and maintain roads.

So far, only a few Member countries have attempted to use energy taxation explicitly as an instrument to promote energy conservation and interfuel substitution. In Sweden, the Parliament decided in 1983 that, in order to maintain the economic incentive for conservation and fuel switching, consumer oil prices should be kept high through taxation even if the world market price decreased, and that taxes on coal and natural gas would follow proportionately. In line with this decision, Sweden decided in 1986 to raise taxes on fuel oil and coal from 1st January 1987. Denmark revised energy taxes in March 1986 to neutralise the effect of falling oil prices. The government now reviews energy taxes twice a year. In the first half of 1986, as oil prices fell, the Portuguese Government phased out subsidies on some oil products. For other products, notably gasoline, taxes were increased.

Since the accelerated decline in world crude oil prices after late 1985, a number of IEA countries besides those mentioned in the preceding paragraph have enacted or proposed changes in energy taxation, mainly for budgetary reasons. Between December 1985 and June 1986, tax increases which had immediate effects were enacted for various fuels — mainly oil — in Australia, Denmark, Ireland, Italy, Switzerland and the United Kingdom. Legislation for increases in taxation had been proposed by the former government in Norway. In July 1986 the Dutch cabinet proposed a broad increase in excise taxes on the sale of petroleum products.

In many countries coal use by industry and for electricity generation is either not taxed or the tax levied is rebated. Where a levy is applicable, coal is often favoured by having a lower level of taxation than other fuels. On the other hand, there are cases where countries try to reflect the cost of the environmental damage caused by coal use by applying an additional levy. Again, Sweden is an example. The tax on coal has been increased 50% to 75% above the tax on oil (on a heat content basis).

Despite these examples, there is no country where taxation and energy objectives are brought together in a systematic and well-balanced fashion. IEA countries could do more to improve integration of general taxation and energy objectives and to develop the necessary internalisation of externalities related to the supply and consumption of energy.

There are some instances of distortions in tax policies which act against energy conservation. Examples include tax regimes that do not give equal treatment to supply and conservation options, for example in the United Kingdom, where a value added tax (VAT) is charged on insulating material, but not on electricity and natural gas. Another example is the tax regime for company cars in Sweden, the United Kingdom and — to a lesser extent — other countries, which favours the purchase of larger cars and more profligate car use than would be the case without such tax incentives. Other examples of tax regimes which could be improved to enhance energy efficiency are motor car purchase and ownership taxes that collect vehicle charges on the basis of technical car characteristics (e.g. weight or engine size), regardless of vehicle use. These exist in most IEA countries [1]. Such charges could be collected as a tax on fuel use, thus making the user pay and promoting energy conservation.

F. Conclusions

Economic energy pricing is a prerequisite for effective energy conservation policies. As a general rule, energy prices should reflect the real costs of supply, give the right signals to producers and consumers in order to facilitate an optimal allocation of resources, and enable energy suppliers to finance economically sound capital investment and thus to ensure the long-term security of energy supply. There are, however, severe theoretical and practical problems about determining and implementing economic energy prices. Policy on taxation of energy products is inevitably determined by considerations outside energy considerations. It would be unrealistic to expect Member countries to adopt uniform price and taxation policies across the board. Wider considerations make some variation inevitable. It is, however, desirable that adequate importance should be given to considerations of energy conservation in formulating price and tax policies.

Specific conclusions are:

- remaining subsidies or controls on oil prices should be eliminated or reduced as soon as possible;
- the conclusions adopted by the Governing Board on 27th March 1985 on electricity prices should be implemented and developed and their applicability to gas prices should be further considered;
- adequate importance should be given to energy conservation in decisions on taxes on energy consumption;
- distortions in the tax regime which work against energy conservation should be eliminated or reduced.

---

1. For more details see *Fuel Efficiency in Passenger Cars — An IEA Report*, OECD, Paris 1984, Chapter IV.

CHAPTER VIII

Government Conservation Programmes

Since the first major oil price increase of 1973-74, all Member governments have seen a gradual evolution in their policies directed specifically at improving energy efficiency. The first government programmes to enhance conservation were broad efforts to educate consumers about simple techniques to reduce energy use and to motivate them to use such techniques. These information campaigns were sometimes supplemented by large financial incentive programmes designed to encourage conservation investments. These actions were initially taken when oil prices were escalating and fears about security of supply were high. Often these first efforts were not as concerned with efficient energy use as reducing demand for cost and security reasons.

Gradually, governments took a more long-term and cost-oriented approach, stressing programmes and measures to promote efficiency improvements. This approach included introducing new policy measures and revising existing ones (e.g. building codes) to reflect energy concerns more adequately.

Most Member countries now have a range of conservation programmes which encourage consumers in all end-use sectors to use energy more efficiently. However, since 1982, there has been a change in emphasis in some countries from subsidy schemes and broad-based information programmes to ones which are more specific.

This chapter is concerned with three categories of measures which are used by governments to encourage consumers to use energy more efficiently: information programmes, financial incentives and regulations and standards. This division is arbitrary. For example, mandatory energy labelling schemes or energy audit programmes could fall under the first and third headings.

The evaluation of conservation policy measures is complex and sometimes costly and time-consuming. But evaluation is necessary to determine whether the objectives have been achieved and is a valuable ongoing management tool to identify how programmes can be adapted to changing situations. Unfortunately, too often the methods and data for evaluating existing programmes are inadequate. Most evaluations have tended to be rather superficial and issues such as the amount of energy savings achieved, the attribution of energy savings to particular measures, incrementality [1] and cost-effectiveness are seldom fully addressed. Some of the major methodological issues are discussed in Annex F. While the existing programme evaluations have many limitations, they do offer many qualitative and quantitative insights into the effectiveness of past conservation efforts. The chapter summarises the available information.

Information Programmes

Information programmes are the cornerstone of all Members' energy conservation programmes. They are important in their own right to create awareness; motivate consumer action; and educate consumers, decision makers and those who provide energy services. Information programmes also complement all the other policy instruments and, indeed, are essential to their effectiveness.

Information programmes can be aimed either at the public-at-large or at major sectors like industry, or at specific groups of consumers or energy service groups. Immediately after the oil price increase of 1973-74, information was aimed mainly at the general public with the objective of saving energy rather than improving efficiency. But the emphasis has since changed. It has become progressively aimed at more selective

---

1. Incrementality is defined as behaviour which takes place as a result of a programme which otherwise would have not occurred or been delayed. Conversely, the expression "free-rider" is the participant in a programme who would have undertaken the conservation action even in the absence of the programme.

audiences and has gradually emphasized efficiency improvements. There is greater stress on consumer education which provides more precise and usable information and on training programmes to develop better conservation services.

New techniques — such as computers, videotext machines, contests, and local advisory centres — are now also being developed as ways to make information available and more interesting to various audiences. To improve credibility and effectiveness, information is often provided to consumers through local governments, utilities, local groups or conservation service industries which have more direct contact with consumers.

The main types of information programmes which have been developed in Member countries since 1974 are:

— *General publicity campaigns* which are aimed at large audiences. These were popular in the 1970s when governments were trying to introduce the concept of energy conservation. They included national campaigns in many countries, the IEA International Energy Conservation Month in 1979, the annual Energy Conservation Month in Japan and Energy Efficiency Year in 1986 in the United Kingdom. As part of the latter, the British Government undertook a high profile television advertising campaign in the autumn of 1986.

— *Energy Audits.* Many governments provide energy audits for industry, the commercial sector and the residential sector. These were done as much to increase awareness of the conservation opportunities as to conduct a detailed, technical analysis of energy use.

  In the industrial and commercial sectors, energy audits offer specific advice to consumers. The "energy bus" was one of the early forms of energy audits. First pioneered by Canada in 1977, it provides mobile computerised energy audits largely to improve awareness. The European Commission operates a comparable service in five of its Member states, and Turkey and has extended the programme until September 1987. Japan also has a similar service for small- and medium-sized companies.

  There are other types of industrial energy audit programmes. These are usually to subsidise more complex audits by private consultants rather than using government staff. For example, the United Kingdom provides two programmes for industry — the Energy

Efficiency Survey Scheme providing grants of 50% of the cost of qualified consultants and the Industrial Heat Recovery Consultancy Service for high energy users. Germany provides similar grants for small- and medium-sized businesses.

The most ambitious residential energy audit programme has been the United States' Residential Conservation Service (RCS) which has mandated gas and electric utilities to provide audit services (Class A audits) [1] for their residential customers to give consumers information about the availability and advantages of conservation measures. The regulations for the national programme have allowed each state and utility to influence the design and implementation of the audits. Canada has had a free Class B [1] residential audit since 1977. Denmark has heat inspections performed by energy consultants which were an eligible item under the subsidy scheme for residential buildings until the scheme ended in 1984. Inspections are now required upon the re-sale of a building.

— *Labels and guides* have largely been used for appliances and for the road transportation sector. A number of countries have appliance labelling programmes. The European Community's regulatory infrastructure has permitted the introduction of an appliance labelling system. The system is designed to increase market transparency and avoid national schemes which in effect create trade barriers. However, only three EC Member states which are also IEA Members have adapted the EC directives to their own legislation. Canada and the United States also have labelling programmes.

In the transportation sector, almost all Member governments consider fuel consumption information as a cornerstone of their automotive fuel efficiency programmes [2]. Most provide information on some combination of new vehicle fuel efficiency, fuel-efficient driving behaviour and fuel-efficient maintenance practices. The information is mainly in the form of labels, lists, booklets,

---

1. For a Class A audit, a trained energy auditor conducts an on-site inspection of the premises, performs most of the calculations for the resident and provides recommendations for conservation measures. For a Class B audit, the resident collects the data himself, transmits the data to the auditing agency which performs the calculations and makes recommendations.
2. *Fuel Efficiency of Passenger Cars - An IEA Report,* OECD, Paris 1984.

guides and advertisements. For most countries, the programmes were instituted to reduce confusion for the consumer because no consistent data existed among different available sources. For example, in the United States there had been 15 separate measurements of fuel economy available before the federal fuel efficiency information programme was instituted. Six Member countries have specific vehicle labelling procedures and they are mandatory in Japan, Sweden, the United States and the United Kingdom. The labels usually include both the results of an urban and highway test.

— *Technical handbooks.* Governments have prepared technical materials for consumers, energy managers and conservation services. For the residential consumer, the handbooks are fairly simple and show the consumer how he can add low-cost measures such as insulation and weatherstripping. For the energy manager and conservation services, there are two types of manuals: technical manuals covering such topics as combustion controls and ventilation requirements and manuals detailing how to set up energy management systems.

— *Advisory Services.* Governments operate advisory services to provide information about available conservation programmes or various technologies and techniques. A few countries such as Canada, Switzerland and the United Kingdom have regional advice centres to provide information. Belgium uses videotext machines in municipal offices. Some countries have central telephone advisory services (Canada, Ireland).

— *Training and Education* programmes are provided for current or future consumers, energy managers and service groups. As stated in Chapter V, a lack of information and technical skills has been a major limitation to achieving the conservation potential.

Most governments that promote energy management initiatives usually include training in the form of workshops, seminars, technical manuals and conferences. Switzerland considers training as the major component of its industrial conservation efforts, and the federal government provides funds to private organisations that have training courses. Japan has an extensive training programme for energy managers through the Energy Conservation Centre. Some countries have conservation driver training courses for professional truck drivers.

Increasingly there are consumer education programmes for residential consumers. One of the conclusions of the evaluation of the Canadian Home Insulation Programme found that consumers did not have adequate knowledge of detailed energy conservation possibilities. Thus the Canadian Government started a major consumer education programme. Switzerland provides training courses for concierges. Many countries also stress programmes for elementary and secondary schools. For example, the government of the Province of Alberta in Canada has prepared an entire set of school aids for conservation issues and in the United Kingdom the government provides educational material for teachers.

Training and education for the conservation service industries are also valuable although few countries identify such activities. As was identified in Chapter VI, a strong and mature service industry is essential for increasing conservation activities. Conservation techniques and technologies have changed, and there is a need to have them diffused and applied. Sometimes providing the information is sufficient; however, in other cases, training and seminars may be the most effective means. For example, the techniques used for building the latest generation of super energy-efficient buildings are quite different. Builders who have been accustomed to traditional methods need to learn new techniques. Thus workshops and seminars are very useful. This has been the experience of the Canadian Super Energy Efficient Housing Programme. Switzerland also has a programme to train professional and maintenance staff on new energy-efficient heating technologies.

The Effectiveness of Information Programmes

The effectiveness of information programmes can be judged by two main criteria — their success in increasing awareness of the need for energy conservation and their success in producing results. Generally, information programmes are not designed to save much energy directly. Assessments have only been made in some cases. The judgement about other programmes has to be qualitative rather than quantitative.

For selected information programmes, a more in-depth review of their effectiveness has been undertaken. These programmes are summarised in Table 8.

(a) *Publicity Campaigns.* The level of consumer awareness is generally the only indicator of the success of publicity campaigns although it does not indicate whether action was actually taken. An Irish evaluation for a 1982 campaign showed that 65% of the respondents were aware of the advertising and 55% indicate they took steps to save energy between June 1982 and November 1982. It is not known how many would have undertaken the actions without seeing the campaign. In 1982 about 67% of those surveyed in Sweden said that they had read or looked at energy conservation brochures in the preceding three years. Both of these surveys were taken when energy prices were increasing.

There have been some attempts to measure the link between energy savings and information campaigns. A Belgian evaluation in 1982 showed that consumers who asked for and read the conservation brochures saved about three times the amount of energy between 1980 and 1981 as the average consumer. The evaluation showed that both making a request for a brochure and reading it attentively had a significant influence on consumption. Since the proportion of the audience that uses written material does not exceed 15%, other means of communication have to be used. That is why the Belgian Government has started applying new techniques such as computer-assisted advisory centres to increase the level of awareness.

Germany also tried to show the effects of their information programmes in terms of energy savings. The 1982 German evaluation of all its policy measures estimated that the three main information programmes brought about on average close to 1% of overall annual energy savings. These programmes accounted for about 20% of all information and education activities.

The United Kingdom has had a series of general publicity campaigns. However, as was proven in the 1984 "Lift a Finger" campaign, they are not all successful and that particular campaign had to be terminated because of the poor consumer reaction. The United Kingdom also has a targeted publicity campaign, in the form of breakfast meetings for senior executives. As of September 1986, the meetings attracted over 20 000 senior executives. From surveys of those attending the first twelve meetings:

— 78% subsequently monitored their energy use;
— 44% arranged an energy survey of their premises;
— 34% appointed a manager responsible for energy efficiency.

Much of the success of these briefings has been attributed to the personal attention of the Secretary of State for Energy. The United Kingdom followed on with a general campaign for Energy Efficiency Year in 1986.

(b) *Energy Audits*

(i) Industrial

The energy bus is a very popular type of energy audit which has been well received by both industry and commerce. But it is a most effective tool for increasing awareness, rather than as an audit procedure. There have been some interesting findings from Canadian analyses. First, surveys have shown that there is still a very low level of awareness of energy use and of the technical possibilities for improving energy efficiency, and that the energy bus is very useful in increasing awareness. Second, with the data available, it appears that the energy bus has been reasonably cost-effective. Benefits probably exceeded programme costs by a factor of at least two. Unfortunately, there have not been sufficient return visits to see how much activity was generated by the first audit. Although a full-scale evaluation has not been undertaken, some preliminary calculations showed that about 25% of the recommendations implemented could be attributable to the audit activities.

Between 1980 and September 1985 the European Community's energy buses had made over 10 000 audits. The programme has been targeted to small- and medium-sized undertakings. A common data base in Ispra, Italy, contains the results of 4 576 audits of industrial premises and over 4 600 audits in the service industry. Potential energy savings of 10-20% of normal consumption in a plant for each energy bus visit were identified. This is equivalent to 130 toe per audit. Over 1 million toe were saved per year. To date there have been no follow-up visits to see whether potential savings were actually achieved. The EC has extended the programme until 1988 to have in-depth audits concentrating on areas of high potential (ceramics, abattoirs, dairies, breweries and malt-houses, textile finishing, the leather industry and industrial washeries). For those seven areas, more accurate and specific audit procedures have been prepared in order to get a better understanding of the remaining potential. A more comprehensive evaluation of the programme is planned.

The energy bus programme, both in Canada and Europe, has had many shortcomings. For example, the energy bus is not equipped to analyse process energy which is mainly what industries want to reduce. The audit has not measured the energy used for a company's transportation purposes (e.g. truck fleets) which can be a high proportion of total energy use. Also the audit has looked specifically at energy savings and many felt it has not considered the economics of the recommendations adequately. Because of these concerns, the energy bus is of benefit to fewer firms than had been originally intended. The extended programme in Europe will overcome many of the shortcomings.

(ii) Residential

The results of home energy audits have been mixed. They vary with design and implementation approaches. In the United States, where the programme was undertaken by utilities, those firms that were aggressive and strongly promoted the programme had much better results. In fact, only 40 states implemented the programme, of which 37 conducted evaluations prior to the cut-off date for the federal evaluation. It was widely agreed that the original design of the RCS programme lacked enough flexibility [1]. One problem for utilities was that the audits often contributed to saving oil (in space heating) which was of no concern to the utility, thus there was little incentive for the utility to promote the programme. Cost was also a problem (about $120 per visit on average), with the consumer often paying $10-15 of the cost and the utility paying the rest. Recently revised audit programmes have been more effective because they are more targeted to electricity and/or gas use. Some analysis has shown that audits are more effective if linked with financial incentives [2].

Energy savings attributed to the RCS audits spanned a wide range, from zero to 9%. An evaluation in one of the states showed that participants realised 32% of the identified potential savings for space heating compared to 12% for non-participants. That programme also had one of the highest participation rates — 10% of eligible households for the first

1. For more details see Walker, Rauh and Griffin, "A Review of the Residential Conservation Service Program", *Annual Review of Energy,* 1985, 10: pp.285-315.
2. See, for example, Stern et al., "The Effectiveness of Incentives for Residential Energy Conservation", in *Evaluation Review,* Vol. 10, No. 2, April 1986, pp.147-176.

two years. Only six states audited more than 10% of their eligible customers. They were in the north with high heat loads. Some evaluations have shown that the audit had an impact on the decision of approximately 67-80% of the households. In the evaluation of the Connecticut programme, participants "were more likely to install measures with longer lifetimes than were control households (eleven as opposed to eight years, on average)" [1]. Most of the evaluations showed that mainly middle-income home owners participated and that low-income groups had very low participation rates.

An evaluation of the Canadian Class B audit "Enersave" showed that the programme did influence consumer decision-making. There was positive incrementality for actions to insulate attics and overall, negative incrementality for walls and no significant difference for basements. Analysis showed that payback and cost information was an "overwhelming explanator of incrementality of insulation activity. When Enersave results indicated low payback and/or low costs for a given activity, there was a major stimulative effect ...".

Both the United Kingdom and Canada have had pilot programmes for full Class A audits. The United Kingdom felt that it was not cost-effective and thus did not follow through with a full programme. Canada is still analysing whether to follow through because of the cost of implementation and the lack of a clear-cut assessment of the cost-effectiveness.

(c) *Labels and Guides*

(i) Appliances

Appliance labelling programmes have been used to upgrade the appliance stock through influencing manufacturers and providing consistent, comparative information to consumers. The Canadian Energuide programme was reasonably successful. Although few consumers used the labels in their purchase decision, the programme was considered incremental and cost-effective in improving the efficiency of

1. Hirst et al., "Connecticut's residential conservation service: an evaluation", *Energy Policy*, February 1985.

appliances, particularly refrigerators and freezers, through introducing competition among manufacturers [1]. An evaluation of the United States' appliance labelling programme concurred with the findings of the Canadian study.

The programme has resulted in significant savings in energy and expenses. Consumers save on energy costs and get a good return on any increased cost of appliances, and Canadian appliances remain competitive with imports. But members of the evaluation team differed in their views on the extent of further economic savings available and on whether improvements to date would be eroded if the programme were terminated.

Currently the Canadian Government is considering options for its programme, including continuation of the programme on a voluntary basis in close co-operation with the appliance industry, electrical utilities and consumer groups.

(ii) Transportation

Fuel efficiency information in transportation was assessed in an earlier IEA study. It concluded that "fuel consumption information is considered a cornerstone of all automotive energy conservation programmes. Without credible information the market cannot be expected to make fuel-efficient choices" [2].

Only the United States has conducted a comprehensive review of the effectiveness of the information programmes [3]. The review was undertaken several times during a period when consumers were more conscious of higher energy prices. The evaluation concluded that:

— About 70% of new car and light truck buyers were aware of labels on vehicles, while less than 20% knew a more comprehensive guide was available;

— Among the aware buyers, about half in each group used the fuel economy information for comparison shopping;

---

1. See Canada, Consumer and Corporate Affairs, *Evaluation of Energuide,* Ottawa, March 1985 and *Energuide Evaluation: Background Study Modules,* 1985.
2. IEA, *Fuel Efficiency in Passenger Cars — An IEA Report,* Paris, 1984, p.104.
3. *Ibid,* pp.99-103.

- New vehicle buyers were very critical of the accuracy of the Environmental Protection Agency (EPA) ratings with 70% of the respondents believing that the EPA ratings overstate actual fuel economy. Nonetheless, 75% of the respondents still believed the ratings were useful for comparisons;
- The lack of credibility of the ratings perceived by consumers was the major cause of non-use by new car and light truck buyers.

As a result of the assessments, the United States took corrective action to improve the programme beginning with model year 1985.

The main conclusions from the assessment of the effectiveness of information programmes are described in Table 8. Some of the conclusions relevant to all the measures are:

- information programmes form the cornerstone of all policy programmes and are valuable in ensuring the success of other types of policy measures;
- energy audits, particularly in the residential sector, can be more effective if combined with financial incentives;
- information programmes can help ensure continuity of conservation momentum, even in periods of changing short-term market conditions;
- information programmes should be adapted to new market situations, consumer needs, and new communication technologies;
- awareness and motivation are not static and need to be reinforced periodically;
- limited, more specific programmes are more effective than broad-based programmes although there is also a role for the latter from time to time.

ii) Financial Investment Incentives

A financial investment incentive has been defined as any measure "designed to influence an investment decision and increasing ... the profit accruing to the potential investment or altering the risks attaching

Table 8

Summary of Information Programmes

Policies/Programmes	Primary Goal	Degree of Use	Market Limitations Addressed[1]	Implementation Environment	General Conclusions
Publicity Campaigns	- awareness	- most Member countries	- lack of information - invisibility	- implemented usually during period of high price increases - many countries have continued them throughout	- valuable for awareness creation
Residential Energy Audits	- awareness - motivation	- Canada, United States and Sweden primarily	- lack of information - invisibility	- implemented during period of high price increases	- valuable to increase awareness on part of consumers and show cost-effective options - problem with cost-effectiveness of Class A audits[2]
Industrial Energy Audits	- awareness - motivation	- Canada, Europe, Japan	- lack of information - invisibility	- initially during periods of high price increases	- valuable to create awareness - problem of degree of sophistication and technical rigour
Appliance Labelling	- awareness - motivation - provide unbiased information to aid purchase decision	- Canada, United States, Europe, Japan	- lack of information - invisibility	- initially during periods of high price increases	- biggest effect on manufacturing industry - has been cost-effective means to produce energy savings - has worked well as voluntary programme
Transportation Fuel Efficiency Information	- awareness - motivation - provide unbiased information to aid purchase decision	- most Member countries	- invisibility - lack of information	- initially during periods of high price increases	- awareness generally high - credibility problems with fuel economy ratings

1. Refer to market limitations described in Chapter V.
2. Refer to page 122 for definition.

to it" [1]. Most conservation incentives are given by governments although in some countries utilities also offer a range of financial incentives — often as a requirement of regulatory bodies.

Most governments offer a wide range of financial incentives for an equally large number of national objectives. This section is concerned with those intended specifically to promote energy conservation. Other incentives, such as general industrial financial incentives, which may help encourage energy conservation indirectly, are not considered.

Initially, financial incentives have been very popular among Member governments as a means to encourage consumers to take action and address some of the market limitations described in Chapter V. But since about 1982 there has been a change in emphasis in some countries away from large incentive programmes [2] because of budgetary constraints, changes in government policy and indications that direct incentive schemes are less effective after a number of years of use. With the energy situation improving, a few countries have perceived less need for financial support; however, other countries recently have started making use of incentive schemes to overcome existing market limitations. These include Italy, which introduced various incentives through its Law 308 in 1982, Norway which now provides incentives as a result of a major policy report to Parliament in November 1984 and Spain which also introduced new schemes under its 1983 National Energy Plan.

Incentives have been used differently in the industrial and residential sectors [3]. Incentives in the industrial sector are available for energy audits or consultants, feasibility studies, CHP and district heating (DH), heat recovery systems, refuse incineration plants, insulating materials, energy-efficient heating systems, heat pumps, control systems and RD&D. In general, industry incentives apply to about 7% and 30% of

1. *Investment Incentives and Disincentives and the International Investment Process,* OECD, Paris 1983, p.10.
2. Since 1982, some programmes that have ended include the United States industrial and residential energy conservation tax credits, the Canadian Home Insulation Programme, the Danish and Swedish grant programmes for industrial energy-savings investments, the Danish residential retrofit scheme, the Swedish residential retrofit scheme for single-family dwellings and the German home retrofit programme.
3. Except for public transport, there are few direct incentives for the transportation sector and, therefore, it has not been included in this part of the study.

the investment. Very few industrial incentive programmes have required a specific payback criterion that eligible conservation investments must meet. In most cases eligibility is determined by lists of acceptable equipment. Some programmes, however, are more selective. For instance, under the Canadian Atlantic Energy Conservation Investment Programme (AECIP), projects with payback periods between three and eight years were eligible for an increasing rate of assistance (10% for short payback and up to 50% for longer payback periods). Other examples are the Danish industrial grants, which were seen after an individual examination and were dependent on the ratio of investment cost and energy savings, and Japan where eligible equipment for a tax credit is required to save at least 30% of the energy used.

For residential programmes, incentives are usually provided for insulating materials and weather-stripping, including labour for installation. Incentives for energy audits, heat pumps, control devices, double- or triple-glazed windows and connections to district heating grids are less common. Generally, incentives apply to 15% to 100% of investment costs, with the average probably being in the 30-40% range. In some cases the percentage varies according to the consumer's financial status [1]. Some residential programmes are specifically targeted to rental accommodation (Netherlands) or to low-income groups (United States). Unlike some more selective industrial grant schemes, the percentage of the grant is usually fixed for the full range of eligible items regardless of the payback. Eligible investments in the commercial sector are similar to those in the residential sector. Eligible investments in the transformation sector resemble those in industry although they have only been available in a few countries.

There are three main types of financial incentives used by Member countries: grants, tax incentives (preferential taxing, e.g. tax credits, and accelerated depreciation), and low-interest ("soft") loans and guarantees. Although they all have the same ultimate objective and can all be adjusted to produce the same financial results, the incentives vary substantially as well as how they are applied. The types of incentives are:

(a) *Grants.* Grants are usually intended to stimulate conservation investment in end-use sectors by covering part of the total investment costs. Grant programmes have been and still are widely

1. For example, the United Kingdom's Home Insulation Scheme gives a much more favourable percentage (almost the entire cost) for the elderly and disabled on low incomes.

used in most IEA countries. Direct grants represent about 60% of all incentive programmes used by IEA Members in 1984. Most grant programmes have been in the residential and industrial sectors although there are a few in the commercial/institutional and the transformation sectors.

Grants have been primarily effective to improve the rate of return on investment, particularly in the industrial and transformation sectors. Grants have helped provide access to capital for some consumer groups, have encouraged the development of a conservation service industry, have provided technical information to consumers and had demonstrated confidence in new products. Grant programmes can be selective but usually have more complex application and approval processes than other types of financial incentives. Therefore, considerable administrative work is often required which can be fairly expensive, especially in the residential sector [1]. Many companies have been reluctant to accept government assistance because of the bureaucratic procedures involved. For small companies, the amount of administrative effort required is often prohibitive.

(b) *Tax Incentives.* Tax incentives represent about 20% of all financial incentives. Among the great variety of tax reductions and exemptions, tax credits appear to be the most widely used form of fiscal incentives. Investment allowances, for example, are less frequently used as are reductions in tax rate of taxable income and accelerated depreciation. Tax incentives for energy conservation investments usually fit easily into the larger array of general tax incentives available to consumers, particularly in the industrial sector. Some unique programmes include: the Netherlands offers a premium of 10%, on top of its tax incentive for general investments, for investments in energy conservation equipment in the industrial/commercial sectors; in Belgium, companies may claim up to 20% of the cost of energy conservation investment (equivalent to about 10% actual tax reduction) or in case of unprofitability for three years, may receive interest free loans of

1. In the Canadian CHIP programme, administrative costs were approximately 7-8% of total costs once the programme was fully operational.

about 6%. Some countries, for example Austria, allow accelerated depreciation for conservation investments in end-use sectors as well as the transformation sector.

From a government's point of view, tax credits and allowances have been generally easier to administer but harder to control in financial terms because of the complexity of accounting within the tax system. They appear to be best used now in conjunction with or integrated within the general tax system. Industrial investors are used to the process and benefits involved and are eager to take advantage of such a system. For large- and medium-sized companies access to capital is less a problem and tax incentives provide a means of reducing corporate taxes. Tax incentives do not, however, help firms which run or have accumulated losses and therefore have no tax obligation to reduce. In a few cases, there has been some confusion because some investments are eligible for tax credits in more than one industrial tax classification. Available residential tax incentives have been used mainly by higher income groups. It is generally difficult to make tax incentive programmes selective.

(c) *Loans*. In the case of soft loans, the interest rate in most cases is just a few percentage points below the market interest rate. This difference can be as low as 0.15 percentage points (as in the loan programme of the Japan Development Bank at the end of 1985). Favourable loans for industry are used by only a few countries. In the residential sector, Sweden has a unique loan system which promotes construction of energy-efficient buildings. In the energy transformation sector, the Netherlands has a system of government guaranteed loans to support expansion of CHP/DH. Austria operates a system of soft loans and fiscal incentives.

In principle, soft loans have some of the characteristics of grants. In both cases, the necessary finance is provided at the investment's start. In practice, the grant element of soft loans is often small due to the small difference from interest rates on commercial loans. The beneficial subsidy effect is, however, spread over the loan period. Therefore, loans are possibly a less attractive form of subsidy than grants for companies with enough liquidity, since companies prefer "quick money". Due to the small grant element in each case, it has been possible to reach more investors with the same amount of subsidy money than with grants.

Effectiveness of Financial Incentives

Although evaluations of financial incentive programmes often received comparably high priority, a rather limited number of programmes has been evaluated. Clearly, the environment in which programmes were initiated has an impact on the programmes' effectiveness and the level of concern about programme evaluations. Many of the programmes were instituted in a period of rapidly escalating energy prices and security concerns. In such periods, governments were less concerned with analysing the incremental effects of programmes than they were in taking quick and strong action to help reduce the negative effects of the "energy crisis".

The fundamental issues concerning the effectiveness of financial incentives include:

— their ability to overcome market limitations;

— their ability to encourage incremental conservation investment;

— the improvement they induce in energy efficiency;

— their cost-effectiveness.

In analysing these factors it is important not to apply stricter standards to financial incentives for energy conservation than to other financial incentives. The problem of incrementality for example arises just as much with tax incentives for energy production as for conservation. Unfortunately, many of the programmes that have been evaluated have not been assessed thoroughly.

Cost-effectiveness has been calculated by the IEA Secretariat using the present value (PV) of the total accumulated energy cost saving during the whole life-cycle of the equipment (real discount rate of 5% per annum, 1984-85 energy price levels). As lifetimes of different efficiency technologies vary widely, rather rough assumptions had to be made. The same procedure has also been applied for the annual government funds spent in order to end up with an indicator for the benefit-cost ratio, excluding external benefits such as environmental gains. As both the shares of genuine free-riders as well as of genuine incremental investors have been rather small, with a large "grey" zone in between, incrementality has been assumed as 50%. However, no real distinction could be made between highly economic "easy" investments (with lower incrementality) and less attractive investments (with higher incrementality) due to the lack of sufficient data.

(a) *Grant Programmes*

Industrial Sector

In industry, grant programmes have generally been more attractive to both non-tax-paying (including new ventures) and tax-exempt companies. They were less attractive to tax-paying companies, although that depended on the specific tax regime.

In general, incrementality ranged from 20% to 80% depending on the eligibility criteria and the percentage of the investment provided. More narrowly defined programmes had better results.

The benefit-cost ratio for each of the programmes analysed [1] was about five to one. This means that for every unit of government funds spent on the programme, there were five units of benefit during the total lifetime of the energy conservation equipment. This indicates very good cost-effectiveness. Recent price declines seem to decrease the benefit-cost ratio by about one-third to about 3.5. The most cost-effective schemes were those which were aimed at specific sectors or in which the percentage of the grants varied depending on the ratio of investment cost and energy savings. For example, the maximum grant allowed under the Danish Industry Scheme was 40% of investment cost up to a certain maximum, but the average was 26%. There was an investment threshold above which energy savings practically doubled.

A Swedish evaluation of grants [2] found, contrary to the other programmes, no systematic correlation between subsidies and improved energy efficiency. The effects of government incentives were difficult to assess because the total grants provided were only a small fraction of total industrial investment due to the high turnover rate of capital stock in Swedish industry. Most of the funds were given to energy-intensive industries in which the motives to invest in energy conservation were strong anyway. This apparently reduced the incremental effect. There is also some evidence that due to the exclusion of the most profitable

1. These include the Canadian AECIP programme, the Danish Industry Incentive Scheme and the German Investment Allowance Programme (Secretariat analysis based on IFO evaluation).
2. *Energihushallnings — programmets effekter* (Effects of the Energy Conservation Programme), Statens energiverk (National Energy Administration), 1984:2.

projects from eligibility, industry actually implemented less economic projects. However, frequent changes of the subsidy programmes actually spread information about their availability and also facilitated contact with sources of finance.

Residential Sector

Grant programmes in the residential sector have been very popular with consumers and government officials largely because they are visible and are useful to motivate consumers who did not invest significantly, even in periods of escalating energy prices. Even though grant programmes had the highest participation rates of all financial incentives for residents, it was estimated that the mean was 7% of eligible participants per year [1]. Thus it takes several years to reach a large segment of the population. Programmes had positive incremental effects, although usually lower than industrial grant programmes. In some cases the low incrementality was possibly due to lax eligibility criteria. The benefit-cost ratio for the total lifetime of the equipment on average is about 3:1, with about 1:1 as the worst case. This is lower than that for grant programmes in the industrial sector but is still beneficial. Recent price declines should lower the ratio by about one-third.

Administration of the programmes can also require inspection and auditing procedures. Many countries use state or local governments as part of the administration. This facilitates having regional variations of the programmes. Because these programmes are often large there can be delays in the administrative process such as approving eligible items or issuing cheques. These delays can affect the overall effectiveness of the programme.

Some programmes did have a positive effect in supporting energy conservation service industries. In Canada, the CHIP evaluation showed that the programme was responsible for increasing sales of attic insulation; and the insulation manufacturing industry grew from nine plants in 1976 with a capacity of 200 000 metric tons per year to 13 plants in 1982 with a capacity of 370 000 metric tons per year. About 50% of the insulation was used for retrofit and it was estimated that 29% of all retrofit activity in Canada was due to CHIP. Many of the retrofit grant programmes were instrumental in improving thermal efficiency by about

1. Stern et al., "The Effectiveness of Incentives for Residential Energy Conservation", *Evaluation Review,* Vol. 10, No. 2, April 1986.

12% per household, about half the 25-30% predicted. There was a large range of efficiency improvements in individual households, partially explained by behavioural changes such as thermostat setbacks.

It is still debatable how effective grants have been in the residential sector. For example:

— Subsidy programmes can make expensive investments more lucrative. Unfortunately, for some programmes, many participants favoured measures such as window replacement, storm doors or attic insulation that did not optimise cost-effective energy savings. This was true in Canada, Germany and the United Kingdom. Consumers were often enticed by grants but were less concerned about saving energy. The list of items eligible for grants needs to be chosen carefully and reviewed periodically, and information programmes such as audits can complement these programmes and better explain conservation opportunities.

— Some programmes encouraged participants to choose eligible items with longer paybacks. For example, conservation measures taken under the Danish Space Heating Scheme had simple paybacks of seven years on average. These would have been marginal investments for consumers without such incentives [1]. The Danish results are very impressive and can only partially be explained by information campaigns and the public's awareness of conservation due to the familiarity with the majority of energy projects under way in the country.

— Programmes have had difficulty penetrating the rental market which is a large segment in many countries. The Netherlands reoriented its programme solely towards rental accommodation to overcome the problem. Although general grants were abolished after 1984, Sweden also kept grants for the rental market.

Commercial Sector

The evaluation of the Institutional Conservation Programme in the United States showed that the average grant recipient saved 13% of his energy consumption or about a total of 5.2 trillion British thermal units (Btus) (0.025 Mtoe). The cost of saving 1 million Btus during the

1. Denmark, Ministry of Energy, *Energy in Denmark: A Report on Energy Planning 1984,* Copenhagen, 1984.

ten-year span of the measures was estimated to be $1.37, compared to government funding of $0.68. Paybacks averaged less than two years, rather than the four years originally predicted.

Transformation Sector

Several government programmes are targeted towards expanding the application of CHP/DH technologies. Financial support is limited to a few European countries [1]. Despite the importance of this sector, only the German Government has evaluated its programmes.

While various evaluations have given conflicting results [2], the German grant programmes have been effective in improving financial viability according to IEA Secretariat analysis. Incrementality has been estimated at one-third, and this has produced a benefit/cost ratio of total cost savings and public expenditures of close to three to one. Total energy savings are expected to be about 25 Mtoe until the year 2000, although this depends on the progress of the expansion of the DH grid. Recent price declines in 1986 lower the benefit/cost ratio to around two and one-half to one — still quite cost-effective.

While there are no specific evaluations, subsidies in Denmark and Sweden for large DH systems (10% for DH pipes) have had an acceleration effect. In both countries the environment has been favourable: practically no gas in the heat market, the existence of rather high levies on oil products, and co-operative utilities. Governments also enhanced the programmes through certain measures. Through the Danish Heat Supply Act, Denmark established price controls for grid-based energy forms to promote DH. Accordingly, for new projects, the total savings in energy costs were related to heat. Over time, however, savings are reduced and even reversed as are efficiency ratios. In mid-1986, Denmark implemented further supportive measures to improve investment conditions by reducing the impact of the high initial outlay through an "index arrangement" with fixed yearly payments based on 1986 Danish kroner and by allowing the total investment to be spread over a period of twenty years.

---

1. The countries include Austria, Denmark, Germany, the Netherlands and Sweden.
2. The four evaluations reviewed were by P. Suding — Cologne, IFO — Munich (two studies; the first, published in 1982, was ordered by the government), and the German Institute for Economic Research, Berlin. They were done between 1982 and 1984.

(b) *Tax Incentives*

Industrial Sector

Tax incentives have generally not been very effective in encouraging incremental investments in North America. In Canada, only 4% of certified capital cost was incremental whereas in the United States there were no incremental projects at all [1]. However, the incentives have had an important symbolic role. On the other hand, once they are in place, it is difficult to remove the incentives without apearing to give a signal to industry that government is no longer interested in conservation.

However, in Japan, a survey of all industrial investors showed two-thirds took the tax credits: close to half expanded or accelerated their investment. Only one-third did not consider the tax credit at all. A macro-economic simulation by the Ministry for International Trade and Industry with the Nikkei-Needs Macro-economic Model suggests tax incentives are highly effective in Japan, with the increase in the investment in eligible equipment about three times the total tax reduction. IEA Secretariat benefit-cost analysis for the Dutch Investment Account Act shows a ratio similar to the analysed industrial grant programmes, i.e. about five to one.

Residential Sector

Although about one-third of IEA Member governments had or still have tax incentives for individual householders, only the United States has evaluated these credits. Some studies in the United States have shown that tax credits had a very low incremental effect (as little as 10%). This was explained in one analysis [2] by the relatively low percentage of the incentive and the fact that the conservation measures usually involved relatively small amounts of money.

1. See Canada, Energy, Mines and Resources, *Class 34 Capital Cost Allowance: Programme Evaluation, Final Report,* Ottawa, November 1984.
2. H. Petersen, "Solar Versus Conservation Tax Credits", *Energy Journal,* Volume 6, Number 3, pp.134-135.

(c) *Loans*

Industrial Sector

The effectiveness of loan programmes in industry has not yet been thoroughly investigated. But in Japan, for example, favourable industrial loans are often given in parallel with tax credits, which forms the largest part of the financial incentive, thereby masking the loan's specific effects.

Residential Loans

In Sweden, favourable loans are largely responsible for the high thermal standards of buildings. Interest rates have been below the inflation rate during the initial years of the projects and reach market levels as late as ten to forty years after the investment. Therefore, these loans are a strong incentive for efficiency investment in new buildings and retrofit projects in residential buildings.

Transformation Sector

In the Netherlands, analysis of financial support (subsidies and risk-sharing loans, on average 15%) for DH projects provided until end-1984, shows about the same benefit/cost ratio as for similar programmes in Germany. Government guaranteed commercial loans were continued because an evaluation pointed out that risk-sharing is considered the most important incentive in government support for DH.

General Assessment of Effectiveness

The main conclusions of the effectiveness of financial incentive programmes are summarised in Table 9. More specifically:

— financial incentives have been generally effective in promoting energy conservation although the incremental effect has varied. Programmes with higher incremental effects usually had specific eligibility criteria, carefully balanced percentage of total investment provided by the measure, and targeted audiences (as in the Canadian AECIP or the Danish industry programmes);

Table 9

Summary of Financial Incentives Programmes

Policies/Programmes	Primary Goal	Degree of Use	Market Limitations Addressed[1]	Implementation Environment	General Conclusions
Industrial Grants	- stimulation of discrete conservation investment	- most countries	- financial attractiveness and access - confidence - lack of information	- largely initiated between two price increases in 1970s - some terminated when energy prices started declining	- expansion and acceleration of investment - introduced new technologies - improved financial attractiveness - good benefit-cost ratio, even given recent price declines - wide range of incremental investment - created awareness - administratively complex - targeting on incremental projects possible
Tax Incentives	"	- North America, Japan, some European countries	- financial attractiveness - confidence	"	- easy implementation - created awareness - application process fairly easy for companies - of little use for non-tax-payers
Loans	"	- Japan, Germany, Austria	- access to capital - confidence	"	- in practice, small interference in market - mainly easing access to capital (companies in poor financial situation) - incrementality difficult to assess

1. Refer to market limitations described in Chapter V.
2. Six countries have programmes for institutional/public sector. Five countries have programmes for the residential sector.

Table 9

Summary of Financial Incentives Programmes *(Continued)*

Policies/Programmes	Primary Goal	Degree of Use	Market Limitations Addressed[1]	Implementation Environment	General Conclusions
Residential/Commercial					
Grants	- stimulation of discrete conservation investment	- about half of Member countries	- financial attractiveness and access - lack of information - confidence - separation of expenditure and benefit	- largely initiated between two prices increases in 1970s - some terminated in early 1980s when energy prices started declining	- popular and visible - created awareness - provided information to consumers - improved financial attractiveness - helped develop conservation service industry - poor results in rental market - poorer benefit-cost ratio than industrial grant programmes - administratively complex
Tax Incentives	"	- Austria, Belgium, Denmark, Germany, Japan, Switzerland, United Kingdom, United States	"	- largely initiated between two price increases in 1970s	- lower government involvement - mainly used by higher income groups
Loans	"	- Denmark, Germany, Japan, Sweden, United States	"		
Energy Transformation Sector					
Grants	- stimulation of investment into CHP and for DH	- Denmark, Germany, Ireland, Italy, Netherlands, Sweden	- financial attractiveness		- subsidies effectively reduced investment risks - benefit-cost ratio similar to industrial programmes - rather high incrementality - often lack of utility co-operation
Tax Incentives		- Austria	"		
Loans		- Austria, Netherlands, New Zealand			

1. Refer to market limitations described in Chapter V.
2. Six countries have programmes for institutional/public sector. Five countries have programmes for the residential sector.

— the grant programmes analysed have had a positive benefit/cost ratio. The ratio has been highest for industrial programmes;

— while the evidence is rather weak, grant programmes have been more cost-effective than other financial incentives;

— financial incentives can be valuable in introducing new energy-efficient technologies depending on how strict the eligibility criteria are. This has been particularly true in the residential sector;

— grant programmes are the most complex and expensive financial incentives to implement;

— grant programmes in the residential sector have been very popular. The average energy savings of a retrofit grant are 12%. But renters have not responded well to these programmes;

— financial incentives need to be closely linked with information programmes, for marketing purposes and to publicise the best incremental investments;

— in the transformation sector there is evidence that risk-sharing, through guaranteed loans, is the most important incentive in encouraging district heating.

iii) Regulations and Standards

Regulations and standards are used to varying degrees in all IEA countries. Standards are efficiency levels established by governments for appliances, buildings or passenger cars. They can be voluntary but most have been mandatory. Regulations refer to controlling or directing conservation actions through government rules or restrictions. Some countries have regulations specifying maximum temperature levels in industrial plants or requiring labels on new automobiles. They have been used in the United States to mandate utilities to offer energy audits to the residential and commercial sectors (see section i). They are sometimes directed towards the energy consumer (e.g. individual billing of heating costs), but more often are directed towards energy service suppliers or industrial manufacturers. The use of government regulations or standards to achieve energy conservation varies considerably by sector and country.

In the industrial sector, there have been fewer standards or regulations because "the huge variety of technologies and organisations in factories, workshops, etc., makes it difficult to design regulations which will not require costly measures of verification." [1] The Japanese Government maintains the right to inspect an industrial site and also requires an energy conservation plan be submitted with an application form for a construction permit for new industrial and residential/commercial buildings larger than 2 000 square metres. In addition, the appointment of energy managers is required by law. In Portugal firms consuming over 1 000 toe per year must implement an energy management service, have their energy-use patterns examined every five years, and develop five-year plans for the rational use of energy which must be approved by the Directorate General for Energy. Italy also has a comparable mandatory programme. The United States requires all firms consuming more than 1 trillion Btu of energy (25 200 toe) per year to report their energy use to the Department of Energy. Norway requires industry to have specific energy conservation plans when establishing large new facilities and/or expansions which call for the allocation of firm electric power supply. A few countries require automated controls for heating systems and at least two countries set efficiency rates for boilers (Austria and Germany), mainly for environmental reasons.

Many regulations and standards have been directed towards residential/commercial buildings, and most have been mandatory. They require minimum thermal efficiencies for new housing by prescribing either particular materials/techniques (prescriptive standards) or levels of performance (performance standards), heating system efficiencies, individual metering according to energy consumption in multi-occupancy buildings, boiler maintenance requirements and restrictions on air-conditioning. Governments sometimes also prescribe increased thermal efficiencies for existing housing stock when retrofitted. Standards for existing buildings may be the only effective way of reaching segments of this sector, such as multi-family rental housing. Both Germany and Sweden initiated model agreements for tenants and home owners. Due to the long life-span of new buildings compared to other energy-related equipment, strict building codes will lead to energy cost savings in the longer term and, if well designed, should require only minor additional investment in the construction phase.

Although many countries also have regulations for appliance labelling which are discussed in section i, few have specific mandatory energy efficiency standards for appliances. The European Commission pub-

1. Commission of the European Communities, *Comparison of Energy Savings Programmes of EC Member States,* COM(84)36 Final, February 1984, p.5

lished guidelines for its Members which include both performance standards and control of servicing of heating systems. Austria has linked tax incentives to minimum efficiency standards. There are few examples of labels showing the thermal characteristics of buildings. Denmark is the only country which requires a heat inspection report with the building's thermal quality when buildings are sold.

In the transportation sector, a number of IEA countries (primarily those with an automobile manufacturing industry) have fuel efficiency standards or targets for new passenger cars, although only the United States (and to some degree Canada, which can invoke mandatory requirements if necessary) has a mandatory programme. Three countries — Germany, the Netherlands and Austria — require mandatory car inspection for environmental and fuel efficiency maintenance on a regular basis. Almost all countries have speed limits on highways and some (e.g. the United States) have lowered them to save fuel. Germany has recommended speed limits for its motorways.

Fuel economy standards have been used to improve the efficiency of new passenger cars. Because of the relatively quick turnover of the car stock, in a few years new car efficiency improvements can have a major impact on fleet efficiency. The standards are directed at the manufacturers and importers to improve the new vehicles. Consumers are involved through parallel aspects of standards programmes — labelling, guides and other information sources.

Effectiveness of Regulations and Standards

Building Codes/Standards

Building codes are an effective means of introducing energy-efficient technologies in buildings. They ensure that minimum levels of efficiency are achieved in new buildings — a sector where the market is especially slow to act. They also permit the construction of more efficient buildings, if demanded by consumers or builders.

The procedure for developing and revising building codes is often laborious because of the multitude of special interest groups involved; also codes are often developed at one level of government and adopted at another and they include other aspects besides. The resulting time-lag has caused some standards to follow trends already started by energy prices, instead of taking the lead. For example, Swedish evaluations of building codes concludes that actual building practices met and sometimes even surpassed the increasingly stringent codes one to two years before they were enacted in 1977 and 1980. To some extent, this acceleration was due to efforts by the government which spurred improvements by distributing information to manufacturers, designers and contractors. In addition, the existence of standards has probably influenced product development. This is especially true for airtightness.

Developing standards is also complicated by the modelling required. But even the most sophisticated models have to assume standard conditions and thus generally cannot accurately predict what actual energy use levels will be using various design assumptions. Savings can be estimated but consumption depends on consumer behaviour such as opening windows and setting thermostats.

In the Danish evaluation, it was estimated that the higher energy standard for new buildings was responsible for approximately a 15-20% decrease in energy use per square metre and saved a total of about 10 petajoules per year between 1975 and 1981.

A Dutch study concluded that, from a benefit/cost point of view, standards in most European countries could be more stringent [1]. Due to the long lifetime of new buildings (fifty to a hundred years), discounted heating costs during the period, which have to be based on probable energy price increases in the medium term, will certainly outweigh the comparatively minor additional expenses for better insulation. Since insulation improvements are often cost-effective only when buildings are substantially modernised, there is a clear need for optimum insulation standards during construction.

According to an evaluation of conservation programmes in the European Community, the difference in the building codes in Europe narrowed between 1980 and 1982 [2]. Nevertheless, the Commission believes building standards can be strengthened, and it is preparing a model

1. Meyer, L.A. "Energiebesparing in de sociale Woningbow", University of Groningen, Groningen, 1981.
2. Commission of the European Communities, *Comparison of Energy Saving Programmes of EC Member States, op. cit.*

"reference" code. The Commission believes that building codes for new buildings are very important because 20% of the buildings still standing in 2000 have not been built yet and insulation is two to four times less expensive during construction than during retrofit [1].

Appliance Standards

No specific studies have evaluated the effect of standards on residential appliance efficiency. The United States ruled in 1982 and 1983 that energy efficiency standards for eight appliances were not economically justified or effective. They did allow states to set their own standards and a number of states, including California, have set efficiency rates for major home appliances. A court decision in 1985 stated that standards should be set by Congress. The Department of Energy will stage new hearings and issue provisional rules. However, legislation is currently being considered which would establish specific national appliance standards while pre-empting existing and future state standards. If enacted, these standards would go into effect more quickly than standards established by the Department of Energy.

Automobile Fuel Efficiency Standards

The IEA reviewed the effectiveness of fuel efficiency standards in 1984. Only three IEA Member countries have attempted to quantify the fuel savings attributable to their programmes — Australia, Canada and the United States. None, however, has been able to clearly attribute improvements to prices or programmes. The United States implemented its mandatory programme in 1977 when energy prices were falling in real terms. Manufacturers were both sceptical they could achieve the targets and unhappy that the standards were based on fleet sales rather than individual car performance which is more predictable. But manufacturers did improve efficiency of their vehicles and they were better prepared when the second major oil price increase came in 1979-80. Once gasoline prices rose dramatically, manufacturers anticipated surpassing the standards and thus saw no need for them. In the early 1980s gasoline prices again declined and two of the major car producers in the United

1. Commission of the European Communities, *Towards a European Policy for the Rational Use of Energy in the Building Sector,* COM(84)614 Final, Brussels, 13th November 1984.

States were not able to reach the fleet average of 27.5 miles per gallon (mpg). The National Highway Traffic Safety Administration has reduced the 1986 standard from 27.5 mpg to 26 mpg in response. As shown in Table 4 in Chapter III, many of the other producers' programmes, however, did meet their target. As shown by the United States, fuel economy standards can be important. They initially were responsible for improved efficiency in new cars and it maintained rates during subsequent declines in gasoline prices. Even though the 1985 target had to be reduced, the improvement since 1978 is commendable.

The main findings of the review of the effectiveness of regulations and standards in Table 10 are:

— standards and regulations have been most useful in the residential/ commercial and transportation sectors where there is more standardization of equipment and where there are special market segments such as rental housing;

— building codes/standards have been mainly limited to new buildings; due to the much larger stock of existing buildings and the ineffectiveness of market forces for major categories of existing buildings, standards should also be considered in this area;

— transportation vehicle standards have been effective in improving fuel efficiency of new vehicles, especially when combined with information programmes. There is no evidence if mandatory or voluntary programmes are more effective;

— regulations and standards ensure that minimum efficiency levels are met, but can discourage producers from exceeding these minimum requirements and thus achieving the full economic potential;

— they have been useful to provide long-term continuity through periods of energy price fluctuations;

— standards and regulations need to be enforced, and governments have backed these measures with small amounts of funds.

General Conclusions

A range of policy measures are used by governments to complement and strengthen market signals that stimulate conservation actions and address many of the market limitations described in Chapter V.

Table 10

Summary of Regulations and Standards

Policies/Programmes	Primary Goal	Degree of Use	Market Limitations Addressed[1]	Implementation Environment	General Conclusions
Building Codes	- upgrade efficiency of new building stock	- all IEA countries	- invisibility of consumption - lack of information - separation of expenditure and benefit	- energy efficiency aspect of existing building codes added after major price increases - have been maintained even in periods of declining energy prices	- very effective in overcoming market limitations - low cost means of upgrading thermal quality of new building stock - provide long-term signals - easy to adapt to regional/local conditions
Appliance Efficiency Standards	- upgrade efficiency of new appliances	- Japan, United States	- invisibility of consumption - lack of information - separation of expenditure and benefit	- initially implemented when energy prices increasing	- insufficient information to draw conclusion - most countries more interested in appliance labelling programmes than efficiency standards
Fuel Economy Standards for New Passenger Cars	- upgrade efficiency of new passenger cars	- nine countries - only United States has mandatory programme	- lack of information prices increasing for	- initially when energy and importers specific period - some kept after target period	- directed towards manufacturers - work in parallel with transportation information programmes - attribution of effects is difficult yet countries have maintained momentum to improve efficiency even when energy prices declining - both mandatory and voluntary programmes have achieved targets

1. Refer to market limitations described in Chapter V.

Different policy measures can address these limitations although in different ways. Their use depends on such factors as complexity and cost of implementation, government policy, legislative requirements and timing required. Often the combination of more than one policy measure or the offering of an array of programmes to suit different market segments can be even more effective. For example, residential energy audits are most effective when combined with financial incentives, and fuel efficiency standards are best when accompanied by guides, labels and brochures. Therefore, policy measures should not be considered in isolation but as part of an integrated approach.

The main conclusions from the analysis of policy measures used by governments are:

— most policy measures have been implemented to either address the lack of information and technical skills or the lack of access to financing or the economic attractiveness of conservation activities;

— financial incentives have shown to have positive benefit-cost ratios with the best results in the industrial and transformation sectors;

— there is a wide range of incremental effects due to a variety of factors including percentage of financial incentive, range of eligible items, implementation and programme design;

— more focussed programmes are more effective than general ones;

— awareness and motivation are not static and need to be reinforced periodically or through feedback mechanisms to give the consumer a better understanding of his energy use;

— standards and regulations have been most useful in the residential/ commercial and transportation sectors where there is more standardization of equipment and where there are special market segments such as rental accommodation;

— standards and regulations ensure minimum efficiency levels are met;

— effectiveness of programmes depends on good implementation, requiring human and financial resources, co-ordination within administrations and with other levels of government;

— in the industrial sector, a combination of programmes encourages businesses to develop good energy management. These programmes include energy audits, training, monitoring and targeting, technical materials and technology transfer programmes;

— there is still too little known about the effectiveness of programmes, and evaluations that have been undertaken have generally not been sufficiently comprehensive.

The specific conclusions are:

— thorough analysis should be made of the remaining opportunities for energy efficiency improvements, the obstacles to their achievement and which decision makers will need to act;

— upon identifying areas of economic potential which are unlikely to be achieved by the market, governments should assess the full range of policy instruments in order to determine the most appropriate and cost-effective programme mix for each situation;

— in designing and implementing programmes, every effort should be made to ensure their effectiveness and maximise the incremental conservation action that results;

— conservation programmes should be rigorously evaluated periodically to ensure they are meeting policy objectives and maximising effectiveness;

— information programmes should be the cornerstone of every conservation strategy: they can motivate and create awareness, explain conservation opportunities, improve technical skills, and publicise other government programmes;

— financial incentives should be used selectively to support the operation of the market by providing access to needed capital; motivating consumers to undertake conservation efforts; helping the introduction of new technologies; and helping to develop conservation services;

— regulations and standards can be valuable to keep the long-term momentum and to reach special market segments (e.g. the residential sector which is the least price responsive end-use sector, rented buildings and markets heavily influenced by style and advertising (e.g. automobiles)). They ensure minimum levels of effort, are useful during periods of energy price fluctuations and should be reviewed periodically;

— energy efficiency objectives should be carefully integrated with industrial, social, fiscal and other policies that affect energy use.

CHAPTER IX

Research, Development and Demonstration

Government support for the research, development and demonstration (RD&D) of more energy-efficient technologies has been a major element of the conservation efforts of many IEA countries since the mid-1970s. These efforts have resulted in the development or demonstration of a large number of new technologies and conservation techniques, some of which have already begun to have significant effects on end-use demand. Some countries, such as the United States and Japan, have emphasized long-range research aimed at achieving major advances in basic technologies, while other countries have focussed on demonstrating, evaluating and supporting the introduction of existing, under-utilised technologies.

The main contribution to promoting energy efficiency over the rest of the century is likely to be made by the commercialisation and diffusion of existing and new technologies rather than research and development. Demonstration — the trial of newly developed technologies or applications under normal working conditions on a large enough scale to determine with relative assurance the economic and technical feasibility of a full commercial application — is thus of particular importance as one of the major means of transferring the knowledge and experience learned about new and existing technologies. Further R&D is required to develop new technologies into the next century. Socio-economic research also has an important contribution to make to the formulation of energy conservation policies. The following chapter examines the government role in the support of such research efforts, the effectiveness of past government efforts and some possible future directions [1].

---

1. See Stern et al., *Energy Use: The Human Dimension,* National Research Council of the United States, 1985.

I. Technologies

A. *Focus of RD&D activities*

RD&D programmes over the last decade focussed on and continue to advance such technologies as heat pumps, heat exchangers, microwave industrial heaters, microelectronic controls, energy storage, municipal solid waste systems, DH and CHP. Utilisation of waste heat, new building materials, building design, modelling of heat loads, ceramic engines and control systems are further examples of the myriad of projects which governments and industry have undertaken.

A recent IEA study [1] identified end-use technologies where it is important to pursue further RD&D. They are:

Industrial Sector

— energy-efficient production systems;
— heat recovery systems (e.g. heat pumps);
— recycling of energy-intensive products.

Residential/Commercial Sector

— building design;
— heating and cooling systems (e.g. district heating, heat pumps);
— total energy management systems.

Transportation Sector

— more efficient vehicles (e.g. ceramic engines).

Energy Transformation Sector

— more efficient power plants (e.g. combined cycle, fuel cells);
— cogeneration systems (organic rankine cycle (ORC)).

---

1. IEA, *Energy Technology Policy*, OECD, Paris, 1985. See Annex G for a review of the identified technologies.

Microelectronic sensor and control systems have an important role in many of these individual technologies. By applying these systems, energy consumption can be precisely adapted to actual requirements, thereby increasing energy efficiency. The same also holds for complete systems such as an electricity distribution system, where the deployment of these systems in conjunction with more effective storage systems could revolutionize electricity demand management.

B. The Roles of Government and Industry

The level and type of government activity in RD&D vary by industry and country. In the transportation sector, the automobile industry for the most part undertakes its own RD&D work relating to more efficient motors and vehicle design, although governments are involved to some extent in basic fuel efficiency research and play a major role in aviation RD&D. Some of this RD&D for automobiles has probably resulted from government targets for fuel-efficiency. Efficiency improvements are incorporated into new vehicles after internal demonstration work but without the educational demonstration stage for consumers.

In the industrial sector, companies often do their own R&D to improve their own efficiency. This is particularly important for certain energy-intensive industries such as cement and pulp and paper. In addition, manufacturers who make equipment (e.g. boilers) for use by other industrial consumers conduct R&D work to improve the efficiency of their products, sometimes followed by demonstrations to test products and educate the potential consumers. Small- and medium-sized firms generally do not have the funds or the expertise to undertake R&D.

The building industry performs much less R&D work, and faces major technology transfer problems when new technologies or processes are developed. Extensive demonstration efforts are necessary to reach the other decision makers in the building sector, including builders, architects, building owners and tenants.

Whether government funds are applied to R&D projects — conducted by government, universities or industry — or to demonstration projects — usually conducted by industry but often in collaboration with government — depends in large part on the philosophy of the government and the economic and industrial structure of the country. Certain countries such as the United States support and conduct R&D,

leaving industry to take over in following through with demonstration. In practice, however, R&D efforts in the United States are sometimes defined in a way which covers projects which in other countries would be regarded as demonstration. In specific technology areas, government funding for all RD&D can be greater than that of industry. In other cases, where industry has undertaken the basic R&D work on its own and is then faced with funding difficulties at the technology transfer stage, governments have concentrated their efforts on technology transfer. The United Kingdom, for example, although involved in R&D work, puts more emphasis on demonstrations which have significant replication potential through the Energy Efficiency Demonstration Scheme. Japan also has achieved notable success in technology transfer through extensive co-operation between government agencies and industry.

Governments also have a role to play in international conservation RD&D activities. International co-operation is generally more beneficial for expensive and universally applicable technology developments, but it can provide advantages for conservation RD&D, such as limiting duplication and allowing more to be done with total available resources; permitting the early adoption of technologies to specific national or regional circumstances; and permitting a wider range of approaches and optimal use of available scientific talent.

C. Funding for RD&D Programmes

Government spending on RD&D on conservation technologies has increased since 1977. While government energy RD&D budgets actually declined 7.6% between 1977 and 1985, conservation RD&D budgets increased 23.3%, although from a low base (see Table 11). In 1984 conservation represented 6.2% of total energy RD&D budgets compared to 4.7% in 1977. Although industry figures are difficult to obtain and IEA aggregate figures are therefore incomplete, indications are that expenditures for conservation RD&D in the private sector are substantially higher than government expenditures [1].

---

1. Nine countries (Australia, Austria, Italy, the Netherlands, Norway, Spain, Turkey, United Kingdom, United States) submitted estimates of industry energy conservation RD&D expenditures in 1984 totalling $3 052.13 million. Total government energy conservation RD&D expenditures for those nine countries in 1984 were $376.03 million.

Table 11

IEA Government RD&D Budgets for Energy Conservation[1]

($ millions)

Country	1977	%[2]	1979	%	1980	%	1981	%	1982	%	1983	%	1984	%	1985	%
Australia	5.3	(18.0)	3.9	(8.1)	5.8	(9.4)	6.8	(9.7)	N.A.	—	10.2	(12.5)	N.A.	—	N.A.	—
Austria	5.3	(27.9)	5.9	(27.4)	7.2	(33.0)	6.0	(29.6)	6.9	(31.2)	5.3	(25.4)	7.6	(33.3)	6.5	(31.0)
Belgium	5.5	(6.6)	3.6	(6.9)	5.4	(6.2)	6.0	(7.9)	4.5	(6.4)	6.0	(9.5)	7.1	(10.60	4.3	(7.0)
Canada	13.4	(5.5)	19.5	(6.9)	25.5	(9.1)	37.2	(11.3)	43.4	(12.7)	54.6	(14.0)	54.8	(12.4)	50.4	(13.4)
Denmark	1.9	(11.4)	2.1	(7.8)	3.4	(17.2)	3.9	(30.0)	3.4	(27.2)	3.8	(31.2)	3.2	(29.1)	2.7	(26.0)
Germany	13.9	(2.2)	32.4	(4.4)	38.2	(5.1)	43.5	(5.4)	28.7	(2.8)	22.1	(3.6)	12.9	(2.1)	14.2	(2.6)
Greece	.0	(0)	.0	(0)	.1	(0.3)	.1	(0.3)	.6	(10.0)	.5	(9.1)	.1	(1.4)	.2	(2.2)
Ireland	.3	(12.5)	.4	(8.0)	.9	(15.0)	2.3	(32.9)	2.1	(35.6)	1.6	(43.2)	.7	(46.7)	.8	(38.1)
Italy	13.4	(6.7)	13.5	(4.9)	15.6	(5.4)	11.6	(2.4)	14.9	(4.0)	21.6	(4.9)	25.9	(4.4)	19.3	(3.4)
Japan	55.9	(8.0)	62.1	(6.5)	31.6	(2.3)	18.2	(1.3)	10.7	(1.0)	11.7	(0.9)	11.9	(0.8)	12.3	(0.8)
Netherlands	11.2	(10.8)	13.6	(12.1)	17.0	(14.9)	17.3	(14.6)	13.8	(14.1)	19.0	(19.5)	16.6	(19.8)	19.2	(16.7)
New Zealand	.8	(21.0)	.8	(12.5)	1.3	(11.9)	1.3	(14.0)	1.2	(14.3)	.9	(11.4)	1.7	(18.1)	1.4	(14.0)
Norway	2.3	(8.2)	5.5	(14.0)	6.2	(17.2)	5.4	(17.5)	4.7	(17.2)	4.4	(19.4)	3.9	(18.0)	3.5	(16.1)
Portugal	N.A.	—	N.A.	—	.1	(2.7)	.2	(6.9)	.2	(6.9)	.3	(8.3)	1.7	(29.8)	1.5	(27.2)
Spain	2.5	(6.4)	2.3	(4.2)	2.2	(3.4)	3.3	(5.4)	8.1	(14.5)	25.2	(19.8)	38.0	(26.1)	34.4	(27.7)
Sweden	22.9	(36.7)	23.5	(22.7)	29.9	(30.2)	32.3	(23.5)	36.6	(33.2)	35.4	(35.9)	28.0	(30.6)	22.3	(27.1)
Switzerland	2.6	(8.6)	4.0	(8.5)	5.8	(11.5)	5.7	(12.1)	4.9	(11.0)	4.7	(10.2)	6.6	(14.6)	7.0	(14.5)
Turkey	N.A.	—	N.A.	—	.2	(7.4)	.2	(16.7)	.4	(21.1)	.9	(32.1)	0.6	(25.0)	0.7	(30.4)
United Kingdom	21.2	(6.4)	22.9	(5.7)	20.2	(5.0)	21.5	(5.0)	40.1	(10.3)	44.3	(10.6)	32.6	(8.4)	37.1	(10.2)
United States	140.1	(3.4)	185.3	(3.6)	380.9	(7.5)	266.8	(6.8)	162.5	(5.4)	226.7	(8.0)	169.7	(7.2)	173.6	(7.7)
TOTAL	318.7	(4.7)	397.3	(4.6)	597.3	(6.5)	489.6	(5.8)	387.8	(5.2)	499.3	(7.0)	423.7	(6.3)	411.5	(6.2)

1. In 1985 U.S. dollars.
2. Percentage of total government RD&D budgets.

Source: Country submissions.

These aggregate IEA figures are greatly affected by United States expenditures which since 1977 have represented as much as 63% of total IEA government RD&D budgets in 1980. In 1985 the United States represented 40% of the IEA total. Netting out the United States from the IEA total shows for the most part a steady increase in spending on conservation from 1977 to 1983, except for a dip in 1979, and a slight decrease since 1979. When the United States is included, RD&D conservation peaked at $597.3 million in 1980, dropped dramatically over the next two years, increased again in 1983 to $499.3 million and declined slowly since.

There is a wide range of perceptions among IEA countries regarding the need for government support for RD&D. As shown in Table 11, support to conservation RD&D as a percentage of total government support for energy RD&D ranged from 0.8% in Japan in 1985 to 31% in Australia and 38% in Ireland.

Government RD&D programmes vary according to the needs of the country, the industrial infrastructure, and the philosophy of the government. No matter what emphasis is decided on, RD&D programmes interact with and are part of other government programmes. They almost always make use of other tools such as financial incentives, information programmes and regulations and standards for both the basic R&D work, and the integration of new technologies into the market. Examples of some innovative energy conservation RD&D government programmes in IEA Member countries include:

— The Canadian Super Energy Efficiency Housing Programme is a demonstration programme which promotes energy-efficient construction techniques in the residential sector. The programme provides consumer education, builder training, product development and monitoring of homes designed to satisfy the "R-2000" energy consumption standard. The programme takes advantage of the builders' knowledge and skills by allowing the builders to design and construct the buildings as long as the energy consumption of the finished product does not exceed a specified amount.

— In Japan, the "Moonlight Project" is designed to develop and commercialise energy-efficient technologies. The programme synthesises efforts applying to all end-use sectors. The government funds large-scale R&D projects (on a cost-sharing basis if the companies are actually able to apply such technologies). Financial assistance has also been provided to one or two private companies on an equal cost-sharing basis to develop efficient household appliances. A number of basic energy conservation research projects are conducted in National Research Laboratories through the Moonlight Project.

— In Sweden, the part of the government's research programme which concerns "new energy systems in buildings", focusses primarily on heat storage and heat pumps. The work is conducted through the co-operation of manufacturing industry, contractors, consultants, building operators and administrators, local authorities, government bodies, research organisations and the Institute of Technology.

— The United Kingdom stimulates the development and uptake of energy-efficient technologies and designs in two ways. First, the Energy Efficiency Demonstration Scheme provides financial assistance to organisations which are able to demonstrate new ways in which energy can be used more efficiently. Independent monitoring of the performance of demonstrated technologies, coupled with a co-ordinated promotional programme, are two key elements of the United Kingdom's approach. Second, the United Kingdom makes available, on a selective basis, funds to assist R&D into energy-efficient technologies and designs. Promotion and dissemination of the results are regarded as an important feature of the programme.

— In the United States, the government sponsors various industrial R&D projects in specific areas such as efforts to reduce energy requirements in aluminium and steel processing, recovery of energy from industrial waste, improved process efficiency and others. R&D programmes in the United States also include advanced propulsion technologies, ceramic materials and alternative fuel utilisation for the transportation sector, advanced energy storage technologies and basic research in materials science, biocatalysis and tribology. The approach taken to transfer new developments to industry is based on proving the performance of new technologies through testing and evaluation in the industrial environment and informing industry of the results. This transfer of information on new technologies is accomplished through close co-operation between the government and trade associations and professional societies, as well as technical reports, workshops, seminars and conferences.

There have also been extensive international efforts in the energy conservation RD&D area:

— the European Community has both R&D and demonstration programmes. For R&D, the Community's role has been to encourage co-ordination, disseminate findings and support R&D of certain problem areas (thermal analysis and improvement of buildings, energy saving and heat recovery in industry, advanced energy-saving technologies in transport, advanced heat pumps, advanced batteries and fuel cells). The demonstration programme offers up to 40% of the eligible cost of the project as a grant which is to be paid back in part if the project is successful. The programme is for projects which create full-sized replicable installations which either use alternative energy sources, save energy or substitute hydrocarbons.

— There are also IEA collaborative projects in energy conservation RD&D, covering a wide range of technologies. Collaborative projects are undertaken on either a cost-sharing basis, a task-sharing basis, or sometimes a combination of both. Results of the projects are shared by all participants. Examples of cost-sharing projects include two information centres: the Air Infiltration Centre and the Heat Pump Centre. The tendency over the years, however, has been towards task-sharing projects such as countries conducting research work in common, or exchanging information on ongoing activities within each country. Consideration is now being given to establishing a separate IEA information centre that would concentrate on end-use technologies.

D. Assessment of government programmes

Assessing the effectiveness of government involvement in RD&D activities is as important as for other types of policy instruments. For both R&D and demonstration, assessments must examine the design, management and results of the programme itself in light of its goals. In addition, assessments of R&D programmes must analyse the type of technology work being done in terms of whether funded projects offer significant potential savings, and how much and whether the technologies are marketable. Demonstration programme assessments must also look beyond a micro analysis of the programme to assessing whether it is attracting useful technologies and whether it is increasing the market penetration of the new technology. The market limitations which can most effectively be overcome by effective RD&D programmes are lack of information and technical skills, and lack of confidence in new technologies. While the results of programme evaluations depend on each country and programme, the lessons learned can be valuable for other governments.

Selected country programme evaluations include:

Sweden — Heat Pump RD&D

— In Sweden, an evaluation by the Energy Research Commission (ERC) of government support on research for heat pump technologies found that the results led to a substantial development of the technology and that there has been good transfer of the know-how. Direct effects in terms of the near-market technical development were concluded to be small, although the assessment was made against a rather short perspective. Therefore, energy-savings assessments were considered to be very difficult. It was concluded that, for further work, co-operation with the manufacturing industries and energy producers and distributors should be strengthened.

Conservation Demonstrations in the United Kingdom

— The United Kingdom's Energy Efficiency Demonstration Scheme is periodically evaluated in terms of its established targets for the level of energy savings, government expenditure and industry expenditure. Assessment of how well these goals are being reached results in the readjustment of interim targets, of how the Scheme is carried out and in what areas. For example, the initial emphasis was on industry, but this has since been expanded to include both domestic and non-domestic buildings. Some R&D work, in addition to demonstration projects, was added while the programme was under way. The fact that the programme is so closely monitored provides valuable guidance for future government expenditure in this field and ensures that national energy policy objectives continue to be served.

Conservation R&D in the United States

— The United States has conducted an extensive analysis of energy conservation R&D activities in order to determine areas that need to be addressed by government. Separating the effects, in terms of energy savings, of government and private sector activities has not been attempted, nor has the role of government in specific R&D projects been evaluated separately. Effort has been made, however, to identify developments which were accelerated through Federal participation.

The study found that government and private sector cost-shared projects in industry have supported over 200 technologies, of which 23 are now complete and being commercialised by industry. Accumulated energy savings from those 23 technologies are estimated to be 116 trillion Btus. Government efforts have also resulted in improvements in the transportation sector due to ceramics and automotive battery improvements.

The study also identified twelve technology areas within the industrial sector which need to be developed to improve energy efficiency [1]; four areas within the transport sector [2]; and three technology areas within the residential/commercial sector [3].

Conservation Demonstrations in the EEC

— The European Community's Demonstration Programme has been working continuously since 1979 and will continue at least until 1989. By the end of 1986 and since the beginning of the programme, some 4 200 proposals have been introduced of which some 1 200 have been selected for financial support. The overall

---

1. Process electrolysis, carbothermic processing; catalysis research; comminution; materials processing; sensors and controls systems; separation/concentration of chemical components; coatings and adhesives; energy cascading; waste heat recovery; combustion efficiency; and industrial waste utilisation.
2. Batteries and propulsion systems for electric vehicle commercialisation; new and more efficient heat engines for automotive applications; heavy-duty diesel engine technology, including more fuel-efficient, fuel-flexible and cost-effective systems; ceramics, ceramic composites, and ceramic coating development.
3. Building subsystems, systems integration, building retrofit and standards and guidelines; building equipment; and community systems, including district heating and cooling, community energy planning, development and management.

budget for the eleven-year programme is approximately 850 million ECU [1], which means that this programme is the biggest of its kind in the world. All areas are covered:

— energy savings (industry, buildings, transport, energy industry);
— alternative energy sources (biomass and energy from waste, solar, wind, hydro, geothermal energy);
— substitution of hydrocarbons (solid fuels, electricity and heat);
— liquefaction and gasification of solid fuels.

Some 150 projects have been completed with a "contractual success" rate of over 50%. In some 40 cases, the success has been repeated through replication elsewhere in the Community.

The programme has been evaluated twice, once in 1982 and again in 1984-85 by an external group of experts. The overall conclusion from both evaluations was that the programme fulfilled a specific need in the area and showed very positive results.

II. Socio-Economic Research

The formulation of effective conservation policies — like the formulation of an effective marketing plan for a product — requires an understanding of the many non-technical factors which influence efficiency improvements and which can make policies (or marketing strategies) more effective. There are four main areas of research:

— Consumer behaviour research. This is important because the factors influencing energy-consuming behaviour are complex and little work has been undertaken. Most of the work has been done in the United States although there is evidence that this has tapered off [2]. Some utilities in the United States which are putting heavy emphasis on conservation are still doing research. Europe started research much later than North America and in the past year

1. In September 1986, 1 ECU = US$1.
2. See Stern et al., op. cit.

activity has started to wane. However, the Swedish State Power Board has just started a seven-year project to study how to use energy more efficiently. The reason for starting the project is that the remaining surplus of electricity generating capacity is small and that options for expansion are few. The project will also study consumer behaviour, the use of energy today and how consumers react to increasing prices and changes in the price structure of electricity.

— Micro-economic research for improving the evaluation of policy measures. As was evident in Chapter VIII, there are still many methodological problems in evaluating the effectiveness of policy instruments. Research has been undertaken to improve the quality of evaluation methodologies and to increase their use in the evaluation of conservation.

— Macro-economic research to determine broad trends, impacts and influential variables, such as the impact of conservation on employment or environment, the development of various energy scenarios or the attribution of aggregate energy efficiency improvements to energy prices or government policies.

— Techno-economic research to have a better understanding of how energy is used, to develop better assessment of conservation potential and to develop better statistical indicators of improvements in energy efficiency. For the last, the European Commission has a two-year research project under way.

It is essential that socio-economic research should be developed appropriately in order to provide a better understanding of the factors which influence consumer behaviour and of the effectiveness of policy measures. More government support for socio-economic research as appropriate is needed to achieve this result. The IEA could help this development by providing a clearing-house for the exchange of information on the results of socio-economic research in Member countries.

Research in all these areas is hampered by the inadequacy of data about energy use. Consideration needs to be given at both national and international levels to improvement of the quantity and quality of these data to the extent that government expenditure and staffing constraints permit.

III. Conclusions

RD&D on energy conservation is widely spread over governments and industry. Substantial progress has been made in the last ten years. More needs to be done, however, to demonstrate new technologies which are technically viable, to improve the assessment of government programmes and to develop social research relevant to the formulation of energy conservation policies and programmes. Specific conclusions to these ends are:

— those Member governments which do not support demonstration programmes or other appropriate methods of technology transfer should look again at their position in the light of the success of such programmes in other IEA countries and in the European Community;

— the design, management and results of RD&D programmes should be carefully and regularly assessed; in the case of demonstration and technology transfer programmes, evaluations should examine the effectiveness of the effort in overcoming market limitations to the adoption of new technologies;

— the lessons learned from assessments should be made available to other governments directly and through the appropriate international organisations;

— socio-economic research should be continued or expanded as appropriate to ensure better formulation and assessment of conservation policies and programmes;

— efforts should be made to improve the quantity and quality of data on energy consumption to improve international comparisons of energy efficiency;

— there is a need for closer collaboration within government (since RD&D are often in ministries that do not handle energy policy) and between government and industry to avoid unnecessary duplication and to optimise use of resources.

CHAPTER X

The Exemplary Role of Governments as Energy Consumers [1]

Governments use energy directly and have the opportunity to set an example of good conservation practices. In doing so they will encourage others to use energy more efficiently. But failing to do so, will suggest that energy conservation is unimportant.

Governments also have a responsibility to manage their resources well. Like industry and commerce, many governments have developed energy management programmes. They have come to recognise that energy costs, constituting a significant proportion of operating costs, can be controlled and provide many financial benefits.

Governments face most of the same obstacles that confront other energy consumers. For example, many governments rent accommodation and face the same landlord-tenant concerns described in Chapter V. In many ways, the task is more complicated than for many industries because of the autonomy of various government organisations and the lack of incentives for cutting costs. While many governments have one central group responsible for conservation policy, the responsibility for energy use is dispersed [2]. For example, in IEA countries, responsibility for

1. This chapter deals with energy use directly controlled by governments and not with the entire public sector.
2. For a description of government organisation, refer to Chapter VI.

conservation in government buildings commonly falls to Ministries of Public Buildings or Administrative Affairs, in the defence estate to Defence Ministries and in schools to Education Ministries.

Government units have varying levels of autonomy in implementing programmes to achieve their energy efficiency goals. There remains, nevertheless, a need for the central co-ordination of government-wide programmes. Common functions of the co-ordinating body are to provide advice, guidance and assistance to the various participating organisations, to undertake training programmes, to report on total public sector energy consumption and to initiate measures to improve energy management. Often the point of co-ordination is within the ministry responsible for energy. In Canada, the Federal Energy Management Programme is co-ordinated by the Department of Energy, Mines and Resources which must make an annual progress report to Parliament. Nevertheless, it is useful to have each government administrative unit responsible for implementing conservation programmes. In the United Kingdom, the Energy Efficiency Office, which takes the lead in Government for co-ordinating energy efficiency policy, includes action within the government estate within its campaign. The Prime Minister has instructed all government departments to appoint energy managers. In the Australian state of New South Wales, a centralised payment of gas and electricity accounts for some departments through a "group vote" system had operated for many years. The decision of the NSW government to introduce individual departmental votes for these cost items from July 1982 clearly assigned management responsibility and accountability for all energy use to the individual departments.

Some countries have comprehensive mandatory programmes. This is valuable in requiring the large energy users (public works, defence) to undertake conservation efforts. In Canada, the Internal Energy Conservation Programme was created by the Cabinet in 1976 with specific targets and requiring each government organisation (departments, agencies and Crown Corporations) to develop its own energy management plan. In Denmark all public buildings (municipal, regional, state) must as far as possible be brought up to a reasonable energy economic standard by the end of 1989. The United States also has a mandatory programme. Among its specific objectives are reducing energy consumption, using energy more efficiently, and transferring applicable energy management experience and technologies among federal agencies and the private sector. Many governments have mandatory temperature settings or speed limits for their vehicles.

Results of Energy Conservation Programmes in the Government

The governments of many IEA countries have introduced several measures to conserve energy in their public sectors. Some of the results include:

— The state Energy Management Programme in New South Wales is the oldest in Australia. It was established in 1979, and the first two years of its operation saw a 16% reduction in government energy consumption. The main features of the programme include maintaining the government's total energy consumption (excluding that required for electricity generation and public transport) at the 1978-79 level for five years; applying energy conservation concepts in the design and construction of all government buildings and projects; and purchasing lighter, more fuel-efficient vehicles.

— In Austria government activities initiated in the mid-1970s led to a decrease of 2.5% in total energy consumption in federal buildings between 1978 and 1983. This decrease occurred despite a 25% increase in the size of the federal building estate and was achieved by the appointment of officials responsible for energy efficiency in all ministries, a high level of capital investment in energy conservation equipment (starting around 1978, increasing up to around 1 billion schillings, between 1981 and 1983), better training of boiler staff and increased use of district heat systems.

— By 1980, the goal of the Canadian programme had already been surpassed and the programme was redesigned. The Federal Internal Retrofit Programme and the Federal Off-Oil Programme were added to the IECP to form the Federal Energy Management Programme (FEMP). The FEMP shifts emphasis and responsibility for energy savings target setting towards individual organisations. Total government direct energy consumption in 1982-83 was nearly 5% lower than in the previous year and more than 21% lower than in 1975-76, the year before the IECP began. Energy savings in government accommodation were even more pronounced — 9% over the previous year and more than 24% over 1975-76.

— In Germany, it was found that the Federal Ministries reduced their energy consumption by 25% between 1979 and 1983. Similar decreases were noted for state and local governments.

- In the Netherlands, 0.18 billion guilders [1] were invested in the period 1980-83 for efficiency measures in central government buildings. Energy savings are estimated as 75 million cubic metres gas per year, indicating rather low payback periods between three and five years.
- In the United Kingdom, there is a target of 70% reduction in the energy consumption level of government buildings between 1972-73 and 1988. The energy bill in the government estate is now £100 million lower than it would have been otherwise.

However, not all programmes have achieved their desired results. For example, in the United States total energy consumption in fiscal year 1984 was 1.6% below that in 1975 and the total expenditure was approximately $1.0 billion less than in 1983. However, there is a requirement to reduce the amount of energy used per square foot by 20% between the fiscal years 1975 and 1985. As of fiscal year 1984 a reduction of only 5.4% had been achieved. The Secretary of Energy has written to all federal agencies to encourage better results for 1985.

Conclusions

Governments need to set a positive example of good energy management to all consumers. In addition, officials need to publicise and explain conservation strategies. Specific conclusions are:

- socio-economic research should be continued or expanded as appropriate to ensure better formulation and assessment of conservation policies and programmes;
- efforts should be made to improve the quantity and quality of data on energy consumption;
- there is a need for closer collaboration within government (since RD&D are often in ministries that do not handle energy policy) and between government and industry to avoid unnecessary duplication and to optimise use of resources.

1. On the average in 1986, 1 guilder = $1.

Annex A

IEA Principles for Energy Policy

In 1977, IEA Energy Ministers adopted twelve Principles for Energy Policy to guide Member countries in planning and carrying out energy programmes:

1. Reduce oil imports by conservation, supply expansion and oil substitution;

2. Reduce conflicts between environmental concerns and energy requirements;

3. Allow domestic energy prices sufficient to encourage energy conservation and development of energy supplies;

4. Slow energy demand growth relative to economic growth by conservation and substitution;

5. Replace oil in electricity generation and industry;

6. Promote international trade in coal;

7. Concentrate natural gas on premium users and expand its availability;

8. Steadily expand nuclear generating capacity;

9. Emphasize research and development through international collaborative projects and more intensive national efforts;

10. Establish a favourable investment climate to develop energy resources, with priority for exploration;

11. Plan alternative programmes should conservation and supply goals not be fully attained;

12. Seek appropriate co-operation with non-Member countries and international organisations.

Annex B

Lines of Action for Energy Conservation and Fuel Switching

Adopted by the Governing Board at Ministerial Level
on 9th December 1980

GENERAL

1. Ensure that governments are, and are seen to be, energy efficient in the operation of their own buildings, transportation fleets and other activities, and that other fuels are substituted for oil in government operations wherever possible.

2. Encourage market confidence in new energy conserving equipment and processes, for example by demonstrating the use of such products and technologies and purchasing them for use within government facilities.

3. Give high priority to assessing the results achieved by existing programmes, as a basis for stronger and more effective action.

4. Develop training programmes for skilled labour to expand technical expertise in energy management, and make parallel efforts to stimulate energy management wherever practical, in line with the IEA Energy Management Initiative announced in October 1980.

APPROPRIATE ENERGY PRICING

5. Allow energy prices to reach a level which encourages energy conservation, movement away from oil, and the development of new sources of energy.

INDUSTRIAL SECTOR

6. Actively support industry efforts to increase energy efficiency and fuel substitution, by ensuring an overall economic climate that encourages investment and by working with industry to provide advice and, where necessary, fiscal, financial and legislative means to encourage rapid adoption of modern equipment and technologies. Target setting and monitoring of progress by industry itself or in co-operation with government can lead to a more rapid move towards efficient processes and substitution away from oil.

7. Investigate urgently the potential for the productive use of waste heat and encourage its use by providing incentives or removing institutional barriers. In particular, governments should encourage the development of combined heat and power facilities, by ensuring that satisfactory and reasonable commercial arrangements exist for linking such facilities to existing electricity grids or in the case of surplus heat, to link such facilities to heat distribution networks.

ROAD TRANSPORTATION

8. Carefully assess and, as appropriate, strengthen and extend through the 1980s current policies to improve automobile fuel efficiency. Countries without fuel economy standards should consider their introduction where necessary.

9. Ensure that automobile testing procedures reflect actual road use and continue efforts to examine the possibility of developing tests that will allow the introduction of standards for vans, trucks and energy-intensive recreational vehicles.

10. Review the level and structure of fuel taxes and purchase and road taxes for automobiles with a view to encouraging oil savings and improving fuel efficiency.

11. Ensure that automobile owners are well informed about the financial and energy savings from better maintenance and driving habits. The regulation of maintenance should be considered along with upgrading of automobile mechanics' knowledge of energy-efficient tuning procedures.

12. Introduce stronger measures to encourage and support the use of public transportation systems.

RESIDENTIAL AND COMMERCIAL BUILDINGS

13. Examine the rate of insulation of present homes and commercial buildings as well as the potential for improved efficiency of heating and cooling systems, and, where necessary, stimulate change by regulatory means or by providing incentives. In particular, the training of installers and builders should be upgraded.

14. Encourage solar heating and cooling technologies where they are economic.

15. Consider the introduction of mandatory codes for new buildings which cover all energy use.

16. Examine urgently the particular constraints inhibiting improved energy efficiency in rented accommodation, and develop solutions to these difficulties.

17. Support and encourage the substitution of non-oil fuel, used either directly (including district heating) or converted to electricity, in residential and commercial use wherever infrastructure exists or can be provided.

ELECTRICITY GENERATION AND TRANSMISSION

18. Reduce oil-fired generation as rapidly as economically and technically possible by substituting other fuels so that oil-fired capacity is used primarily to meet middle and peak loads. No new oil-fired plants should be authorised except in particular circumst-

ances where there are no practical alternatives. Particular efforts should be made to facilitate reconversion of oil burning plant to coal or other solid fuels, wherever possible.

19. Examine the potential for reducing transmission losses by upgrading electricity grids.

Annex C

Extract from Conclusions of the Meeting of the IEA Governing Board at Ministerial Level on 9th July 1985 pertaining to Energy Conservation

Ministers adopted the following Conclusions.

I. Conservation

Ministers noted that the IEA have under way a wide-ranging study designed to help governments assess which conservation programmes are likely to be most cost effective. While detailed policy proposals must await the completion of this study, on the basis of the work so far done, Ministers concluded that in order to further reduce the energy intensity of IEA economies, government conservation policies should be actively pursued and should focus on the following types of action which, depending on national circumstances, could assist in achieving greater energy efficiency:

(a) Ensuring that the energy pricing and tariff systems give the right signals to consumers.

(b) Ensuring that information programmes are well directed towards the removal of the obstacles to energy conservation.

(c) Identifying what financial barriers exist, helping to improve access to financial resources and encouraging where appropriate the use of innovative financing schemes by the parties concerned.

(d) Improving the skills of the conservation service industry.

(e) Developing more effective evaluations of their conservation programmes and a better understanding of the factors which influence consumer decisions.

(f) Standards and regulations.

(g) Well designed programmes of research, development and demonstration.

Annex D

Developments in Energy Demand and Energy Efficiency, 1973-1985; Supporting Data and Analyses

This Annex includes the additional data and analyses underlying the summary material included in Chapter III of the study. For an introduction to the basic trends and concepts examined in this Annex, refer to Chapter III. In order to avoid duplication, the Annex contains a number of references to tables or figures contained in the body of the report.

The main purpose of this Annex is to describe recent trends in energy efficiency and how they have affected overall energy demand. One statistical indicator of energy efficiency improvements is the ratio of the energy use to GDP of Member countries, called energy intensity. The energy intensity of IEA Member countries decreased by about 20% from 1973 to 1985. This decrease in energy intensity, however, reflects both efficiency improvements and structural changes in IEA Member countries and therefore is only a rough indicator of the rate of progress toward improved efficiency. For individual countries, structural changes and other factors can have major effects on energy intensity. As a result, while efficiency gains were made in all Member countries over the past ten years, some countries actually increased their energy intensity. Table 1 (see Chapter III) provides a more detailed breakdown of energy intensity levels and trends.

Response to Movements in Energy Prices

One explanation for the change in energy intensity, particularly after 1979, is the change in the relative prices of production factors (labour, capital and energy). As shown in Figure 2 (see Chapter III), energy

Figure D.1

IEA — Indices of Real Energy End-Use Prices
1978 = 100

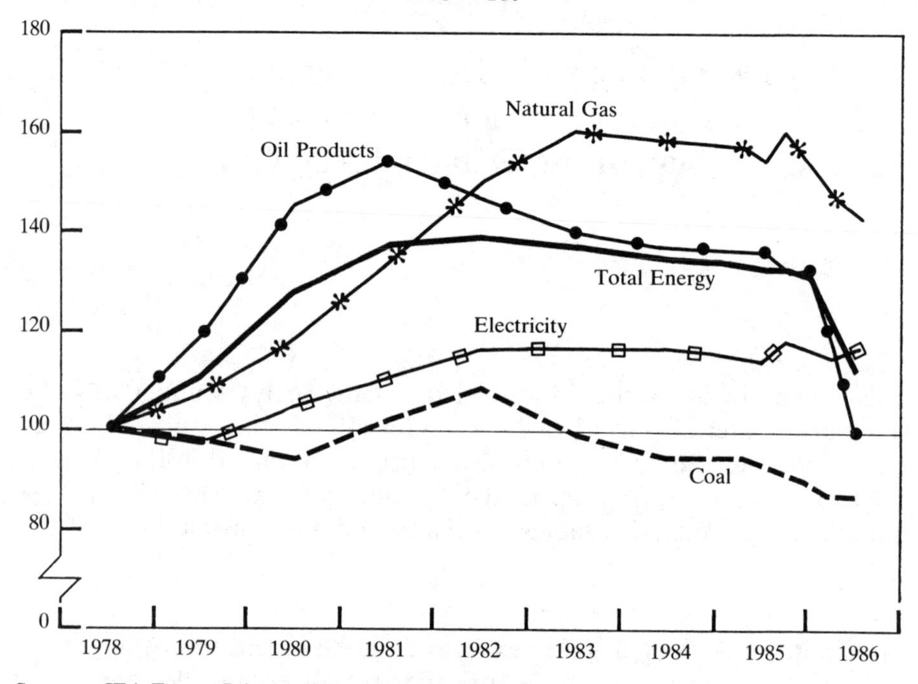

Source: IEA Energy Prices and Taxation.

prices increased roughly in line with the prices of other production factors between 1973 and 1978 following the steep rise in the relative price of energy in the 1973-74 period. Absolute energy prices in the total OECD rose by 78% from 1973 to 1978, while the absolute prices of capital (measured by the deflator of fixed capital formation [1]) and labour (measured by unit labour cost) increased by 57% and 53% respectively

1. From a theoretical point of view, cost of capital should be defined in more rigorous ways reflecting the institutional factors (such as depreciation method, and tax rules) and interest rates in the capital market. The fixed investment deflator adopted here is only a rough proxy to the true "user cost of capital".

during the same period. However, after 1979, energy prices rose much faster than in the 1973-78 period in both absolute and relative terms. Energy prices rose by 80% in just four years from 1978 to 1982, while the cost of capital rose by only 36% and that of labour by even less.

Figure D.2

North America — Indices of Real Energy End-Use Prices
1978 = 100

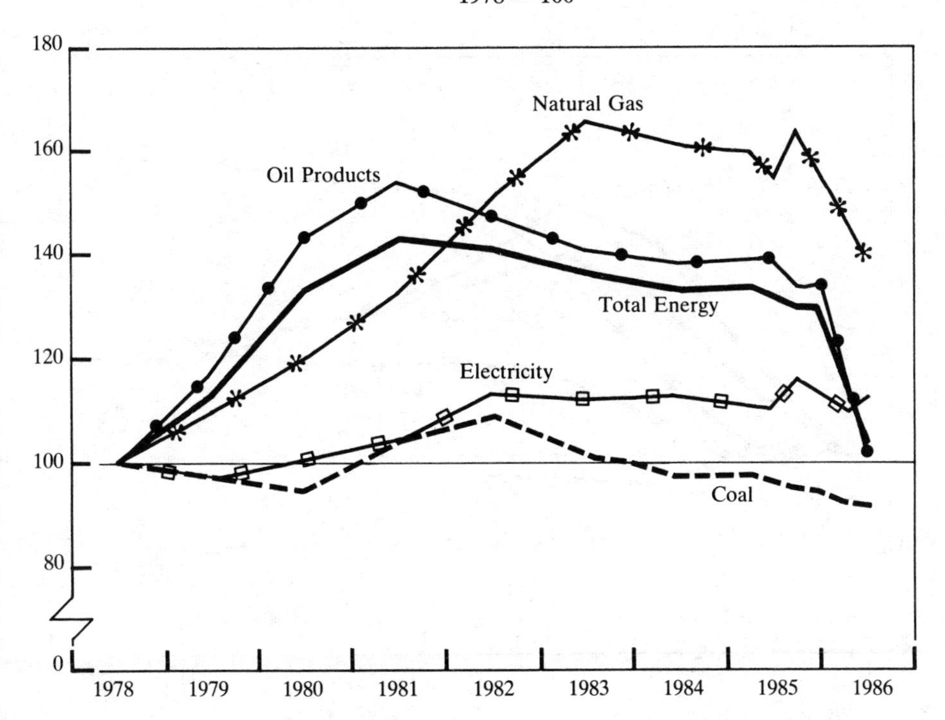

Source: IEA Energy Prices and Taxation.

Figures D.1-D.4 show changes in real prices for the major energy forms for the IEA as a whole and by region for the period 1978-1986. Note that average real end-use energy prices (based on price indices in national currencies) reached their peak in 1982 for IEA as a whole. They remained relatively stable until late 1985 and fell sharply during the first six months of 1986 to approximately the same level that existed in late 1979. Oil and gas prices have been most affected by the recent price decline, while coal has continued its gradual trend toward lower prices and electricity has been virtually unaffected.

Table 2 in Chapter III summarises the results of an IEA Secretariat analysis of the responsiveness of energy demand to increasing energy prices that have been experienced during several periods within the past thirty years. The analysis attempted to exclude the effects of income

Figure D.3

Europe — Indices of Real Energy End-Use Prices
1978 = 100

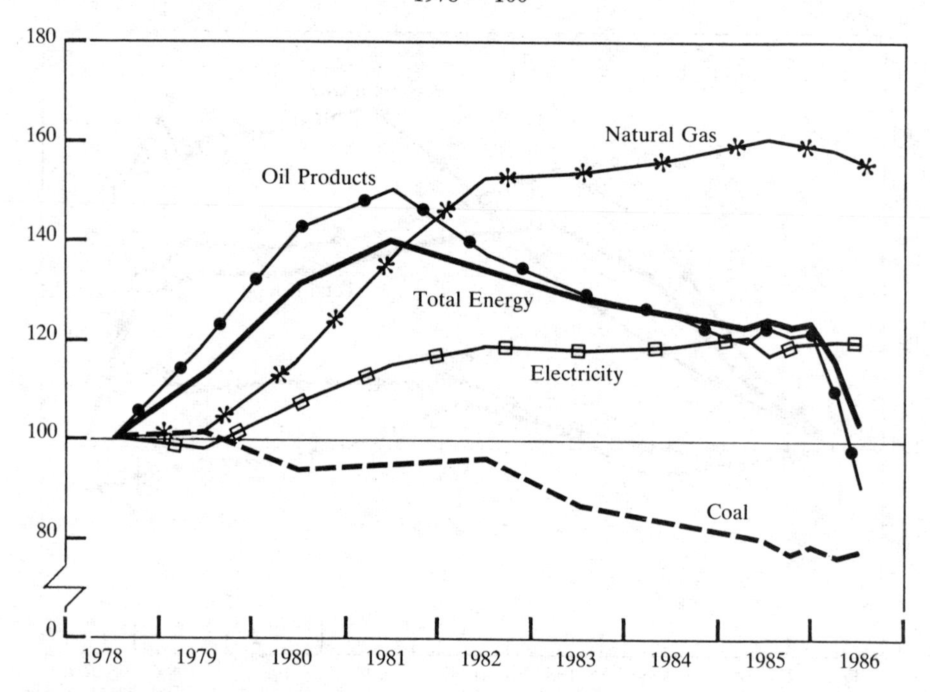

Source: IEA Energy Prices and Taxation.

growth and major structural changes that were not primarily or directly dependent on energy prices. The table indicates that energy demand has apparently responded strongly to rising retail energy prices, especially over the long term (up to ten years), in every sector and region. On average, a 10% increase in end-use energy prices results in about a 5% decrease in demand (an own price elasticity of 0.5) and, as other analyses have shown, most of this reduction is achieved through improved energy efficiency. A similar analysis concluded that energy demand was generally less responsive to declining energy prices; that is, demand generally increased *less* than 5% in response to price decreases of 10%. The analysis was based on data from periods of declining prices for specific fuels which occurred over the past thirty years.

Figure D.4

Pacific — Indices of Real Energy End-Use Prices
1978 = 100

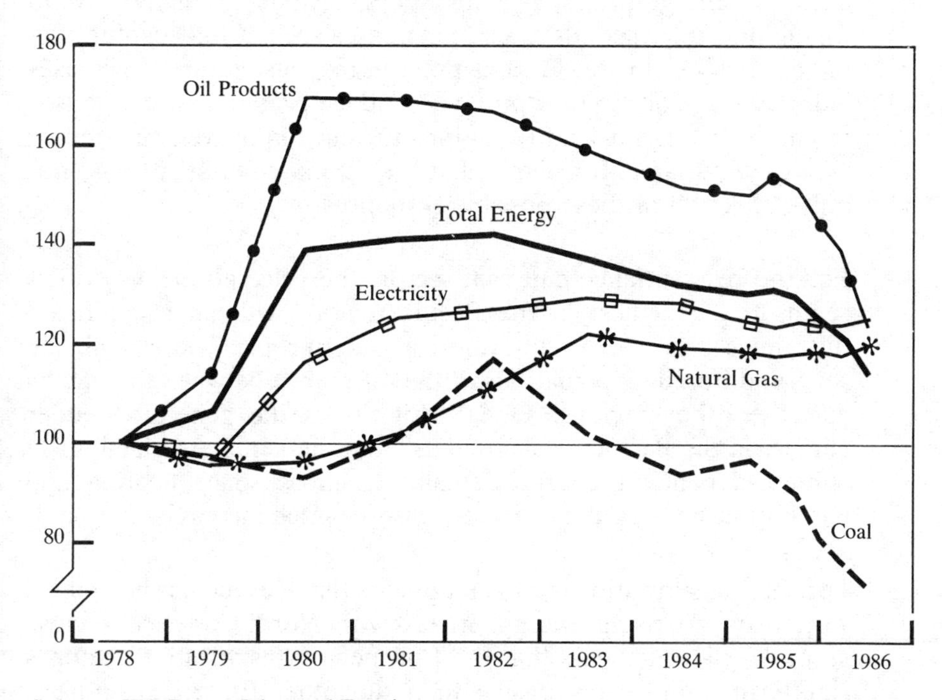

Source: IEA Energy Prices and Taxation.

Table 2 (see Chapter III) also reveals some interesting differences among sectors and regions. While small variances (less than 0.1) are not statistically significant, there are a few regional or sectoral elasticities that appear sufficiently different to warrant comment:

— the average elasticity for industrial fuels appears to be below those for the transportation and residential/commercial sectors. This suggests that industry may be less price elastic than other sectors, even though industry generally appears to be sensitive to energy price changes and the observed reductions in energy intensity have been greatest in this sector. These apparent contradictions have several possible explanations. Industry's lower average price elasticity may be because industrial energy demand is more strongly determined by technological and structural changes that are not directly dependent on energy price movements. Other analyses have indicated that there are long-standing industrial trends towards improved energy efficiency (and reduced energy intensity) and that these trends have continued even during periods of declining energy prices. It is also possible that the many ways in which industry responds to price increases are not fully captured in the elasticity estimate. For example, in response to price increases industry may shift to new products, reduce production, or sponsor research that might lead to major efficiency improvements ten or twenty years later. It is likely that the estimated elasticities do not fully reflect all of these possible responses.

— For the residential/commercial sector, the price elasticity of fuels seems to be highest in the Pacific region, with an insignificant difference between North America and Europe. There might be several different explanations of the difference between the Pacific and the other regions. One possibility is the generally milder climate in the Pacific, which makes larger percentage reductions in climate-dependent energy demand (such as space heating and cooling) easier to achieve in response to price increases.

— For the transportation sector, it appears that Europe has been most responsive to rising energy prices, with North America a close second. It also appears that the European transportation sector is significantly more responsive than the other European end-use sectors. In contrast, the price elasticity of transportation fuels in the Pacific (which is dominated by Japanese demand) is clearly lower than in the other regions and lowest among its other end-use sectors. These various differences are less easily explained. They may reflect the differing effects of economic growth on the transportation sector and the major differences in transportation patterns among the three regions.

Table D.1 shows the ratio of percent changes in energy intensity in terms of GDP to percent changes in real energy prices. The table provides a different perspective on the relationship between energy prices and

Table D.1: **Real Energy Prices and Energy Intensity** (Cumulative percentage change)

	Changes in Real Energy Prices[1]			Changes in Energy Intensity[2]			Ratio of Intensity Changes to Price Changes		
	1973-78	1978-82	1973-82	1973-78	1978-82	1973-82	1973-78	1978-82	1973-82
Residential/ Commercial[3]									
United States	57.7	40.6	121.7	-10.7	-11.4	-20.8	-0.18	-0.28	-0.17
Japan	10.1	69.8	86.9	- 8.3	-19.2	-25.9	-0.82	-0.28	-0.30
Germany	14.0	23.5	40.9	- 5.7	- 4.0	- 9.4	-0.41	-0.17	-0.23
France	47.1	64.3	141.6	-17.5	-18.2	-32.5	-0.37	-0.28	-0.23
United Kingdom	36.7	28.2	75.4	- 3.4	- 2.2	- 5.5	-0.09	-0.08	-0.07
Italy	59.2	67.4	166.4	- 5.4	-12.8	-17.5	-0.09	-0.19	-0.11
Canada	4.6	78.9	87.2	- 8.8	- 9.0	-17.0	-1.90	-0.11	-0.20
Total OECD	43.6	46.2	109.9	-12.9	-14.1	-25.2	-0.30	-0.31	-0.23
Industry[4]									
United States	96.8	42.1	179.6	-17.4	-23.4	-36.8	-0.18	-0.56	-0.20
Japan	39.9	44.8	102.6	-27.4	-16.3	-39.2	-0.69	-0.36	-0.38
Germany	56.0	32.0	106.0	-24.2	-24.8	-43.0	-0.43	-0.78	-0.41
France	73.5	65.9	187.9	-12.7	-23.7	-33.4	-0.17	-0.36	-0.18
United Kingdom	92.6	36.3	162.5	- 1.6	- 7.9	- 9.3	-0.02	-0.22	-0.06
Italy	52.2	80.6	174.9	-20.6	-22.5	-38.5	-0.39	-0.28	-0.22
Canada	70.3	77.7	202.7	-22.4	-29.5	-45.3	-0.32	-0.38	-0.22
Total OECD	79.8	49.9	169.5	-21.1	-24.9	-40.7	-0.26	-0.50	-0.24
Transportation									
United States	9.9	32.8	45.9	- 0.01	-18.6	-18.6	0	-0.57	-0.41
Japan	0.9	24.3	25.5	3.0	- 7.0	- 4.1	3.22	-0.29	-0.16
Germany	3.0	34.5	38.5	- 1.4	- 2.8	- 4.2	-0.46	-0.08	-0.11
France	22.9	8.9	33.9	- 1.2	- 2.1	- 3.2	-0.05	-0.23	-0.10
United Kingdom	9.5	40.7	54.2	- 1.0	-16.1	-17.0	-0.11	-0.40	-0.31
Italy	37.6	8.4	49.2	0.5	1.3	1.8	0.01	0.15	0.04
Canada	-4.3	34.9	29.2	- 6.5	-23.6	-28.5	1.52	-0.68	-0.98
Total OECD	9.5	30.8	43.3	- 3.7	-18.3	-21.4	-0.39	-0.59	-0.49

1. Change in end-use prices.
2. Change in the index of final energy demand divided by the change in the index of real GDP.
3. Includes public and agricultural use.
4. Includes non-energy use.

Source: IEA, Energy Price Data and Energy Balances of OECD Countries.

intensity. It clearly indicates that industry has experienced the largest percentage increases in energy prices and achieved the largest reductions in energy intensity.

Table D.1 also indicates the comparatively small increases in end-use transportation prices since 1973 and radical differences in the rates of change in transportation energy intensity among the countries for which data are provided. The very substantial reduction in energy intensity in the United States, United Kingdom and Canada compares to virtually no change in Japan and Germany, among others. These trends are discussed in more detail in the transportation section below.

Analysis of End-Use Sectors

While the efficiency of energy use has increased substantially since 1973 in almost all sectors, the rate of change in efficiency, as well as in the other factors affecting energy demand differs greatly from one sector to another. Table D.2 provides energy demand data for all end-use sectors for selected years from 1973 to 1985. The following sections review the trends in energy demand and efficiency for each end-use sector, and the transformation sector.

Industry

Industry remains the most energy consuming end-use sector in the IEA even though between 1973 and 1985 industry's share of total final energy consumption (TFC) dropped from 41% to 38% while TFC decreased by 9% to 951 Mtoe [1] (see Table D.2). This decrease has primarily taken place since the second major oil price increase in 1979-80. The decline in consumption has largely been at the expense of oil which decreased 20% or 87 Mtoe for the IEA countries as a whole, the share of oil in industrial energy consumption dropping from 42% in 1973 to 37% in 1985. Only electricity use increased between 1973 and 1985 in absolute terms.

1. Including non-energy use.

Table D.2

IEA Total Final Energy Consumption by Sector[1]

	1973		1979		1983		1985		Average Annual % Change		
	Mtoe	Share of Total (%)	Mtoe	Share of Total (%)	Mtoe	Share of Total (%)	Mtoe	Share of Total (%)	1973-79	1979-83	1983-85
TFC											
Total	2 502.9	100.0	2 692.0	100.0	2 388.4	100.0	2 507.9	100.0	1.2	-2.9	2.5
Oil	1 434.2	57.3	1 537.4	57.1	1 269.6	53.2	1 305.6	52.1	1.2	-4.7	1.4
Solid Fuels	300.5	12.0	285.6	10.6	265.2	11.1	289.2	11.5	- .8	-1.8	4.4
Gas	479.3	19.2	508.6	18.9	474.9	19.9	501.8	20.0	1.0	-1.7	2.8
Electricity	288.9	11.5	354.6	13.2	371.1	15.5	403.6	16.1	3.5	1.1	4.3
Industry[2]											
Total	1 045.8	100.0	1 061.4	100.0	860.0	100.0	950.6	100.0	.2	-5.1	5.1
Oil	434.4	41.5	461.5	43.5	311.6	36.2	347.2	36.5	1.0	-9.4	5.6
Solid Fuels	230.1	22.0	215.6	20.3	197.5	23.0	219.2	23.1	-1.1	-2.2	5.3
Gas	244.2	23.4	222.1	20.9	196.6	22.9	215.9	22.7	-1.6	-3.0	4.8
Electricity	137.1	13.1	161.5	15.2	154.0	17.9	167.9	17.7	2.8	-1.2	4.4
Transport											
Total	641.3	100.0	748.2	100.0	717.6	100.0	730.5	100.0	2.6	-1.0	.9
Oil	635.5	99.1	743.0	99.3	712.3	99.3	724.9	99.2	2.6	-1.0	.9
Resi/Comm.[3]											
Total	815.8	100.0	882.5	100.0	810.8	100.0	826.8	100.0	1.3	-2.1	1.0
Oil	364.2	44.6	332.8	37.7	245.7	30.3	233.5	28.2	-1.5	-7.3	-2.5
Solid Fuels	68.5	8.4	69.6	7.9	67.4	8.3	69.7	8.4	.3	- .8	1.7
Gas	234.9	28.8	286.1	32.4	277.9	34.3	285.5	34.5	3.3	- .7	1.4
Electricity	148.1	18.2	188.7	21.4	212.6	26.2	230.7	27.9	4.1	3.0	4.2

1. Shares may not add up to 100% because direct consumption of heat is not shown.
2. Includes non-energy use.
3. Includes public and agricultural use.

Source: Energy Balances of OECD Countries.

i) *Changes in Energy Intensities*

Industrial energy consumption declined on average by 0.8% per year between 1973 and 1985; and industrial output of IEA Member countries increased at an annual rate of about 2%. This resulted in an overall reduction of industrial energy intensity (the ratio of industrial energy use to value added) of about 28%. The development of industrial energy intensities for Member countries of the IEA is shown in Table D.3.

For all IEA industry, the decreases in energy intensity were markedly different between 1973-79 and 1979-84. Whereas overall energy intensity decreased at a rate of about 1.7% per year between 1973 and 1979, the annual decrease in the period 1979 to 1984 was about three times higher (about 5% per year). According to preliminary data, energy intensities increased in 1985. This can be attributed primarily to the differences in the rate of growth of energy prices during these two periods, but may also reflect necessary delays in the response of industry to the initial oil price increases of 1973-74. This accelerated trend can be found in most IEA Member countries, with a few exceptions. On a country-by-country basis, Japan had by far the steepest decline, followed by Luxembourg, Belgium, Denmark and Germany. It should be noted, however, that due to different starting levels in 1973 and variations of industrial structure, the possibilities for relative changes of energy intensities were quite different in the various countries. Countries which were still in the process of industrialisation (Greece, New Zealand and Portugal), as well as several other countries (Ireland, Netherlands and Norway) actually increased their intensities.

The trend of decreasing energy intensities did not start in 1973, but has been occurring in the industry of most IEA Member countries for more than thirty years. For manufacturing industry in all OECD Member countries it has been estimated that in the period 1960-73 energy intensity decreased each year at a rate of about two-thirds of the rate experienced between 1973-79. The decline in the 1960s was partly due to substitution of oil for coal.

Besides increasing energy efficiency, the major cause of decreasing industrial energy intensity is structural change in the types of industrial products. Since 1973, structural changes primarily involved shifts away from energy-intensive industries to less energy-intensive industries, but another major shift was from the industrial to the service sector. In

Table D.3: **Energy Intensity in Industry** (Energy Consumption[1]/Value Added[2] Ratio)

Country	Ratio[3]				Average Annual % Change			Average Annual % Change Based on National Currencies
	1973	**1979**	**1984**	**1985[4]**	**1973-84**	**1973-79**	**1979-84**	**1973-85[4]**
Australia	0.51[6]	0.55	0.51	0.51	0.1	1.2	-1.4	—
Austria	0.24	0.22	0.21	0.21	-1.3	-1.1	-1.4	-1.3
Belgium	0.45	0.38	0.32	0.30	-3.1	-2.8	-3.5	-2.6
Canada[7]	0.74	0.68	0.70	0.70	-0.5	-1.5	0.8	-0.5
Denmark	0.21	0.22	0.15	0.16	-2.9	0.7	-7.0	-3.0
Germany	0.25	0.20	0.18	0.18	-3.0	-3.8	-2.0	-3.0
Greece	0.35	0.39	0.40	0.37	1.3	2.0	0.4	1.2
Ireland	0.31	0.32	0.34	0.35	0.9	0.7	1.2	1.1
Italy	0.32	0.27	0.24	0.23	-2.7	-2.7	-2.7	-2.7
Japan	0.35	0.30	0.21	0.16	-4.4	-2.3	-6.8	-4.7
Luxembourg	1.70	1.42	1.05	1.0	-4.2	-2.9	-5.8	-4.1
Netherlands	0.32	0.38	0.34	0.37	0.6	3.1	-2.3	0.9
New Zealand	0.32	0.35	0.40	0.40	2.0	1.3	2.8	2.3
Norway	0.52	0.50	0.55	0.48	0.5	-0.4	1.6	0.6
Portugal	0.34	0.38	0.38	0.45	1.1	1.8	0.3	1.6
Spain	0.31	0.28	0.23	0.27	-2.6	-1.9	-3.3	-2.4
Sweden	0.38	0.37	0.30	0.29	-2.1	-0.7	-3.7	-1.9
Switzerland	0.11	0.10	0.10	0.08	-0.7	-1.0	-0.3	-0.9
Turkey	0.41	0.42	0.37	0.43	-0.7	0.4	-2.0	-0.5
United Kingdom	0.33	0.30	0.24	0.24	-2.9	-1.6	-4.5	-3.0
United States	0.61	0.57	0.46	0.43	-2.4	-1.2	-3.9	-2.4
IEA Total	0.42	0.38	0.29	0.30	-3.2	-1.7	-5.0	—

1. Final energy consumption excluding non-energy use and feedstocks.
2. Value added of manufacturing, construction and mining/quarrying industry but excluding coal/oil/gas production; calculated based on industrial shares of GDP.
3. Expressed in toe per thousand 1980 United States Dollars.
4. Preliminary data; therefore change rates until 1985 not yet calculated.
5. Average annual intensity changes based on the industrial value added on a 1980 National Currency basis. This is another proxy for absolute energy intensity which is designed to eliminate the distortions resulting from exchange rate fluctuations compared to the U.S. dollar.
6. Because of a break in time series of energy consumption, feedstock levels for 1974 had to be used.
7. Canadian data under revision (including about 6.9 Mtoe consumption in 1973, which is not yet accounted for in the OECD data base).

Sources: Energy Balances of OECD Countries;
OECD - National Accounts.

addition, the product mix within industrial branches has shifted towards products with a higher value added ("intra-industrial"). Due to limited data quality, this factor's influence often cannot be separately identified.

The rate of structural change also accelerated after 1979, similarly to the rates of overall intensity changes. From 1973 to 1983, the iron and steel, textiles, wood and non-metallic minerals industries all grew more slowly than overall industry, whereas chemicals, pulp and paper, food, metal products, telecommunications and engineering all grew more rapidly. Even though chemicals, as well as pulp and paper, are energy-intensive industries, a net trend towards less energy-intensive industries took place due to the dominating decrease of iron and steel. As a result of the economic recession, this structural shift was more pronounced in the period from 1979-83 — a cyclical effect which has been partially offset with the economic upturn in 1984. When assessing individual country developments, another factor to be considered is that energy-intensive industries (e.g. aluminium) sometimes shifted their production to countries with large, low-cost energy resources (e.g. Canada, Australia), thereby affecting industrial energy intensity trends in those countries.

ii) *Efficiency Improvements*

Secretariat analysis for the six major European IEA countries indicates that energy efficiency improvements within single industries have been the most important contributor to the decline in total industry's energy consumption in the 1973-83 period. The rates of efficiency gains during the period 1979-83 were also comparatively higher than the previous six-year period 1973-79. These improvements are particularly important against the background of decreasing industrial investment between 1979 and 1982 which adversely affected industrial energy efficiency. The dominant factor behind the decline in overall industrial energy consumption was the acceleration of energy efficiency improvements, a consequence of much higher energy prices and other market-related factors, but probably also of government conservation programmes (see Chapter VIII). These results are in line with some studies on a national level. For example, a Swedish study [1] showed that, apart from lower production growth in the period 1974-83, efficiency improvements contributed to almost 60% of lower final energy consumption (structural effects to slightly more than 40%). For Germany, it was found for the

1. Source: National Energy Administration, Stockholm.

period 1979-83 that, apart from changes in the level of industrial activity, genuine efficiency gains were twice as high as effects of inter-sectoral structural changes [1]. These conclusions were recently reenforced by an independent study, the results of which are summarised in Table 3 (see Chapter III).

Table D.4 summarises the results that have been achieved in Japan according to an analysis done by the Ministry of International Trade and Industry (MITI) for the period 1979-85 in the manufacturing and mining industry. The table indicates that efficiency gains were the most important factor.

In the United States, industrial energy use (including conversion losses) in 1983 was down 456 Mtoe from 1960-1972 energy use trends; that is, if pre-1972 energy intensities had prevailed, industrial energy consumption would have been 456 Mtoe higher in 1983 than it actually was. Of this total decline, a 25% reduction in industrial output (relative to pre-1972 trends) accounted for about 250 Mtoe and a 31.6% reduction in energy use intensity (also relative to pre-1972 trends) accounted for 206 Mtoe. Improved equipment efficiency accounted for about 70% of the latter, while structural changes accounted for approximately 30%. It should be noted that due to the different treatment (inclusion of conversion losses, extrapolation of previous trend), the growth term is much larger than in Table D.4 and the analysis on Europe.

Improvements in industrial energy efficiency have been achieved by several means (which were depicted generally in Figure 1; see Chapter III):

— *Integrated conservation investments,* which were part of general technological progress in industrial processes and equipment, probably made the single largest contribution to improved energy efficiency. In such cases, improved energy efficiency is an integral part of general plant and equipment investments.

— *Operational* improvements ranging from sophisticated control systems, to better maintenance ("housekeeping") had a major effect on industrial efficiency during this period.

— *Discrete conservation investments* intended primarily to improve energy efficiency were another major contributor. Such invest-

1. Source: ISI - Fraunhofer Institute, Karlsruhe.

ments (in heat recovery equipment, etc.) or other modifications to existing facilities were especially evident after 1979.

A method widely used by industry to achieve efficiency gains during this period was the institution of integrated energy management programmes. A central feature of such programmes has usually been the designation of a senior manager with a broad range of energy-related responsibilities, including the operation and maintenance of energy-using systems, the identification and evaluation of major efficiency investments and decisions on fuel choice. There has been a general trend toward the establishment of such energy management programmes in large, energy-intensive industries over the past ten years. This trend has been accelerated in several countries through both voluntary programmes, such as the current U.K. efforts and mandatory requirements, such as the Japanese requirement that industries whose energy consumption is over a fixed threshold must appoint energy managers.

Two other factors have had important influences on industrial efficiency, especially in certain industries or during particular periods.

— Short-term changes in production levels, as a result of fluctuations in economic activity, have affected industrial efficiency. These effects differ, however, depending on the industry and facilities

Table D.4

Determinants of Japanese Industrial Energy Consumption in the Period 1979-85[1]

(Mtoe)

	79-80	**80-81**	**81-82**	**82-83**	**83-84**	**84-85**	**Period 1979-85**
Total change in energy consumption	-10.1	-13.8	-9.8	+ 5.3	+ 7.2	+ 0.5	-20.7
by factor:							
Change in production level	+ 0.2	+ 1.0	-2.6	+10.4	+14.9	+ 4.6	+28.5
Change in industrial structure	- 7.4	- 4.8	-2.1	- 1.7	- 4.1	- 3.5	-23.6
Change in energy efficiency coefficient (energy consumption/unit of output)	- 2.9	-10.0	-5.1	- 3.4	- 3.6	- 0.6	-25.6

1. All data are based on Japanese Fiscal Years, 1st April to 31st March.

Source: Ministry of International Trade and Industry, Tokyo. Because of the high quality of the Japanese data base, individual factors add up to total changes.

involved. Reduced production levels usually result in the closing of less efficient facilities. In some cases, however, low capacity utilisation can introduce inefficiencies not usually present. Similarly, very high capacity utilisation during an economic upswing can reduce efficiency because less efficient production capacity is utilised. Some recent data suggest that, on balance, rapid increases in overall economic activity tend to slow overall industrial energy efficiency gains, but that steady growth results in more rapid efficiency improvements.

— Energy or fuel substitution has also affected industrial energy efficiency. Switching from oil or gas to coal, for example, usually reduces overall efficiency levels (often caused by increased deposits on heat transfer surfaces), but also reduces total energy costs. The trend toward greater industrial use of electricity has increased industrial energy efficiency in terms of TFC. Electricity has often permitted the adoption of radically different, economically more productive and sometimes more energy-efficient industrial processes — especially when energy costs are concerned. If, however, the primary energy required to generate the electricity is considered, overall efficiency may decrease.

For better indicators of the actual efficiency improvement realised in the industrial sector, data on individual industries or industrial products must be examined. The following figures list several different examples of industrial efficiency improvements. In every case, substantial progress was made since 1972-73 — a reflection of both long-term trends towards improved productivity, as well as more vigorous conservation efforts.

On a product-specific level, Japan for example improved energy efficiencies of selected energy-intensive industrial products in the period 1979/85 compared to 1973 as shown in Figure 3 (see Chapter III). In particular, Japanese steel producers adopted energy-efficient continuous casting processes in close to 90% of its plants, compared to about 40% in the United States and the Netherlands, about half in the United Kingdom and only close to 30% in Australia. One of the reasons for the lack of improvement in the Japanese aluminium industry may be the large production decreases that occurred in this industry over the past ten years. As a result of this decline there were no major investments in new production facilities. But despite the lack of improvement, the Japanese industry is still among the most efficient of IEA producers.

In the United Kingdom, specific energy consumption for the production of crude steel decreased between 1970 and 1981 as shown in Figure D.5.

Figure D.5

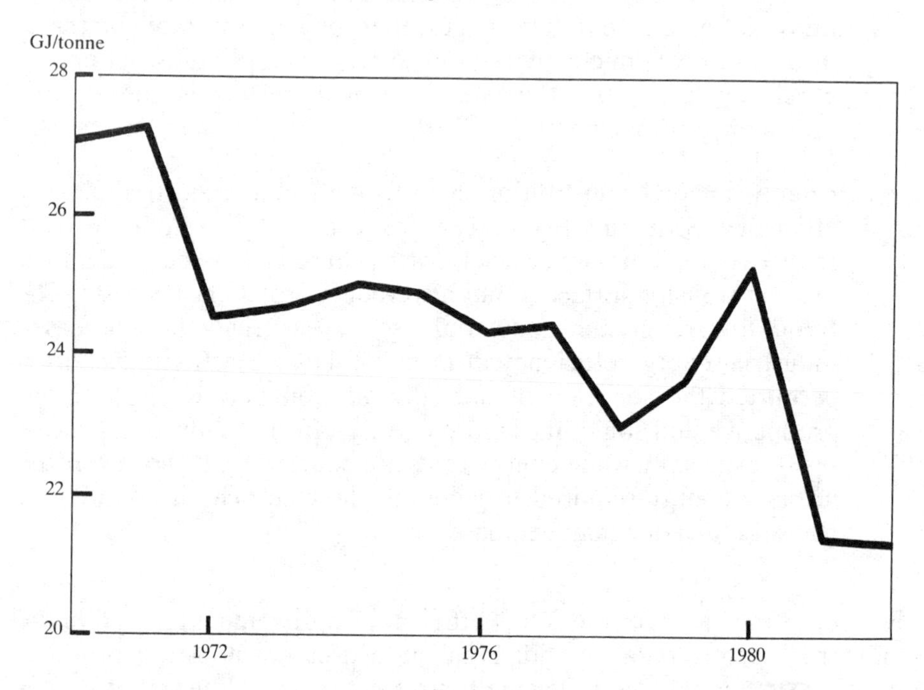

Source: Energy Use and Energy Efficiency in the U.K.Manufacturing Industry up to the Year 2000. Energy Efficiency Office, 1984, London.

Residential/Commercial Sector [1]

The residential/commercial sector is complex to analyse because it comprises several unrelated subsectors which have been grouped together for convenience — from home heating to farming. Figure D.6 shows the change in the energy use of the different subsectors since 1973. The residential subsector represented over half of the total consumption for the sector in 1984.

---

1. This sector includes residential, commercial, agriculture, public and "non-specified" subsectors.

As for all end-use sectors, there is no statistical indicator of total energy efficiency improvements. In addition, because of the structure of the sector there is no single indicator of energy intensity which adequately describes the changes for the entire sector. Population data are the most readily available although for some subsectors (e.g. agriculture) per capita energy consumption is not a suitable measure of energy intensity.

Figure D.6

IEA Residential/Commercial Sector Energy Consumption by Subsector from 1973 to 1984
(in Mtoe)

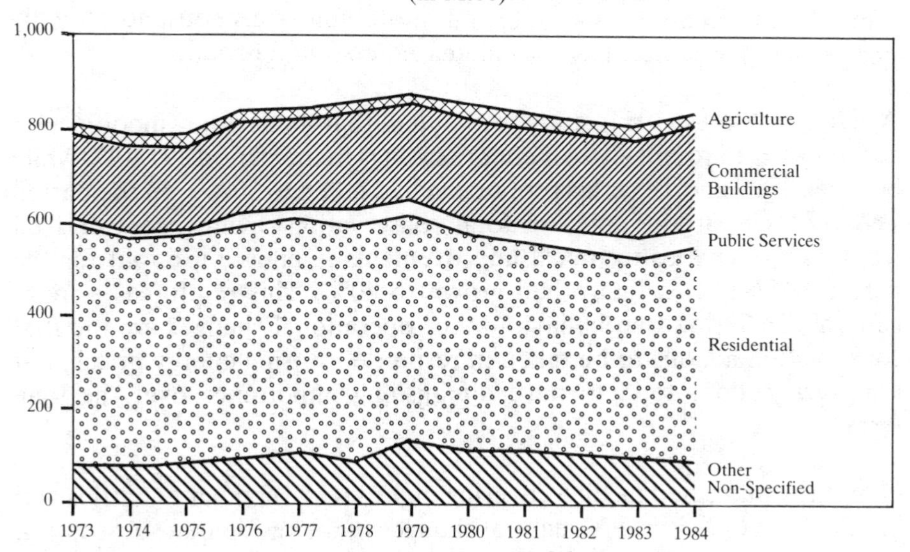

Source: Energy Balances of OECD Countries.

For the IEA as a whole, per capita energy use in the sector increased by 2.5% between 1973 and 1979, declined by about 12% between 1979 and 1984, and increased by 2.1% from 1984 to 1985. In Europe and the Pacific, energy intensity increased about 10% between 1973 and 1979. After 1979, European energy intensity decreased about 15% through 1983, then increased again during both 1984 and 1985. For the Pacific region, residential/commercial energy use per capita experienced both annual increases and decreases from 1979 to 1985, but on average there was no change. In North America, energy intensity fell during both periods although the fall was more pronounced from 1979 to 1985. These differences in the rate of change in intensity can be attributed largely to regional differences in the rate of growth of the penetration of major space conditioning equipment and appliances.

(a) *Residential Subsector*

In the residential subsector, energy use per capita decreased continuously in most IEA Member countries from 1973 to 1983, but increased slightly from 1983 to 1985 (Table D.5). Increased energy prices and government policies and programmes have influenced behavioural changes and consumer investments which have resulted in major energy efficiency improvements. Changing demographic patterns and lifestyles have also introduced major structural changes to the residential sector over the past twelve years. Increasing penetration of central heating and major energy-using appliances have boosted residential energy use. Similarly, the creation of smaller, but more numerous households, with more space per capita, has also increased energy demand.

A United States 1984 study estimated that behaviour modification accounted for around 45% of energy savings in the period 1973-77, while between 1977 and 1982 this proportion had fallen to 30% [1]. In the period 1973-77, only around 15% of total conservation savings were derived from investments to technical improvements in energy use, but for the period 1977-82 this figure had grown to 40%. This reinforces a more general conclusion that behavioural or operational changes are the first conservation actions to be taken in response to rising energy prices, but that such actions tend to have diminishing importance over the long term.

One study published in 1985 covering eight [2] IEA Member countries and France found significant structural changes in the residential subsector in the period 1972-82 [3], including an increase in the living area per household, an even larger decrease in the number of people per dwelling, a notable increase in the penetration of central heating in those countries of the eight with low levels (except Japan), and the continued growth in the stock of electrical appliances. The study calculated that these structural changes would, in isolation, have increased energy use per dwelling between 1972 and 1982 by about 10-15% in North America, Denmark and Sweden, 15-20% in Norway, the United Kingdom and Germany and more than 20% in Japan and Italy. In most countries,

1. U.S. Department of Energy, *A Retrospective Analysis of Energy Use and Conservation Trends: 1972-1982,* Washington, D.C., February 1984.
2. The eight are: Canada, Denmark, Germany, Italy, Japan, Norway, Sweden and the United States.
3. Schipper, Ketoff and Kahane, "Explaining Residential Energy Use by International Bottom-Up Comparisons", *Annual Review of Energy* 1985. 10:341-405.

Table D.5

IEA Final Energy Consumption, Residential Subsector[1]

(toe per capita)

	1973	1979	1983	1985	Average Annual % Change 1973-79	1979-83	1983-85[2]
Canada	1.08	1.10	0.96	1.00	0.2	- 3.3	2.4
United States	1.31	1.12	0.98	0.94	-2.6	- 3.3	-2.2
NORTH AMERICA	1.29	1.12	0.98	0.94	-2.4	- 3.3	-1.8
Australia	0.33	0.35	0.40	0.40	0.6	3.9	-1.2
Japan	0.16	0.23	0.24	0.25	6.4	0.4	2.4
New Zealand	0.30	0.34	0.36	0.45	2.2	1.3	11.5
PACIFIC	0.18	0.25	0.26	0.27	5.2	1.1	2.1
Austria	0.85	0.93	0.80	0.90	1.7	- 3.7	5.9
Belgium	1.14	1.04	0.79	0.91	-1.5	- 6.8	7.8
Denmark	1.41	1.21	0.82	1.09	-2.6	- 9.2	15.6
Germany	1.04	0.71	0.54	0.81	-6.1	6.8	22.4
Greece	0.17	0.22	0.21	0.23	4.0	- 0.7	4.4
Ireland	0.61	0.60	0.58	0.57	-0.3	- 0.7	-1.4
Italy	0.24	0.17	0.10	0.11	-5.4	-12.2	2.3
Luxembourg	1.35	1.30	1.02	1.22	-0.6	- 5.8	9.0
Netherlands	0.99	1.03	0.84	0.83	0.6	- 4.9	-0.6
Norway	0.74	0.80	0.78	0.89	1.2	- 0.6	7.2
Portugal	0.09	0.09	0.12	0.12	0.0	5.5	1.2
Spain	0.13	0.15	0.14	0.15	2.6	- 1.0	2.7
Sweden	1.51	1.27	0.89	1.13	-2.9	- 8.3	12.7
Switzerland	1.18	1.10	0.81	0.85	-1.1	- 7.4	2.4
Turkey	0.26	0.27	0.28	0.29	0.2	0.8	2.9
United Kingdom	0.60	0.67	0.63	0.68	1.8	- 1.5	3.6
IEA EUROPE	0.59	0.52	0.43	0.50	-2.3	- 4.8	8.6
IEA TOTAL	0.76	0.68	0.59	0.61	-1.9	- 3.4	2.2

1. Changes in the definitions of the residential and commercial sectors made between 1973 and 1979 mean that the percentage reductions for the 1973-79 period are sometimes overstated.
2. 1985 data are preliminary. The large annual changes observed, especially from 1983 to 1985, may be partially the result of changes in weather conditions.

Source: Energy Balances of OECD Countries;
OECD — National Accounts.

Figure D.7

Efficiency Improvements of New Electric Household Appliances

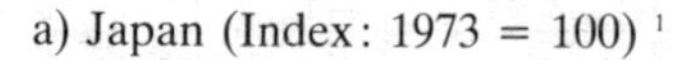

a) Japan (Index: 1973 = 100) [1]

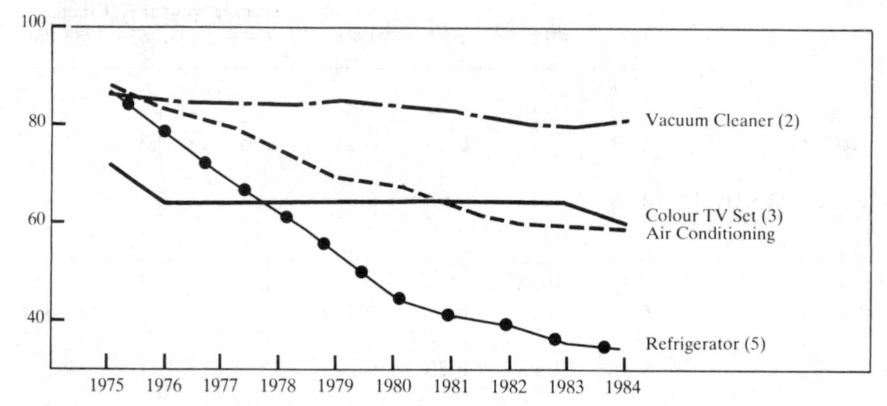

Specific Electric Power in 1973

(1) All figures by Japanese fiscal years which is 1st April to 31st March.
(2) 628 W.
(3) 148 W — 19.2 inch type.
(4) 847 W — 1 600 kcal/h class.
(5) 79.6 kWh/month — 170 litre type.

Source: "Energy Conservation in Japan", The Energy Conservation Center, Tokyo.

b) Germany (Index: 1970 = 100)

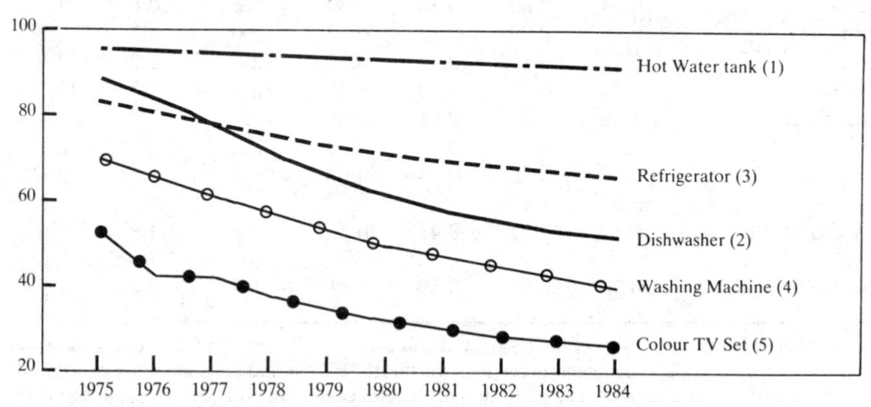

Specific Electricity Consumption in 1970.

(1) 1.14 kWh/kWh of hot water.
(2) 2.7 kWh/cycle.
(3) 1.2 +/- 0.5 kWh/100 litre and day.
(4) 0.8 kWh/kg of clothes washed.
(5) Big screen

Source: Prof. Schäefer, Technical University of Munich, Germany

however, energy use per dwelling did not grow this quickly because of large improvements in energy efficiency which offset the effects of the structural changes in housing characteristics.

As an example, Figure 4 (see Chapter III) shows efficiency improvements in specific useful energy for space heat in the period 1970-82.

Another example, Figure D.7 shows efficiency improvements of electric household appliances in Japan and Germany compared to the 1973 level.

(b) *Commercial, Agriculture and Public Subsectors*

Neither the IEA nor most of its Member countries keep comprehensive and consistent statistics on final energy demand for these subsectors. Analysis is complicated by their use as a statistical "catch-all", representing an aggregation of energy demand for a variety of dissimilar economic and social activities. The public subsector is rarely defined and can be difficult to distinguish from the commercial subsector, which includes a wide array of activities related to the provision of services [1].

The first comprehensive study of the service sector in seven IEA countries was not done until 1985 [2] largely because of the complexity and lack of end-use data. The study estimated that space heating accounts for 60-65% of total delivered energy consumption in commercial subsector buildings in central Europe and the United States. Only in the United States is air-conditioning significant (at 13%). Lighting accounts for an estimated 9% in Germany and 12% in the United Kingdom and the United States.

Again, there are no good statistical measures of energy efficiency or intensity in this sector. The study uses two types of indicators: per square metre and per employee. Each indicator has elements of both efficiency changes as well as some structural and income changes.

---

1. Analysis of the sector is further complicated by the different categorisations used in various IEA countries. For example, Japan reports no energy consumption in the public subsector in 1983 but a very large consumption in the "other (non-specified)" subsector.
2. Lee Schipper, Steve Meyers and Andrea N. Ketoff, "Energy Use in the Service Sector. An International Perspective", *Energy Policy*. Volume 14, No. 3, June 1986, pp.201-218.

The per square metre indicator showed:

	Percentage Change	Period [1]
Denmark	-27	1972-82
Sweden	-17	1972-82
United States	-18	1970-80

(1) Those years were used because they had comparable weather.

Source: Schipper, Meyers, Ketoff, *Energy Use in the Service Sector: An International Perspective,* Lawrence Berkeley Laboratory (19443), June 1985.

See Chapter III for trends in the energy use per employee indicator.

Energy demand in the commercial subsector is being shaped by a number of factors — including the rate of growth of the overall building stock and of particular businesses within the commercial subsector and the energy intensity of each process that occurs in a building. The transfer of residential activity to the service sector, such as increased use of restaurants, hotels and care for the elderly is one cause of increased commercial subsector energy consumption (although there have also been shifts of commercial activities, such as laundry and entertainment, to residences).

Increased energy efficiency in this sector has taken many forms, from technological improvements to building design, materials and equipment to operational changes that improve building maintenance and energy use control.

Transportation Sector

Oil consumption accounts for close to 99% of energy consumption in this sector, and road transport is responsible for 80% of this oil use. Energy efficiency has been improving since 1973. While overall energy use rose between 1973 and 1979, it declined from 1979 to 1982. Since 1983, total transportation demand has gradually increased, but in 1985 it still was below 1979 levels. Although the technical fuel efficiency of passenger cars has improved, better economic conditions act as a stimulus to overall consumption, particularly in certain European countries and by commercial vehicles.

Passenger Cars and Commercial Vehicles

Table D.6 shows the oil consumption in 1973 and 1983 by country for the two major categories of transportation: passenger cars and commercial vehicles. Passenger cars accounted for almost 67% of consumption in the road transport subsector in 1973, and 61% in 1983, while consumption by commercial vehicles increased from 33% to 39%.

Changes in consumption are mainly determined by the number of vehicles in the fleet, the average distance travelled per car, and fuel efficiency. Secretariat analysis based on technical data, estimates and historical trends provides a broad indication of changes in road transport, the largest subsector. Of the 19.6% reduction in consumption per passenger and commercial vehicle observed between 1973 and 1983, 65% was due to efficiency improvements (45% was price induced efficiency changes, 10% other efficiency improvements, and 10% attributable to shifts from gasoline to diesel), and 35% to reduced average distance travelled per vehicle.

Table D.6

IEA Oil Consumption for Road Transport

(Mtoe)

	Passenger Cars			Commercial Vehicles		
	1973	1983	% Change	1973	1983	% Change
Germany	17.2	23.2	34.9	10.7	12.4	15.9
Italy	9.5	10.7	12.6	7.6	11.5	51.3
United Kingdom	15.3	18.2	19.0	8.7	9.1	4.6
Netherlands	3.1	4.4	41.9	2.1	2.9	38.0
Sweden	2.7	3.3	22.2	1.6	1.9	18.8
Spain	4.0	6.3	57.5	3.3	5.1	54.6
Other Central Europe[1]	6.7	8.4	25.4	4.2	5.0	19.0
Other Northern Europe[2]	2.5	2.8	12.0	1.4	1.6	14.3
Other Europe[3]	3.1	6.0	93.6	3.1	5.8	87.1
United States	233.3	221.7	-5.0	108.3	148.4	36.8
Canada	18.7	17.2	-8.0	8.2	9.1	11.0
Japan	18.8	26.2	39.4	9.7	12.0	23.7
Australia	7.8	9.5	21.8	2.7	5.1	88.9
New Zealand	1.4	1.3	-7.0	0.5	0.6	0.2
IEA Total	344.1	359.2	4.4	172.1	230.5	33.9

1. Central Europe: Austria, Switzerland, Belgium, Luxembourg.
2. Northern Europe: Denmark, Norway.
3. Other Europe: Portugal, Greece, Turkey, Ireland.

Source: Secretariat analysis and International Road Federation.

Demographic trends are the most significant determinant, along with gasoline and car prices and disposable income, of passenger car consumption. The average distance travelled per passenger car is dependent upon economic activity, gasoline prices, the number of cars per family, the age of the car, the degree of urbanisation, the level of public transport, and motoring habits.

The economic cycle has an important role in commercial subsector consumption: increasing industrial output increases the total demand for freight transport, which in turn increases demand for commercial road transport. Population density, location of cities, and government expenditure on transport influence energy use among buses and coaches. Unlike passenger cars, there has been an increase in the average distance travelled per commercial vehicle, probably as a result of increased economic activity, which has played a greater role than higher oil prices.

Table D.7

Annual Gasoline Consumption per Passenger Car in IEA Countries

(Litres per Car)

	1973	1979	1985	% Change in litres per Car[1] 1973-85	% Change in Number of Cars 1973-85
North America	3 806.0	3 571.9	2 987.4	-21.5	30.6
Pacific	1 994.4	1 726.0	1 457.7	-26.9	83.7
Europe	1 512.4	1 359.1	1 200.3	-20.6	56.3
IEA Total	2 866.7	2 594.2	2 151.8	-24.9	44.8

1. For statistical reasons this figure includes gasoline consumption of light commercial vehicles.

Sources: Energy Statistics, IEA; International Road Federation, World Road Statistics, Editions 1977 to 1986; Motor Vehicle Manufacturers' Association of the United States; World Motor Vehicle Data, 1983, 1984 and 1986 Editions.

Average vehicle fuel efficiency in the passenger car subsector has been improving since 1973, which is important since turnover of the car fleet is very rapid (about ten years). Table D.7 shows the changes in annual gasoline consumption per passenger car between 1973 and 1984. For the IEA as a whole, the biggest change occurred since 1979, mainly due to the progress in North America. Voluntary and mandatory standards in nine countries have played a role — substantial improvement has been noted in the United States, Germany, Japan and the United Kingdom in the adoption of technical efficiency improvements in new cars, such as improved engine efficiency and more aerodynamic design. Table 4 (see Chapter III) describes the improvements made in new car fuel economy since 1973. Fuel economy derives from weight, engine power and design and is therefore greatly affected by the size and power of individual cars. There is no single accepted method of defining the technical efficiency of automobiles, which ideally would be independent of car size and power. Overall fleet fuel economy is also dependent on the proportion of urban versus highway driving and consumer driving patterns. There have been improvements in all weight categories, and there are now smaller differences between countries within each category. For commercial vehicles, technological progress in vehicle engine design and aerodynamic rather than fuel prices may be the key to affecting energy demand. Fuel efficiency improvements have been slower because units were already more technically efficient than most passenger cars and they have longer lifetimes. Goods vehicles have shown more improvement than buses, and also the relative use of diesel versus petrol has been responsible for different results across IEA countries.

Maritime, Aviation and Railway Subsectors

The maritime, aviation and railway subsectors are all very dependent on petroleum products for energy. (Only in Europe are electric railways widespread.) Together they accounted for 17.4% of oil consumption in the sector in 1973, and only 17.1% in 1984. Higher fuel prices and other factors have encouraged efforts to improve efficiency in each of these subsectors over the past decade.

Factors contributing to improved efficiency levels in the maritime sector include: a gradual shift from steam turbine ships to motor diesel powered ships; improvements in the design of diesel engines and propellers; and slow steaming. Figure D.8 depicts the trend in maritime energy efficiency as measured by world bunker oil demand divided by laden ton-miles.

In recent years, new commercial jet aircraft have been introduced which are substantially more energy-efficient that existing commercial jets. Measures taken in the aviation sector to improve energy efficiency in addition to these technological advances include: removing paint to reduce the weight of planes, steeper angle of descent in landing and take-off to reduce air resistance, and lower cruise speeds and better taxiing procedures.

An energy efficiency measure used for the aviation sector is the ratio of fuel consumption to ton-kilometres for both passenger and freight traffic. Figure D.9 shows the trends in this factor since 1973.

Figure D.8

Energy Intensity in World Shipping [1]

(Index: 1973 = 100)

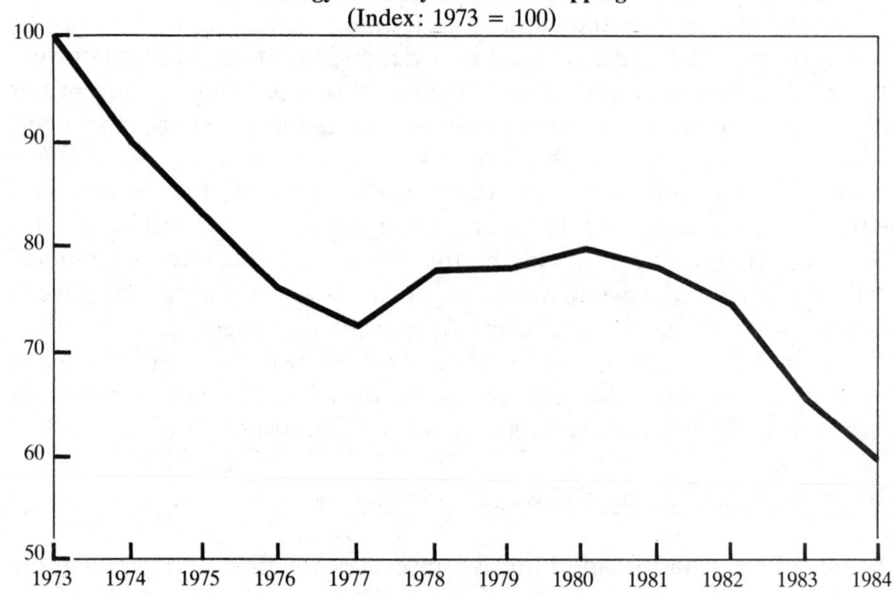

(1) Index based on world bunker oil demand divided by laden ton-miles.

Source: Fearnley's Limited, Norway;
World Energy Statistics Yearbook, United Nations.

Figure D.9

Energy Intensity in Aviation [1]

(Index: 1973 = 100)

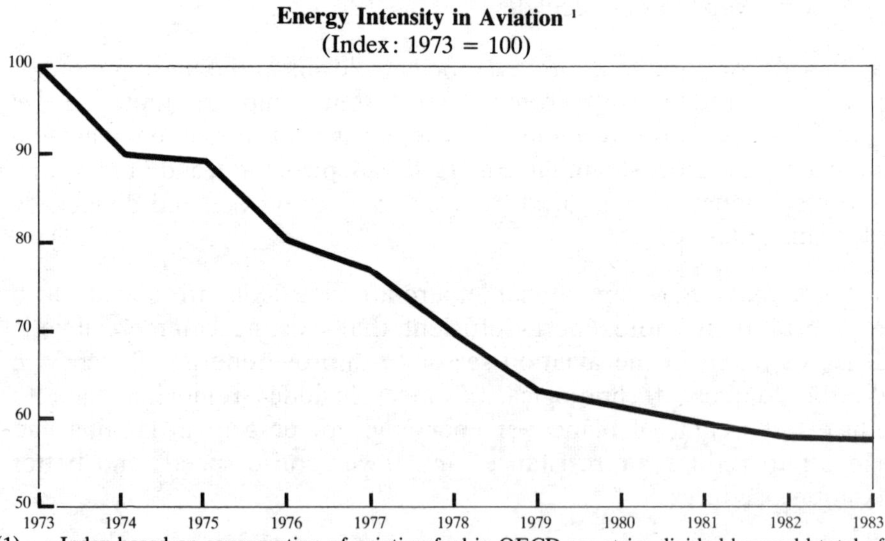

(1) Index based on consumption of aviation fuel in OECD countries divided by world total of ton-kilometre flown.

Source: Energy Balances of OECD Countries;
International Civil Aviation Organization, Montreal.

The electrification of railways has been the primary measure used to improve the efficiency of railway transport. However, the ratio of energy consumption to passenger kilometres has not shown any consistent trend. The ratio fluctuates primarily as a result of changes in load factors (i.e. passenger/kilometre transported) and may go up or down depending on the country or period examined.

Figure D.10 shows specific energy consumption and respective efficiency improvements for different means of transport in Germany between 1978 and 1983.

Energy Transformation Sector

The transformation sector encompasses the conversion of primary fuel into more useful energy forms, as well as the distribution of energy to end-users. In the IEA energy balances, the energy demand of the transformation sector is the difference between the input of primary energy (TPER) and the secondary energy forms produced (TFC). It therefore represents exclusively the losses in the conversion processes and in the operations of extraction and distribution of energy. In 1985, this sector was larger than any of the individual end-use sectors. It can be divided into four components:

— losses in central station electricity/heat generation (equivalent to the energy content of the fuels and/or other energy sources used to generate electricity/heat minus the energy content of electricity/heat produced);

— consumption by power plants for their internal use including electricity consumed by pumped storage hydro plants;

— fuel consumption and losses in petroleum refineries and by other energy sectors such as coal mines, oil and gas extraction;

— distribution losses (electricity grids; oil, gas and district heat pipelines).

The transformation sector is dominated by the electricity sector (close to 80%), with most of the rest accounted for by the energy use and losses in oil refineries. This sector merits attention due to its absolute size and growth rate. Between 1973 and 1985, the transformation sector grew by 19% to slightly above 1 000 Mtoe. While 1985 TFC in all of IEA was about the same as the 1973 level, TPER was 170 Mtoe higher than 1973 due mainly to growing demand for electricity (the generation of which usually results in greater energy losses than direct use of the primary

Figure D.10

Improvement of Specific Energy Consumption of Different Modes of Transport in Germany between 1978/83
(grams hard coal equivalent per passenger-km or metric ton-km)

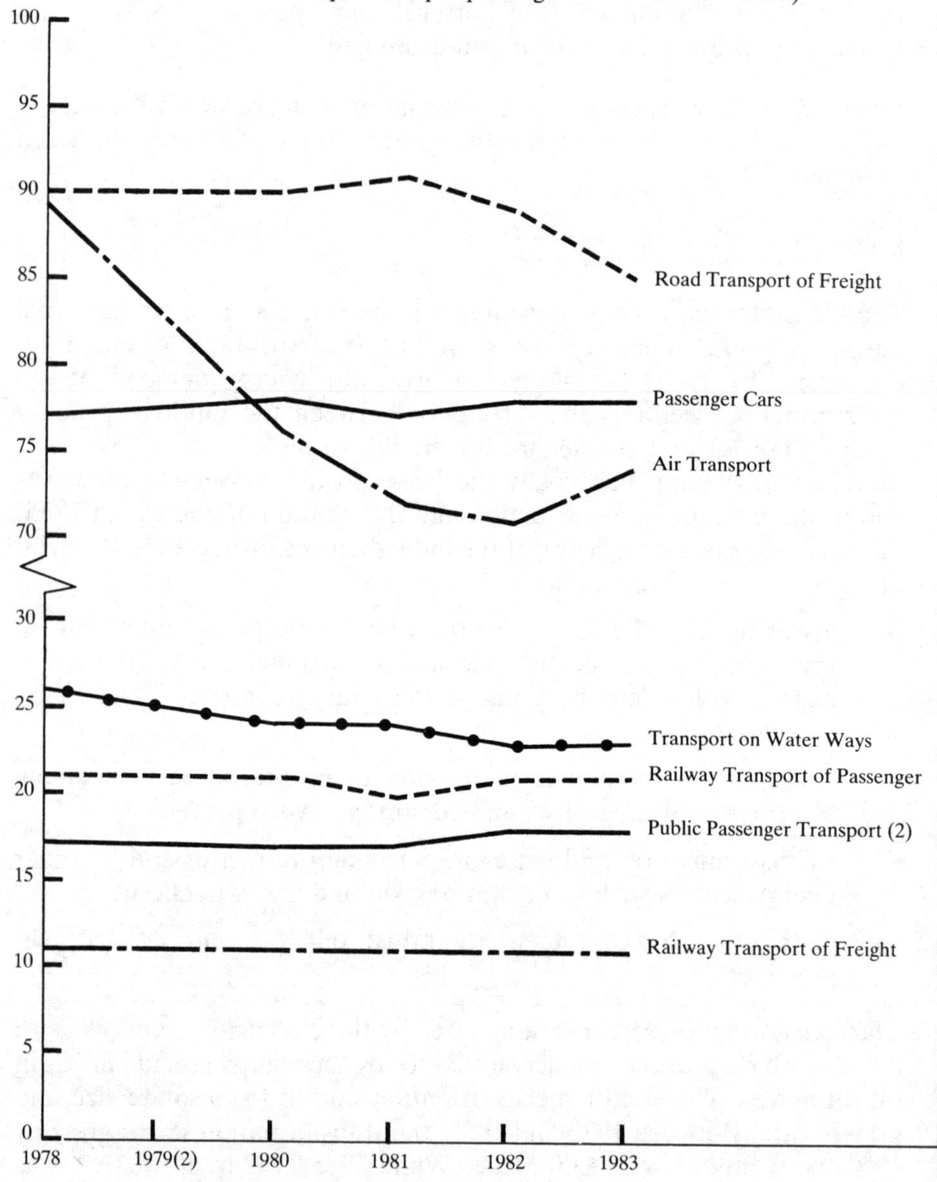

(1) No data available for 1979: trend thus extrapolated.
(2) Mainly road transport by buses, etc.

Sources: German Institute for Economic Research (DIW), Berlin.

fuels). During that period, electricity's share of TFC increased from 12% to about 16% with total fuel inputs increasing from 930 Mtoe to about 1 250 Mtoe in 1985. The shift to electricity, because of its high end-use efficiency and convenience, has resulted in a shift of energy losses previously experienced in energy end-use to the energy transformation sector.

The total own consumption and losses of oil refineries accounted for 124 Mtoe (or about 7.7% of total refinery input) in 1973 and about 90 Mtoe in 1985 (about 6.4% of refinery input); a 17% reduction over this twelve year period. Refinery fuel use and losses decreased in comparison to overall oil input because of major efficiency improvements and the shutdown of both smaller and less efficient units. These factors outweighed the effects of lower overall utilisation rates of refineries (83% in 1984 compared to 90% in 1973) and the increased proportion of fuel consuming upgrading/cracking units, both of which decrease efficiency. The solid line of Figure D.11 indicates that the United States petroleum refining industry reduced overall energy consumption by 25.7% in 1983 compared to the 1972 level, corrected to account for changes in processing schemes, lead phase-out, reduced feed quality, etc. The dashed line of Figure D.11, which presents actual fuel consumption per barrel of refinery input uncorrected for any changes since 1972, shows that efficiency did, in fact, increase until 1979. It then decreased (probably due to increases in the production of lighter petroleum products compared to total output, decreases in crude oil qualities, and reductions in feed rates). During the 1979-1981 period, continued conservation efforts were not able to offset these effects, but the long trend toward improved efficiency resumed in 1982.

Absolute conversion losses in electricity generation were placed at about 590 Mtoe of TPER in 1973 and about 780 Mtoe in 1985. This increase in "losses" reflects the increasing demand for electricity and not a change in the efficiency of electricity generation. Figure 5 (see Chapter III) shows the development of average efficiency for steam-electric power plants in the United States since 1925. It clearly indicates that generating efficiency has not significantly changed since the 1960s. The transformation sector's own consumption and losses — except refineries — were 146 Mtoe in 1973 and about the same level in 1985. Electricity transmission and distribution losses decreased slightly to about 7% of total final electricity in 1985 (from about 8% in 1974). Own energy use in power plants increased to about 5.5% of total final electricity in 1985 (from 4.6% in 1974) due in particular to more coal-fired plants (coal grinding). Pumped storage plants consumed 0.5% of total final electricity in 1985 (0.4% in 1974). While pumped storage plants do use energy, they increase system-wide efficiency by replacing low efficiency (below 25%) gas turbine power plants normally used to meet peak demand.

Figure D.11

United States Refining Industry Fuel Consumption

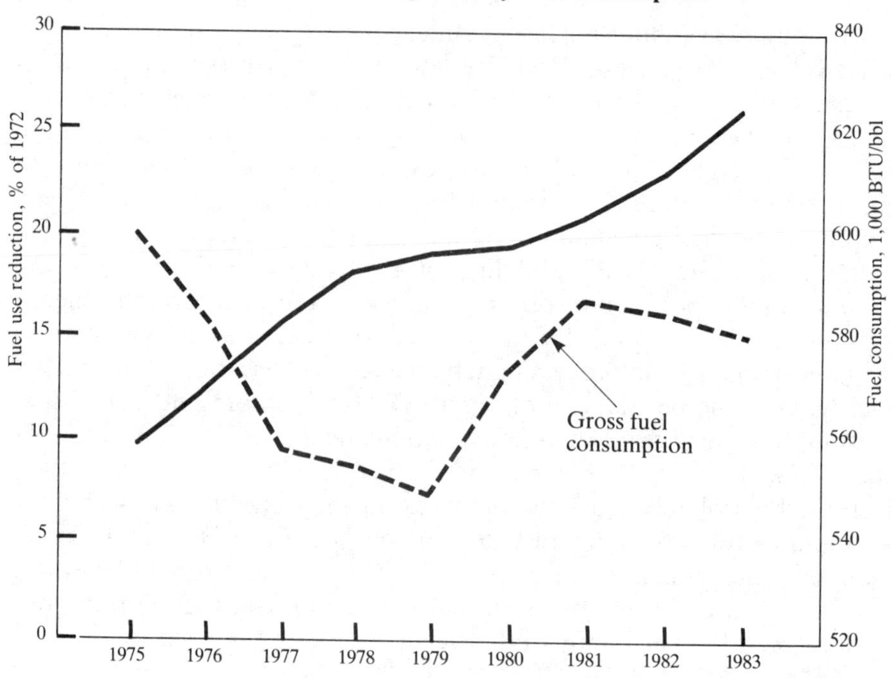

Source: Energy Efficiency Improvement Report, API, United States.

To account for the energy losses in hydro and nuclear electricity production in terms of fuel equivalent (that is the theoretical amount of oil necessary to produce the same amount of electricity), some fixed assumptions about the efficiency of these processes are used in the TPER statistics of IEA and most Member countries. Given that hydro actually encounters only insignificant heat losses due to turbine friction and nuclear energy has limited alternative uses [1] the most important

1. In the case of nuclear, real losses of about 125 Mtoe are released into the environment in the form of waste heat. Even so, because of special siting and design concerns, there are few cases where the heat from nuclear plants is used (e.g. in an industrial plant in Switzerland).

opportunities for efficiency improvements are in fossil fuel-fired power stations (440 Mtoe in 1973, about 500 Mtoe in 1985). Most of the energy losses in electricity generation are released as low temperature heat from oil-, coal- and gas-fired thermal power stations. The size of this unused resource has grown because of increased electricity production and the slightly lower efficiency of coal plants compared to gas- or oil-fired electricity generation. Largely in response to the substitution of oil by coal-fired generation, the thermal efficiency of electricity generation decreased by less than 1 percentage point to about 36% in 1985.

One of the existing electricity generating technologies capable of increasing efficiency (by 30-40%) is combined heat and power (CHP), also called cogeneration. In industry, CHP played a major role in the past in most countries, e.g. in the United States where in 1920 about 60% of industrial electricity consumption was autoproduced. In many countries industrial CHP application declined as a result of deteriorating economics (decreasing ratio of electricity to fuel prices and process heat savings) but also structural and tariff reasons. Actual market penetration of these technologies is now very limited, except in a few countries. In 1985 in Europe, heat from CHP systems contributed only about 6 Mtoe to end-use heat demand in industry and the residential/commercial sector. This is about 5% of total final electricity consumption. When heat from district heating (DH) plants is included, about 10 Mtoe were produced. For total IEA, total heat supplied from CHP plants was about 8 Mtoe (about 2% of total electricity consumption). Since the early 1980s in the United States, the Public Utility Regulatory Policies Act (PURPA), which required utilities to offer to purchase excess electricity produced by CHP at marginal prices, and special tax incentives reversed the downward trend and led to a real boom in industrial CHP applications.

Market penetration of CHP/DH technologies is already very high in some northern countries, notably Sweden and Denmark. In Sweden, heat from CHP amounted to about 5% of TFC with DH providing nearly one-half of all the heat required for multi-family dwellings and about 40% of the requirements for non-housing premises in 1985. In Denmark, heat from CHP plants contributed about 17% to the total net heat consumption for space heating and hot tap water of about 4 Mtoe (an additional 26% from pure DH plants) in 1985. In Germany and Austria, DH contributed close to about 2.5% to total TFC in 1985.

Annex E

Selected Studies on Energy Conservation Potential; Data Sources and Supporting Analyses

This annex provides the sources of the data in Tables 5 and 6 contained in Chapter IV (referred to as Tables E.1 and E.2 in this annex) and a brief description of the development and purpose of the tables.

The data presented in Tables E.1 and E.2 do not represent a comprehensive or precise review of the potential for conservation in IEA Member countries. Such an assessment would require the gathering of detailed and current data on end-use technologies in all Member countries, which is far beyond the scope of this study. Although comprehensive assessments of conservation potential are extremely useful in the design of the specific conservation policies and programmes to be undertaken by Member governments, such a detailed assessment was not necessary to address the broader policy issues examined by this study. To fulfil the IEA requirements, a selection of recent technology and sectoral studies of conservation potential were sufficient.

Table E.1 provides an assessment of the economic potential for conservation by major end-use or technology categories. Column (a) indicates the percent of IEA's total primary energy requirements accounted for by the specific energy end-use or technology category for which data are provided. The column is intended to provide a rough indication of the magnitude of the potential savings indicated in columns (c) and (e). The estimates were derived from IEA data on energy end-use, as well as from a variety of other sources. Ranges are provided to indicate that these values do vary over time and by major region; however, they do not reflect the full range of the differences among IEA Member countries. Column (b) contains estimates of the average efficiency of the existing stock of the technology in question. The estimates are usually based on actual surveys of end-use efficiency conducted by Member governments or other organisations, but they sometimes have been derived from data on the efficiency of new products over time. Data on the average efficiency of existing physical stocks are often difficult to obtain but they are essential to a reliable assessment of conservation potential. Column (c) almost always provides data on the average efficiency of new products sold in the most recent period for which data are available. For the industrial sector, however, it contains data on the most efficient existing industrial facility in the category. It is assumed that the most efficient facility approximates the efficiency of the newest plants for the industry in question. The figure in parentheses in column (c) is the percentage reduction of the column (c) efficiency compared to the column (b) efficiency. The data on the Best Available Technology are intended to represent the savings that could be achieved through the application of the most energy-efficient technologies which are proven, commercially available and economically viable. Column (d) provides an estimate of efficiency and column (e) is the percentage reduction of column (d) compared to column (b). Column (f) contains an estimate of the average useful life of the technology and therefore is a rough indicator of the time span required to achieve the full potential savings (indicated in column (e)). Column (g) provides the reference numbers for the sources listed in the annex, as well as other relevant information. Finally, it sould be noted that the technology categories and the percentage savings indicated in columns (c) and (e) sometimes overlap. For example, the estimates of potential savings in all electric homes in the United States and Sweden include estimates of potential savings for electric appliances, although independently derived estimates for certain electric appliances are provided elsewhere. Also, efficiency improvements in electric motors, gas turbines or other cross-cutting technologies would result in energy savings in all sectors. The basic conclusion drawn from Table E.1 is that the average potential for savings (column (e)) is at least 25%.

Table E.2 lists the estimates made by independent studies of the potential energy savings that could result from cost-effective efficiency improvements. Column (a) indicates the end-use sector to which the estimate applies and column (b) the year of the study. Column (c) indicates the potential reduction in energy demand that could be achieved compared to the energy demand levels that would have resulted if no further efficiency improvements were made. For example, such estimates usually include the savings that will result from the gradual replacement of existing appliances and automobiles by the more energy-efficient products that are now being sold. Column (d) lists estimates of the energy demand reductions possible in comparison to the energy demand levels that are expected to result from current trends in product efficiencies and energy prices. For example, these estimates exclude the savings that are already expected because of the production of more energy-efficient appliances and automobiles and they usually exclude the savings that are anticipated to result from the further improvements in product efficiencies which are likely to result from market forces and current government policies. For studies that contain both types of potential estimates, column (d) would usually be lower than column (c) and the difference between the two would approximate the efficiency gains expected to result from market forces and current government policies alone. But this is not true for at least one study, which took into account the effects on potential savings of rising energy prices. All potential estimates are based on existing technologies, although in some cases the technology may not be widely available at present. Column (e) indicates the year in which the savings indicated might be achieved and column (g) provides reference numbers and other relevant notes. The two basic conclusions drawn from Table E.2 are that the average potentials for savings among all sectors and countries are: at least 25% compared to levels that would have been realised if no further efficiency improvements were made (column (c)); and at least 10% compared to levels that would exist in the year 2000 as a result of market forces and current government policies alone (column (d)).

On the whole the data provide convincing support for the general conclusion that a substantial potential for further energy efficiency improvements exists. While the studies examined do have some analytical weaknesses and use different economic and energy price assumptions, they all represent serious analytical efforts to estimate the further potential for energy efficiency improvements. There are, however, two major weaknesses of the data that should be addressed specifically. Because almost all of the studies were performed prior to 1986, they generally did not take into account the dramatic fall in oil prices that has occurred in 1986 nor do they reflect the efficiency improvements that were instituted after the time of each study's

Table E.1

Energy-Efficient Technologies and the Economic Potential for Conservation

Energy End Use/ Technology	(a) % of IEA TPER	(b) Existing Stock Average Efficiency (Units)	(c) New Stock Average Efficiency (% Savings)	Best Available Technology (d) Efficiency	(e) % Savings	(f) Average Useful Life of Technology	(g) Notes
RESIDENTIAL	20-25%						
- U.S. (All electric)		1 501 (Watts per capita)		328	-78%	Over 30 years	(1)
- Sweden (All electric)		1 242 (Watts per capita)		266	-78%		(1)
Heating and Cooling	8-12%						
- Building shell thermal efficiency						Over 30 years	
- U.S. (winter)		160 (KJ per m² per degree day)	100 (-37%)	50	-70%		(1)
- Sweden (winter)		135 (KJ per m² per degree day)	65 (-52%)	35	-74%		(1)
Heating	8-12%						
- Oil/Gas - System Efficiency						10-20 years	
- U.S.		65-70% (% of TPER converted to useful heat)	75-80% (-13%)	84-94%	-23-26%		(2)
Cooling	1-2%						
- Central a/c -						10-20 years	
- U.S.		7 (Energy Efficiency Rating)	9 (-22%)	14	-50%		(3)(4)
Refrigerators/freezers	2%					10-15 years	
- U.S.		1 500 (kWh/year)	1 300 (-13%)	750	-50%		(3)(4)

Table E.1

Energy-Efficient Technologies and the Economic Potential for Conservation *(Continued)*

Energy End Use/ Technology	(a) % of IEA TPER	(b) Existing Stock Average Efficiency (Units)	(c) New Stock Average Efficiency (% Savings)	Best Available Technology (d) Efficiency	(e) % Savings	(f) Average Useful Life of Technology	(g) Notes
- Germany		About 400 (kWh/year)	(-20%)		At least -20%		(5)
- Japan		35 (kWh/month)	28 (-20%)		At least -20%		(6)
Water Heating	3-5%					15 years	
- U.S.		4 000 (kWh/yr)	3 600	1 700	-57%		(3)(4)
COMMERCIAL	15-20%						
Heating and Cooling	10-12%					30+ years	
- U.S.		1.31 (GJ per m²/yr)	0.73 (-44%)	0.32	-75%		(1)
- Sweden		1.04	0.76 (-27%)	0.25-0.46	-55-75%		(1)
Large Office Buildings	5%						
- U.S.		270 (KBtu/ft²year)	200 (-26%)	100	-63%	30+ years	(7)
Lighting	3-5%					1-10 years	
- U.S.							
Ballast/Tubes		64 (lumens/Watt)	73 (-12%)	86	(-26%)		(8)(9)
Controls					(-20-30%)		
Total					-40-50%		
TRANSPORTATION	20-25%						
Automobiles	10-13%					10 years	(1)(10)40 mpg may be technically possible at cost of less than $1/ gallon saved.

Table E.1

Energy-Efficient Technologies and the Economic Potential for Conservation *(Continued)*

Energy End Use/ Technology	(a) % of IEA TPER	(b) Existing Stock Average Efficiency (Units)	(c) New Stock Average Efficiency (% Savings)	Best Available Technology (d) Efficiency	(e) % Savings	(f) Average Useful Life of Technology	(g) Notes
- U.S.		19.0 (miles per gallon)	26.1 (-34%)	31.5	-46%		
- Japan		[11] (km/1)	13 (-15%)				
Other road transport	7-10%						
Air Transport	2-3%	[25] (passenger miles/gallon)	30+ (-20%)	40+	-40%	15-30 years	(1)(11)(18)
- All							
Rail/Marine/Other	2-3%						
INDUSTRY	35-40%						
Chemicals	6-8%						
- U.K. (Inorganic)					-13%		(12)
Iron and Steel	5%						
- U.S./Japan/U.K./Neth.		22-24 (GJ/tonne)	17-18 (-20-25%)	N.A.	At least -20-25%	10-30 years	(13)(14)
Non-ferrous metals	3%						
- OECD (Aluminium)		15-17 (mWh/tonne)	13.5 (-10-20%)	N.A.	At least -10-20%	20-30 years	(8) Technology being developed could reduce consumption 30-40%
Paper	3%						
- U.K. (Paper & Board Making)					-30%		(12)
Stone, Clay and Glass	2%						
- U.S./France/Switz./U.K. (Bricks/Pottery)		2.5 (MJ/kg)	1.5-2.0 (-20-40%)	N.A.	At least -20-40%	10-30 years	(15) 0.5 MJ/kg theoretical minimum

Table E.1

Energy-Efficient Technologies and the Economic Potential for Conservation *(Continued)*

Energy End Use/ Technology	(a) % of IEA TPER	(b) Existing Stock Average Efficiency (Units)	(c) New Stock Average Efficiency (% Savings)	Best Available Technology (d) Efficiency	(e) % Savings	(f) Average Useful Life of Technology	(g) Notes
- France/U.K./Switz./Germany (Cement)		3.6-3.8 (MJ/kg)	3.3 (-8-13%)	N.A.	At least -8-13%	10-30 years	(15) 1.5 & 3.0 MJ/kg theoretical minimums for dry and wet processes
Food	1%						
Space Heating, Cooling, Water, Heating, Lighting	2-3%						
ALL SECTORS							
Electric Motors	20%	75-90% (% converted to motive power)	80-92% (-2-7%)	85-93%	(-3-12%)	10-20 years	(8)(11) More efficient motors.
					(-10-20%)		(16)(17) Variable speed controls
					-15-30%		Total Potential Electric Motor Savings.
Central and On-Site Electricity Generation	35%						
- U.S. (Gas Turbines)	N.A.	30% (% converted to electricity)	35% (-15%)	39-41%	-25%		(18) Best combined cycle and steam injected gas turbines. Projected 1990 best: 44-48% efficient; total savings: 32-37%.

Table E.2

Sectoral Studies of Conservation Potential

Country	(a) End-Use Sector	(b) Year of Study	(c) Estimated Economic Potential for Demand Reductions %	(d) Remaining Potential After Projected Effects of Market Forces %	(e) Year in which Economic Potential could be Achieved	(f) Notes
United States	Residential/Commercial	1981	-50%		2000	(11)
United States/Texas	Residential/Commercial	1986	-50%		2000	(19) Electricity only
United States	Residential/Commercial	1985	-27%	-17%	2000	(20)
United States	Transportation	1985		-30%	2010	(20)
United States	Industry	1984	-35-40%		2000	(21)
		1985		-18%	2010	(20)
United States	All Sectors	1983		-22%	2000	(22)
Canada	Residential	1986	-33%	-21%	2000	(23)
	Commercial	1986	-22%	-24%	2000	(23)
	Industrial	1986	-15%	-9%	2000	(23)
	Transport	1986	-20%	-18%	2000	(23)
	All	1986	-22%	-17%	2000	(23)
Japan	All	1983	-15%		1990	
		1984	-more than 20%		1995	(6)
United Kingdom	Industry	1984	-21-29%		2000	(12)(24) United Kingdom study estimates that 21-25% efficiency gain is likely to occur 29% estimate based on "technical potential"
Netherlands	Industry	1985	-21%		2000	(25)
	Residential	1985	-21%			
	Commercial	1985	-38%			

Table E.2

Sectoral Studies of Conservation Potential *(Continued)*

Country	(a) End-Use Sector	(b) Year of Study	(c) Estimated Economic Potential for Demand Reductions %	(d) Remaining Potential After Projected Effects of Market Forces %	(e) Year in which Economic Potential could be Achieved	(f) Notes
Austria	Industry	1981	-10%			Heat pumps only
	Residential	1984	-25%			
Norway	Industry	1984		-12%		(26)
	Residential/Commercial	1984		-10%	2010	
Sweden	Residential	1985	-50%		N.A.	(27)
	Commercial	1985	-40%		N.A.	(27)
	Industry	1983		-20-50%	1990	(28)
European Community	Residential/Commercial	1984	-30%		1995	(29)
	Small-/Medium-sized Industry		-10-20%		N.A.	(30)
	Industry	1986	-25%		2000	(31)
Western Europe (EUR-9)	All Sectors	1983		-19%	2000	(22)
	Industry	1985	-30%		2000	(32)

Data Sources

1. Goldenberg, J., Thomas, J., Aulya, K.N.R., and Williams, R.H., "An End-Use Oriented Global Energy Strategy", *Annual Review of Energy,* Volume 10, pp.613-88, 1985.

2. U.S., Department of Energy, Office of Buildings and Community Systems (direct communication).

3. American Council for an Energy-Efficient Economy for Pacific Gas and Electric Company, *Residential Conservation Power Plant Study, Phase I - Technical Potential,* February 1986.

4. Geller, H.S., "Energy Efficient Residential Appliances: Performance Issues and Policy Options", *IEEE Technology and Society Magazine,* March 1986.

5. Schipper, L., et al., *Changing Patterns of Electricity Demand in Homes: An International Overview,* Lawrence Berkeley Laboratory, 1985. (Estimates derived by IEA Secretariat.)

6. Japan, the Energy Conservation Center, *Energy Conservation in Japan,* 1984.

7. Rosenfeld-Hafemeister, "Energy Sources: Conservation and Renewables", *American Institute of Physics Conference Proceedings,* 1985.

8. Moe, R.J., et al., *The Electric Energy Savings from New Technologies* (Draft), Pacific Northwest Laboratory for the U.S. Department of Energy, January 1986.

9 Lighting Group, Lawrence Berkeley Laboratory, Berkeley California (direct communication). Estimated data — existing stock: 80% efficient ballasts, 80 lumen/watt fluorescent tubes; new stock: 85% efficient ballasts, 86 lumen/watt tubes; best available (T12 size): 90% efficient solid state ballasts, 95 lumen/watt tubes.

10. U.S., Department of Energy (direct communication).

11. Solar Energy Research Institute, et al., *A New Prosperity, Building a Sustainable Energy Future, the SERI Solar/Conservation Study,* Brick House Publishing, 1981.

12. United Kingdom, Department of Energy, *Energy Efficiency Office, Energy Use and Efficiency in UK Manufacturing Industry up to the Year 2000,* London, October 1984, p.60.

13. Energie Studie Centrum, Petten, Holland, "Energy and the Iron and Steel Production", WEC-ECIP Draft Paper, November 1985. (Secretariat estimate derived from data on energy consumption for 1 000 kg of hot rolled coil.)

14. Thoreson, R., Rowberg, R., and Ryan, J.F., "Industrial Fuel Use: Structure and Trends", *Annual Review of Energy,* Volume 10, pp.165-199.

15. Giovannini, B., data from CUEPEDE (direct communication.)

16. Hickock, H.N., "Adjustable Speed — A Tool for Saving Energy Losses in Pumps, Fans, Blowers and Compressors", *IEEE Transactions on Industry Applications,* Vol. IA-21, No. 1, January/February 1985.

17. Baldwin, S., "New Opportunities in Electric Motor Technology", *IEEE Technology and Society Magazine,* March 1986.

18. Williams, R.H., and Larson, E.D., "Steam-Injected Gas Turbines and Electric Utility Planning"; *IEEE Technology and Society Magazine,* March 1986.

19. Hunn, B.D., et al., from the University of Texas and Rosenfeld, A.H. et al., from Lawrence Berkeley Laboratory, *Technical Potential for Electrical Energy Conservation and Peak Demand Reduction in Texas Buildings,* Report to the Public Utility Commission of Texas by the Center for Energy Studies, University of Texas, February 1986.

20. United States, Department of Energy, *FY 1987 Energy Conservation Multi-Year Plan,* May 1985, Draft No. 3, p.53, p.136.

21. Sant, Roger W., et al., *Creating Abundance: America's Least Cost Energy Strategy,* McGraw-Hill, 1984, pp.90-92.

22. United Nations Economic Commission for Europe, *An Efficient Energy Future, Prospects for Europe and North America,* 1983, Butterworths.

23. Canada (direct communication).

24. United Kingdom, Department of Energy, Energy Conservation Research, Development and Demonstration, Energy Paper No. 32, HMSO, 1978, p.60.

25. The Netherlands (direct communication).

26. Norway, Ministry of Petroleum and Energy, *Action Plan for Energy Conservation,* Report No. 37 to the Storting (1984-85), November 1984.

27. Ficner, Charles A., *An Overview of Residential/Commercial Energy Consumption and Conservation in IEA Countries, New Energy Conservation Technologies,* IEA Conference Papers, 1981, pp.52-53.

28. Johansson, T.B., Steen, P., et al., "Sweden Beyond Oil: The Efficient Use of Energy", *Science,* Volume 219, January 1983.

29. Commission of the European Communities, Directorate General for Energy, "Energy Efficiency in the EEC".

30. Commission of the European Community, "Findings of the Energy Bus Programme".

31. Commission of the European Communities (direct communication).

32. Albinsson, Harry, "Energy Consumption in Industrial Processes", Report presented to the 13th Congress of the World Energy Conference, Cannes, 1986.

Annex F

Methodological Issues in Programme Evaluation

> "Programme evaluation is one means of providing relevant, timely and objective findings — information, evidence and conclusions — and recommendations on the performance of government programmes, thereby improving the information base on which decisions are taken. In this view, programme evaluation, as part of this decision-making and management process, should not be seen as an exercise in scientific research aimed at producing definitive 'scientific' conclusions about programmes and their results. Rather it should be seen as input to the complex, interactive process that is government decision-making, with the aim of producing *objective but not necessarily conclusive* evidence on the results of programmes. While credible analysis is always required in programme analysis, a strict research model ... is often inappropriate because of timing constraints and an inability to adequately take into account the multiple information needs of the client and users of the evaluation." [1]

Energy conservation programmes have proven to be difficult to evaluate for effectiveness. There are many reasons for this such as: inadequate data collection, poor programme design, lack of measurable objectives,

1. Treasury Board of Canada, *Guide on the Program Evaluation Function,* Ottawa, 1981, p.4.

the decision to evaluate often not taken until after the programme has been operating, evaluations are costly, time-consuming and complex and evaluations are often not done by an objective party.

There is no single ideal evaluation design since different policy measures require different approaches. Over the years, however, much knowledge has been gained in understanding evaluations. The bibliography includes many references explaining various aspects of evaluation techniques.

There are basic methodological issues which are common to all evaluations. This annex describes some of the major issues and how selected evaluations attempted to resolve them.

Incrementality

Incrementality is defined as the difference between what happened and what would have happened in the absence of a programme. It is common to also use the related expression "free-riders". A free-rider would be the participant in a programme who would have undertaken the activity even in the absence of the programme.

Incremental analysis can consider both the effects on conservation activity and on expenditure; however, both are not always done. It is usually the analysis of the effect on expenditure that is done.

Incremental analysis is an essential component in determining the effectiveness of a programme. Unfortunately, it is not always assessed in evaluations undertaken. Often, the estimates that are available, have not been done in a thorough and rigorous manner.

The following excerpt from the Canadian evaluation of its Ener$ave audit programme describes how they used the concept:

"The two-group design ... is based on a specific concept of incrementality. It identifies the significant factors of the decision process through models applied to the "normal", untreated, or control group. The factors are framed as a decision model. Activities are modelled as the consequences of a wide range of independent, or prior, conditions. The model is a predictive one, to be interpreted as saying:

— these are the important characteristics which are related to predicting activity

— the relationship of these characteristics to activities is defined by the model through the use of multivariate analysis.

"Assuming that the decision structure of the two groups is the same, this model will yield an unbiased estimate of what the program group would have done without the program. The differences, if any, between the predicted, *ceteris paribus,* levels of activity and the actual levels can therefore be attributed to that different characteristic, the treatment of one group with Ener$ave. The basic formula is:

Incrementality = (Actual Activity) - (Predicted Activity)

"The two groups do not need to have identical values on the model characteristics. The use of statistical matching, i.e. the multivariate model, will compensate for the differences. There are two prerequisites: data for both groups exist on all salient characteristics, and an assumption that the same decision structure applies. In terms of statistical reliability (as opposed to unbiasedness or validity), the model's range of variables should be similar for the two groups. Otherwise, the standard error of the predictions will be extremely large.

"This methodology should be clearly distinguished from that used to develop the cross-sectional profile ... . The profile identified the differences between the two groups on a wide range of characteristics; it was a single model whose dependent variable was a binary switch indicating participation in Ener$ave. The incrementality models are quite different. For one thing, they predict conservation activity and expenditures, not Ener$ave participation. But more importantly, they use statistical matching to apply models developed for the control group to the Ener$ave users, to produce predictions of what Ener$avers would have done had they used exactly the same decision structure as non-users. That aspect is not produced by the profile model, but is then used as the input to the incrementality estimate where it is subtracted from actual behaviour to determine the effect of Ener$ave (as the only difference between the relevant decision models of the two groups). The profile model shows differences in characteristics; the incrementality models control those differences to model the decision structure.

"An example may be useful. Let's say the average non-Ener$aver spends $500 on his attic while Ener$avers spend $600. The profile will show that Ener$avers spend more on their attics. But these people have some different characteristics. Perhaps Ener$avers are more likely to have moved into old homes that need more attic insulation. Perhaps Ener$avers are generally more enthusiastic about fixing up houses. The incrementality models ask: "What would similar people who didn't use Ener$ave have spent?" For the sake of argument, let's say people with those characteristics would have spent $700. Since Ener$avers spent $600 where similar non-Ener$avers spent $700, incrementality is $600 - $700 or (-)$100. In this example, Ener$ave users would have spent less than a *comparable,* or statistically matched, group of non-Ener$avers."

Attribution

Many evaluations have found it impossible to separate the effects of individual policies from the effect of other influences, such as energy prices, and structural changes in the economy. There are, however, many who feel that research designs can be developed to "control for or rule out alternative explanations for observed energy savings. Because alternative explanations such as increased fuel prices may account for some of the change in consumption, failure to take them into account will result in savings being incorrectly attributed to the programme. This situation is further complicated by the fact that few mechanisms presently exist for gathering information on energy consumption, and all factors that may influence energy use, in a direct, manageable and inexpensive manner" [1].

In a Swedish study of the effects of the energy economy programme, to determine attribution:

1. They removed a number of changes which *cannot* have been affected by the policy instruments.
2. They also tried to note other factors which may have affected the changes.
3. The econometric studies allowed a third kind of analysis to describe the effects. In these analyses, an imagined undirected development is described, based on assumptions about various economic parameters, primarily elasticity of price and incomes.
4. The resulting differences — negative or positive — between actual and hypothetical developments can thus be ascribed to government policy.

(1) Jon Soderstrom, "Measuring Energy Savings", *Energy Policy,* March 1984, p.104.

In work undertaken at the Oak Ridge National Laboratory on measuring the energy savings of U.S. Department of Energy conservation programmes, the following comment was made regarding the State Energy Conservation Programme (SECP) and its estimated saving due to thermal efficiency standards (which accounted for about 66% of the estimated savings):

> "...The standard algorithm for estimating savings is a function of the annual square footage of new building construction multiplied by an estimate of energy saved per square foot, times the number of years the thermal efficiency code has been in effect. Separate estimates are made for each of nine building types and summed for a total savings estimate."
>
> "Recent work suggests that this algorithm may attribute too much of the estimated savings to the SECP. Critical examination of each component of this algorithm reveals a number of flaws which could contribute to an overestimate of savings. For example, building construction rates were based on 1973 data, producing estimated rates considerably higher than are actually occurring. Further, this algorithm ignores the influence of factors other than the code. These other factors include code-compliance rates, the effect of local codes, construction lag time, and changes in the price of energy. Thus, there is some doubt about how much energy savings should be attributed to the SECP, and the actual savings could be less than reported in the evaluation" [1].

The evaluation of the impact of the Belgian information campaign also appreciated the complexity of separating the effects [2]. Because the impact study was not started at the same time as the actual information campaign, it was difficult to establish a proper control and thus impossible to accurately distinguish between the effects of the campaign and other effects. They undertook an approach of sampling two groups of households that used gas, one group of which had used the government's information brochures. These groups were surveyed and then the data was analysed using multidimensional regression analysis to evaluate the relative importance of each cause.

---

1. Ibid, p.105.
2. Institut Interuniversitaire de Sondage d'Opinion Publique, *Mesure de l'impact de la campagne d'information en faveur des économies d'énergie*, Brussels.

Cost-Effectiveness

Most evaluations have made some attempts to determine cost-effectiveness. For some it has simply been to take the difference between the gross energy savings and the government (and sometimes only departmental) costs without controlling for incrementality. Although there are many methods, it is useful here to describe the methodology used for the evaluation of the Canadian Home Insulation Programme:

> "The evaluation attempted, through sophisticated modelling, to estimate both the incremental number of insulation activities and the dollar value of incremental spending on insulation among CHIP users that could be attributed to CHIP. These incrementality estimates were then used in a cost-benefit analysis. Over the 20-year life of the insulation, the present value of energy savings of 28.1 PJ per year (from all CHIP activities) is $2 100 million at a 7% discount rate. Subtracting $833 million (estimated total spending on CHIP) leaves $1 267 million as net economic benefit [1]. As an estimated 75% of CHIP-related investment was incremental, it follows that $950 million is the incremental net economic benefit. (At a 10% discount rate, the incremental net benefit would be $652 million.)"
>
> "These calculations assume that the incrementality was permanent, as the benefits from the incremental actions were measured for the full 20 years. However, the methods used in the evaluation were able only to determine what proportion of activities was incremental during 1977-82; they could not distinguish activities that might have occurred later and were simply accelerated by CHIP. To the extent that some CHIP users would have eventually acted without CHIP, the analysis overestimates the benefits."
>
> "A sensitivity analysis was performed to demonstrate the impact of various assumptions about accelerations (see Table)."
>
> "Even if CHIP only accelerated the insulation activity by four years, the benefits are still worthwhile. Probably closer to reality is the assumption that half the incrementality was permanent and half was accelerated by four years. This gives a net benefit from the

1. These figures are assumed to be 1980 dollars, as the mid-point of CHIP activity; they are based on reference cost price estimates of the various heating fuels.

programme for 1977-82 of almost $600 million (for comparison, the net cost to the government was $409 million). This analysis excludes any indirect benefits, for example, increased activity by *non*-CHIP participants as a result of publicity about the programme and increased awareness of energy conservation possibilities."

Table. Cost-Benefits under different acceleration assumptions. (Discount rate of 7% is used and where acceleration is noted, assumes this would be by four years on average.)

Assumption	Total Incremental	Total Benefits	Net Benefits	Net Benefits/ $1 Grant
1. Incremental CHIP activity is all acceleration	150	380	230	0.56
2. Incremental CHIP activity is 50% acceleration and 50% additional	390	980	590	1.44
3. Incremental CHIP activity	625	1 575	950	2.32

"The high benefit-cost ratio is based both on the relatively high incrementality of the programme and on the high cost-effectiveness of the individual actions taken. From the point of view of the individual home owner and average CHIP participant, an expenditure of $643 ($173 + grant) achieved a saving of about 14% (21 GJ per year) from 1.9 CHIP-eligible insulation activities, plus an additional 2.5% from air-tightening. This is $44 per GJ of annual reduction in energy use, well under the threshold levels for cost-effectiveness cited elsewhere in this report; it is equivalent to a levelized price of energy of $2.93 per GJ at a 7% discount rate (equivalent to about 1¢ per kWh). Individual activities were thus highly cost-effective, on average."

Annex G

Excerpts from IEA Energy Technology Policy Study on Conservation Technologies (pp.74—81)

Table G.1 provides an overview of the main areas of energy-efficiency technology, indicating their potential application and current status, and suggesting measures for encouraging policy decisions to accelerate change. It should be noted that microelectronic sensor and control systems have an important role in many of these individual technologies, as well as in complete systems such as an electricity distribution system, where their deployment in conjunction with more effective storage systems could revolutionise electricity demand management. The total systems approach should be emphasized; the energy efficiency and economic performance of the total system should be considered before specific end-use measures are determined. Another important idea might be to improve communication between and among sectors and consumers, as there are many cases where deployment in the market is delayed by lack of information.

Examples of technologies considered important to pursue are the following:

— *Transportation sector:* Use of liquid fuels will be inescapable as far into the future as can be foreseen, and the development of energy-efficient automobiles will play an important role in determining requirements. More efficient (e.g. ceramic) engines, which promise to halve fuel consumption, may significantly alter the gasoline supply/demand situation. Electric cars are a possibility whose advent could limit the demand for liquid fuels, but present planning should not anticipate their early success, which is

Table G.1

Overview of the Main Energy-Efficient Technologies

	Energy Savings Potential	Timescale for Significant Commercial Applications	International Variations	Impediments to Commercial Applications	Principal Policy Options
Internal Combustion Engines	Large, entirely oil-savings	10-15 years, allowing for introduction of new designs and rotation of car stock	United States, Japan, Germany, Italy, United Kingdom and Sweden, the main IEA automobile manufacturers	Car industry highly competitive. Main impediments would be if petrol prices fell sharply and were expected to stay low in medium term	Government standards for higher miles per gallon (probably combined with environmental standards); direct intervention to stimulate search for radically new automobile designs
Electric Battery Vehicles	Entirely oil-savings	Economic now for some delivery vehicles. 15-25 years for light vans and shorter distance car journey	"	Efficiency of battery which limits payload distance between charges. High costs per vehicle due to small market. Increasing efficiency of petrol and diesel vehicles	Continuing effort to improve battery efficiency. Public sector procurement for organisations with large fleets of light vans
Heat Pumps	Oil and electricity saving, but attractive only in certain countries and regions	Technology available today. 10-15 years where competitive with alternative fuels	Most competitive where - requirements for heating and cooling - air is space-heating medium (ground water, where available is more competitive) - gas and district heat not available	High initial cost Alternative heating fuels	Provision of subsidy for heat pump investment. Support for heat pump R&D, especially to reduce capital cost

Table G.1

Overview of the Main Energy-Efficient Technologies *(Continued)*

	Energy Savings Potential	Timescale for Significant Commercial Applications	International Variations	Impediments to Commercial Applications	Principal Policy Options
Electronic Controls	Large, all sectors; all types of fuel. May improve competitiveness of electricity	Economic now	Smaller and less advanced IEA countries lack strong electronics industries	Lack of consumer awareness and cashflow constraints on investment	Information and advisory services. Assistance for demonstration schemes. Low-interest loans
Process Heat	Oil, gas and coal in industry	Waste heat recovery often economic now	Applicable especially where concentration of heavy industries	Often difficult to sell recovered waste heat (matching loads required close by). Lack of consumer awareness and cashflow constraint	"
Water Heat	Oil, gas and coal mainly in industrial and commercial boilers	Improved boiler designs and controls currently economic		Slow rotation of capital stock, but improved controls and insulation can be fitted to existing boilers and systems	Information and Advisory services. Assistance for demonstration schemes. Low-interest loans
Combined Heat and Power/ District Heating	Primary energy for electricity generation, oil for domestic heating	Economic now in some situations	Viability depends on (a) lack of competition from natural gas, and (b) favourable political and administrative structures	Surpluses and shortages tend to coincide with those on electricity grids. Obtaining a "fair price" from electric utilities. Availability of sites and connections with grid.	Reserve existing power station sites in favourable locations for potential future CHP use. Ensure utilities do not discriminate through unfair pricing for buying and selling. Assistance for demonstration schemes. Low-interest loans.

Table G.1

Overview of the Main Energy-Efficient Technologies *(Continued)*

	Energy Savings Potential	Timescale for Significant Commercial Applications	International Variations	Impediments to Commercial Applications	Principal Policy Options
Urban Waste	Local and minor, except in large conurbations	"		Organisation of refuse collection and of electricity supply industry	"
Building Design	Mainly domestic and commercial buildings	After 2000	Climatic variations and local building materials and design traditions affect what is feasible and economic	Slow rotation of building stock	Higher standards for design and construction and insulation

contingent upon the development of high-charge-density, multiple-charge/recharge, lightweight batteries. Research into battery storage is therefore imperative for widespread use of electric cars to be possible by the turn of the century.

— *Industrial sector:* Energy-efficient production systems, especially in energy-intensive industries such as steel, aluminium and other basic materials, will be decisive factors in determining demand in this sector. Total energy management systems in factories, including the recycling of waste heat, are being developed by the more competition-minded enterprises, but widespread use of these techniques could have a major overall impact. Technologies permitting the recycling of energy-intensive products could be important to pursue.

— *Residential and commercial sectors:* Reductions in energy demand must be achieved either by retrofitting existing buildings or, more efficiently, by building new stock. The technologies to be concentrated on include insulation, heating or cooling systems such as district heating, and/or heat pumps. The development of total energy management systems is essential to efficient energy use in large commercial and institutional buildings. Measurement technologies, among which micro-electronics will have a crucial role, will be a major direction for RD&D.

Energy conversion is an area which requires continued emphasis. In conventional power plants with steam turbines, about two-thirds of the input energy is lost in conversion. Investment in RD&D to increase conversion efficiency is essential, and deployment of more efficient electricity generating technology, even by small percentages, might have great impact on the entire system. For example, use of combined cycles (the combination of gas turbines and steam turbines) provides higher efficiency in producing electricity, with input sources such as low-calorific gasified coal, high-temperature gas-cooled reactors and/or pressurised fluidised bed combustors. Similarly, co-generation systems need to be further developed to utilise high-grade energy for electricity production and low-grade heat for industrial or district heating applications.

Magneto-hydrodynamics (MHD) is one of the technologies being developed for the direct conversion of high-temperature gas-flow energy. Its engineering feasibility has not yet been fully demonstrated and any future commercial application would have to be considered long term. Fuel cells also offer other possibilities for converting energy directly to electricity without requiring the intermediate formation of heat. Experiments have shown that they are highly efficient when compared with conventional technology. The major R&D obstacles seem to be the difficulties of scale-up and low cost effectiveness, but they have the advantage over MHD in that the technological basis is better established.

Principal Findings

— Potential for energy saving technology application is still considered large although priority areas will vary from country to country.

— Raising energy efficiency can be less costly than equivalent expansion of energy production and is an area where environmental benefits may reinforce energy benefits.

— Diffusion of technology already developed is of overriding importance.

— Measurement and control technologies are likely to make a significant contribution to further technical achievements in energy saving.

— Better understanding of end use is essential for determining where R&D is most needed, as well as the development of analytical technologies.

— Total systems approach should be adopted to assist the choice of R&D and other non-technical measures.

— Non-technical promotional initiatives such as fiscal policies, standards and information dissemination are likely to be as important as RD&D itself.

— Efficient direct energy conversion systems require continued RD&D.

Annex H

Members of the Informal Advisory Group of Conservation Experts

Mr. Cameron D. Beers
Director
Administrative Services and Special Projects
The Gillette Company
Boston, Massachusetts, U.S.A.

Dr. W.G. Bentley
Chief Economist
Florida Power and Light Co.
Energy Management Planning Department.
Miami, Fla., U.S.A.

Mr. Evald Brond
Consultant
Copenhagen, Denmark

Mr. David B. Brooks
Marbek Resource Consultants Ltd.
Ottawa, Canada

Mr. John H. Chesshire
Science Policy Research Unit
University of Sussex
Brighton, United Kingdom

Professor B. Giovannini
Centre Universitaire d'Etudes des Problèmes de l'Energie
Geneva, Switzerland

Dr.-Ing. Eberhard Jochem
Fraunhofer Institut für Systemtechnik und Innovationsforschung (ISI)
Karlsruhe, Germany

Dr. Tage Klingberg
Head, Department 5
Impact of Policy Analysis on Buildings
National Swedish Institute for Building Research
Gävle, Sweden

Prof. Dr. F. Moser
Head, Institute for Process Engineering
Technical University
Graz, Austria

Mr. Takashi Niikura
Vice President
The Energy Conservation Center
Tokyo, Japan

Mr. J.G. Potter
Chief Executive
The Freeman Group
Cambridge, United Kingdom

Professor W. Fred van Raaij
Department of Economic Psychology
Erasmus University
Rotterdam, Netherlands

Mr. Roger Sant
Applied Energy Services, Inc.
Arlington, Virginia, U.S.A.

Prof. Dr.-Ing. Helmut Schaefer
Lehrstuhl für Energiewirtschaft und Kraftswerkstechnik
Munchen, Germany

Dr. Paul Stern
Committee on Behavioural and Social Aspects of Energy Consumption and Production
National Research Council
Washington, D.C., U.S.A.

Bibliography

Agence Française pour la maîtrise de l'énergie, *La Maîtrise de l'Energie en France,* Paris, août 1985.

Agence pour les Economies d'énergie, *Une comparaison des résultats des politiques d'économie d'énergie dans six pays industrialisés,* janvier 1982.

Albinsson, Harry, "Energy Consumption in Industrial Processes", Report presented to the 13th Congress of the World Energy Conference, Cannes, 1986.

Alliance to Save Energy, *Industrial Investment in Energy Efficiency: Opportunities, Management Practices and Tax Incentives,* Washington, 1983.

Alliance to Save Energy, *Utility Promotion of Investment in Energy Efficiency: Engineering, Legal and Economic Analysis,* Washington, 1983.

Alliance to Save Energy, *Third Party Financing: Increasing Investment in Energy Efficiency Industrial Projects,* Washington, 1982.

American Council for an Energy-Efficient Economy, *Doing Better: Setting an Agenda for the Second Decade; Proceedings of the ACEEE 1984 Summer Study on Energy Efficient Buildings,* August 1984.

American Council for an Energy Efficient Economy, *Proceedings from the ACEEE 1986 Summer Study on Energy Efficiency in Buildings,* August 1986.

American Council for an Energy-Efficient Economy (for Pacific Gas and Electric Company), *Residential Conservation Power Plant Study, Phase I-Technical Potential,* February 1986.

Association for the Conservation of Energy, *Administering Energy Saving: An Evaluation of the Administration of Energy Conservation Programmes by European Governments,* London, 1983.

Association for the Conservation of Energy, *Jobs and Energy Conservation,* London, 1983.

Association for the Conservation of Energy, *Lessons from America: No. 1 — Home Energy Ratings,* London, September 1984.

Association for the Conservation of Energy, *Lessons from America: No. 2 — U.S. Gas and Electricity Utilities and the Promotion of Conservation,* London, 1984.

Association for the Conservation of Energy, *Lessons from America: No. 3 — Third Party Finance,* London, 1985.

Australian Minerals and Energy Council, *Consolidated Papers on Energy Labelling,* Canberra, July 1983.

Austria, *Energiebericht und Energiekonzept 1984* (Energy Report and Energy Concept 1984), Vienna, November 1984.

Austria, *Energiesparbuch,* 1985.

Baldwin, S., "New Opportunities in Electric Motor Technology", *IEEE Technology and Society Magazine,* March 1986.

Barnes, R.W. et al., *Energy Use from 1973 to 1980: The Role of Improved Energy Efficiencies,* Oak Ridge National Laboratory, December 1981.

Berry, L. and Eric Hirst, "Evaluating utility residential conservation programmes: an overview of an EPRI workshop" in *Energy Policy,* Vol. 11, No. 1, March 1983, pp.77-81.

Berry, Linda et al., *Review of Evaluations of Utility Home Energy Audit Programmes,* Oak Ridge National Laboratory, March 1981.

Bonneville Power Administration, *An Analysis of BPA Conservation Program Levels for Fiscal Years 1984 and 1985 and their Relationship to a Least-Cost Resource Mix,* Portland, Oregon, May 1983.

Bonneville Power Administration, *1985 Resource Strategies,* Portland, Oregon, November 1984.

Bonneville Power Administration, *Scoping Document: For the 1986 Long-Range Conservation Projection,* Portland, Oregon, April 1985.

Bonneville Power Administration, *Technical Report: Conservation Policy in the Pacific Northwest,* Portland, Oregon, May 1985.

Bonneville Power Administration, *An Assessment of Achievable Residential Sector Conservation Potential in the Pacific Northwest,* Portland, Oregon,

Bronfman, B.H. and Lerma, D.I., *Process Evaluation of the Bonneville Power Administration Interim Residential Weatherization Program* (ORNL/CON-158), Oak Ridge National Laboratory, August 1984.

Brooks, David B., *Economic Impacts of Low Energy Growth,* Economic Council of Canada, Discussion Paper 126, Ottawa, Canada, 1978.

Bucy, J. Fred, "Meeting the Competitive Challenge: The Case for R&D Tax Credits", *Issues in Science and Technology,* Summer 1985, pp.69-78.

California Energy Commission, *1983 Electricity Report,* January 1983.

Canada, Consumer and Corporate Affairs, *Evaluation of Energuide,* March 1985.

Canada, Energy, Mines and Resources, *Economics of Energy Conservation in Canada,* Ottawa, Winter 1984.

Canada, Energy, Mines and Resources, *A Comprehensive Analysis of the Atlantic Energy Conservation Investment Programme,* January 1986.

Canada, Energy, Mines and Resources, *Evaluation of the Canadian Home Insulation Programme,* Ottawa, November 1983.

Canada, Energy, Mines and Resources, *Federal Energy Management Program, Seventh Annual Report,* Ottawa, 1983.

Canada, Energy, Mines and Resources, *Home Energy Programmes: Prince Edward Island Energy Audit Demonstration Programme Final Report,* July 1984 (contractor's report).

Canada, Energy, Mines and Resources, *Evaluation Report on the Residential Enersave Energy Audit Programme,* April 1984 (contractor's report).

Canada, Energy, Mines and Resources, *Development and Implementation of a Strategic Plan to Support Savings Financing of Energy Management,* May 1985 (contractor's report).

Canada, Treasury Board, *Guide on the Program Evaluation Function,* May 1981.

Canertech Inc., *Market Potential and Economic Impacts of Energy Conservation in the Canadian Residential, Commercial and Industrial Sectors,* December 1983.

Chesshire, John, *Investment Appraisal in the Energy Sector* (draft), November 1985, to be published in Public Money's Annual Energy Review, 1986.

Claxton, J.D. et al., *Consumers and Energy Conservation: International Perspectives on Research and Policy Options,* Praeger, 1981.

Collins, N.E. et al., *Past Efforts and Future Directions for Evaluating State Energy Conservation Programs, Final Draft* (ORNL-6113), Oak Ridge National Laboratory, October 1984.

Commission of the European Communities, *Comparison of Energy Saving Programmes of EC Member States,* COM(84)36 Final, Brussels, 2nd February 1984.

Commission of the European Communities, *Employment Effects of Energy Conservation Investments in EC Countries,* Brussels, November 1984.

Commission of the European Communities, *Towards a European Policy for the Rational Use of Energy in the Building Sector,* COM(84)614 Final, Brussels, 13th November 1984.

Commission of the European Communities, *Evaluation Report on the Energy Demonstration Programme,* COM(85)29 Final/2, Brussels, 25th February 1985.

Commission of the European Communities, Directorate General for Energy, "Energy Efficiency in the EEC".

Commission of the European Community, "Findings of the Energy Bus Programme".

Commission of the European Communities, "Rational Use of Energy in Road, Rail and Inland Waterway Transport", communication from the Commission to the Council (COM(86)393 final), Commission of the European Communities, Brussels, 22nd July 1986.

Crossley, David and J.D. Kalma, "Inequities in domestic energy use" in *Energy Policy,* Vol. 10, No. 3, Sept. 1982, pp.233-243.

Crossley, David, *Barriers to Household Energy Conservation,* paper presented to International Conference on Consumer Behaviour and Energy Policies, The Netherlands, September 1982.

Currie, W.M., *The Energy Conservation Demonstration Projects Scheme — What it is all about,* Energy Technology Support Unit, Oxon, England, 1981.

de Man, Reiner, "Barriers to energy conservation — the case of the Netherlands social housing sector" in *Energy Policy,* Vol. 11, No. 4, December 1983, pp.363-368

Denmark, Ministry of Energy, *Energy in Denmark: A Report on Energy Planning,* Copenhagen, December 1984.

Eden, R., Posner, M., Bending, R., Crouch, E. and Stanislaw, J., *Energy Economics, Growth, Resources and Policies,* Cambridge University Press, 1981.

Edmonds, J. and Reilly, J.M., *Global Energy, Assessing the Future,* Oxford University Press, 1985.

Eisner, Robert, "The R&D Tax Credit: A Flawed Tool", *Issues in Science and Technology,* Summer 1985, pp.79-86.

Electric Power Research Institute, *Workshop Proceedings: Measuring the Effects of Utilities' Conservation Programmes,* California, July 1982.

Electric Power Research Institute, *Proceedings: International Workshop on Electricity Use in the Service Sector,* EPRI P-4401-SR, January 1986.

Energie Studie Centrum, Petten, Holland, "Energy and the Iron and Steel Production", WEC-ECIP Draft Paper, November 1985.

European Conference of Ministers of Transport (ECMT), Round Table 52, "*Transport and Energy",* ECMT, 1981 (available from OECD, Paris).

Fichter Beratende Ingenieure, Stuttgart, Fraunhofer-Institute (ISI), Karlsruhe, *Der Beitrag ausgewählten neuer Technologien zur rationellen Energieverwendung in der deutschen Industrie* (Contribution of selected new energy conservation technologies in German industry), study ordered by the German Ministry for Economics, March 1986.

Ficner, Charles A., *An Overview of Residential/Commercial Energy Consumption and Conservation in IEA Countries, New Energy Conservation Technologies,* IEA Conference Papers, 1981.

Ford, Andrew, "Uncertainty in the price of gasoline and the automobile manufacturers' 1990 retooling decision", *Energy,* Vol. 9, No. 6, 1984, pp.519-540.

Frieden, Bernard J. and Kermit Becker, The Market Needs Help: The Disappointing Record of Home Energy Conservation", *Journal of Policy Analysis and Management,* Vol. 2, No. 3, pp.432-448 (1983).

Garnreiter, F. et al., *Structural Change and Technological Progress as Influencing Factors of Specific Energy Consumption in Energy-Intensive Industries* (in German), Fraunhofer-Institute (ISI), Karlsruhe, November 1985.

Garnreiter, F., Jochem, E. et al., *Effects of stricter measures to rationalise the energy consumption on environment, employment and income,* Fraunhofer Institute for Systems and Innovation Research (ISI), Karlsruhe, Germany, Erich Schmidt Verlag, Berlin, 1983.

Gates, R.W., "Investing in energy conservation: are homeowners passing up high yields?" in *Energy Policy,* Vol. 11, No. 1, March 1983, pp.63-71.

Geller, H.S., "Energy Efficient Residential Appliances: Performance Issues and Policy Options", *IEEE Technology and Society Magazine,* March 1986.

Geller, Howard S., *Energy Efficient Appliances,* American Council for an Energy Efficient Economy and the Energy Conservation Coalition, June 1983.

German Institute for Economic Research, Berlin (DIW), *Assessment of financial incentives for CHP/DH from an overall economic aspect,* 1985.

German Institute for Economic Research, Munich (IFO), *The Impact of Economic Development on Energy Consumption,* 1985.

Giovannini, B. and Delfosse, A., *Influence sur la consommation d'énergie des scenarios de politique énergétique en Suisse,* Geneva, Université de Genève, 1983.

Goldenberg, J., Thomas, J., Aulya, K.N.R., and Williams, R.H., "An End-Use Oriented Global Energy Strategy", *Annual Review of Energy,* Volume 10, pp.613-88, 1985.

Gorzelnik, Eugene F., "Rebates push sales of efficient appliances", *Electrical World,* January 1984, pp.83-85.

Gruber, E., Garnreiter, F., Jochem, E. et al., *Evaluation of Energy Conservation Programmes in the EC Countries,* Fraunhofer Institute for Systems and Innovation Research (ISI), Karlsruhe, Germany, December 1982 (study prepared for the European Commission).

Hickock, H.N., "Adjustable Speed — A Tool for Saving Energy Losses in Pumps, Fans, Blowers and Compressors", *IEEE Transactions on Industry Applications,* Vol. IA-21, No. 1, January/February 1985.

Hillman, Dr. Mayer, *Conservation's Contribution to UK Self-Sufficiency,* London, Policy Studies Institute, 1984.

Hirst, Eric et al., "Connecticut's residential conservation service: an evaluation", *Energy Policy,* February 1985, pp.60-70.

Hirst, Eric et al., *"Recent Changes in U.S. Energy Consumption: What Happened and Why" in Annual Review of Energy,* Volume 8, 1983, pp.193-245.

Hirst, Eric et al., *Evaluation of the BPA Residential Weatherization Program* (ORNL/CON-180), Oak Ridge National Laboratory, June 1985.

Hirst, Eric et al., *Three Years after Participation: Electricity Savings due to the BPA Residential Weatherization Pilot Program* (ORNL/CON-166), Oak Ridge National Laboratory, January 1985.

Hirst, Eric, "Analysis of Hospital Energy Audits", *Energy Policy,* September 1982.

Hirst, Eric, et al., "Improving energy efficiency: the effectiveness of government action" in *Energy Policy,* Vol. 10, No. 2, June 1982, pp.131-142.

Hofbauer, E., Rogner, W., *"Untersuchung über energiesparende Maßnahmen in der wärmeintensiven Industrie",* Springer Verlag, Wien — New York, 1983.

Hunn, B.D., et al., from the University of Texas and Rosenfeld, A.H. et al., from Lawrence Berkeley Laboratory, *Technical Potential for Electrical Energy Conservation and Peak Demand Reduction in Texas Buildings,* Report to the Public Utility Commission of Texas by the Center for Energy Studies, University of Texas, February 1986.

IEA/OECD, "District Heating and Combined Heat and Power Systems — A Technology Review", Paris, 1983.

Institut Interuniversitaire de Sondage d'Opinion Publique, *Mesure de l'impact de la campagne d'information en faveur des économies d'énergie,* Brussels.

International Chamber of Commerce, "Energy Conservation in Industry: A Background Report by the International Chamber of Commerce", *Twelfth Congress of the World Energy Conference,* New Delhi, September 1983.

International Energy Agency, *Annual Report on Energy Research, Development and Demonstration — Activities of the IEA 1982-83,* OECD, Paris, 1983.

International Energy Agency, *Energy Research, Development and Demonstration in the IEA Countries — 1984 Review of National Programmes,* OECD, Paris, 1985.

International Energy Agency, *Energy Policies and Programmes of IEA Countries — 1984 Review,* OECD, Paris, 1985.

International Energy Agency, *Energy Technology Policy,* OECD, Paris, 1985.

International Energy Agency, *Fuel Efficiency of Passenger Cars — An IEA Report,* OECD, Paris, 1984.

International Road Federation, *World Road Statistics,* 1976, 1977, 1978, 1979, 1983, 1984, 1985, Geneva, Switzerland.

Japan, The Energy Conservation Center, *Energy Conservation in Japan,* Tokyo, 1984.

Johansson, T.B., Steen, P., et al., "Sweden Beyond Oil: The Efficient Use of Energy", *Science,* Volume 219, January 1983.

Karl, H.D., Rammner, P. and Scholz, L., Institute for Economic Research, Munich, Germany (IFO), "Effects of public financial incentives on the economics of CHP/DH and on the total economy", *IFO Study No. 5,* Munich, 1984; summary in *IFO-Schnelldienst 3/84.*

Karl, H.D., Rammner, P. and Scholz, L., Institute for Economic Research, Munich, Germany (IFO), *Quantitative effects of energy conservation policy in the Federal Republic of Germany,* IFO Study No. 3/1, Munich, 1982.

Klingberg, T. and Wickman, K., *Energy Trends and Policy Impacts in Seven Countries,* National Swedish Institute for Building Research, Gävle, August 1984.

Klingberg, Tage, editor, *Effects of Energy Conservation Programmes,* National Swedish Institute for Building Research, Gävle, Sweden, 1984.

Klingberg, Tage, editor, *Energy Conservation in Rented Buildings,* National Swedish Institute for Building Research, Gävle, August 1984.

Klingberg, Tage, editor, *Swedish national and local government programmes for conservation of energy in buildings,* National Swedish Institute for Building Research, Gävle, May 1982.

Kushler, M.G. and Saul, J.A., "Evaluating the Impact of the Michigan RCS Home Energy Audit Program", *Energy,* Vol. 9, No. 2, 1984, pp.113-124.

Maier, W. and Suttor, K.H., (Arbeitsgemeinschaft Kraftwärme-Kopplung, consisting of the consulting companies Fichtner Beratende Ingenieure, Stuttgart and ECH Energieconsulting, Heidelberg GmbH), *Electric Potential and Economics of CHP in industry and commerce,* Stuttgart/Heidelberg, Resch-Verlag, 1984.

Marcus, A.A., et al., "Barriers to the adoption of an energy efficient technology" in *Energy Policy,* Vol. 10, No. 2, June 1982, pp.157-158.

Marlay, Robert C., "Trends in Industrial Use of Energy", *Science, Vol. 226, No. 4680, 14th December 1984.*

Marlay, Robert C., "Trends in U.S. Energy Use", United States Department of Energy (presented at the International Roundtable on Energy, Technology and the Economy, 28th-30th October 1986, International Institute for Applied Systems Analysis, Laxenburg, Austria).

Marsh, Robert L., Practice, Problem-Solving, and Skills Development for Energy Program Evaluation (ORNL-Sub 83-39112), Oak Ridge National Laboratory, November 1983.

McDougall, G.H.G. and Mank, R.B., "Consumer energy conservation policy in Canada: behavioural and institutional obstacles" in *Energy Policy,* Vol. 10, No. 3, Sept. 1982, pp.212-224.

Meier, Alan et al., "Supply Curves of Conserved Energy for California's Residential Sector", *Energy,* Vol. 7, 1981, pp.347-358.

Moe, R.J., et al., *The Electric Energy Savings from New Technologies* (Draft), Pacific Northwest Laboratory for the United States Department of Energy, January 1986.

Moser, F., Schnitzer, H., *"Energieeinsparung durch Wärmepumpen in Industrie und Gewerbe",* Springer Verlag, Wien, New York — 1983.

New South Wales, Energy Authority of New South Wales, *Energy Management in the State of New South Wales,* Sydney, November 1983.

New South Wales, Energy Authority of New South Wales, *Development of the Government Energy Management Programme,* Sydney, July 1981.

New York State Energy Research and Development Authority, *Performance Contracting for Energy Efficiency: An Introduction with Case Studies,* January 1984.

Northwest Power Planning Council, *Evaluation of Utility Sponsored Conservation Programmes* (Volumes I, II and Recommendations), Portland, Oregon, May 1985.

Norway, Ministry of Petroleum and Energy, *Action Plan for Energy Conservation,* Report No. 37 to the Storting (1984-85), November 1984.

Organisation for Economic Co-operation and Development (OECD), *Environmental Effects of Energy Systems: the OECD COMPASS Project,* OECD, Paris, 1983.

Organisation for Economic Co-operation and Development (OECD), *Aluminium Industry — Energy Aspects of Structural Change,* OECD, Paris, 1983.

Organisation for Economic Co-operation and Development (OECD), *Economic Aspects of Energy Use in the Pulp and Paper Industry,* OECD, Paris, 1984.

Organisation for Economic Co-operation and Development (OECD), *Petrochemical Industry — Energy Aspects of Structural Change,* OECD, Paris, 1985.

Pacific Gas and Electric, *Market Research Summary: 1975 to Present* (MR-85-0711), San Francisco, 1985.

Parikh, S.C., co-ordinator, Energy Saving Impacts of DOE's Conservation and Solar Programs, Oak Ridge National Laboratory, ORNL/TM-7690, March 1981.

Peabody, Gerald, *Weatherization Program Evaluation,* United States, Department of Energy, Energy Information Administration, Washington, August 1984.

Peelle, Elizabeth et al., *Reaching People with Energy Conservation Information: From Statewide Residential Case Studies* (ORNL-5984), Oak Ridge National Laboratory, September 1983.

Petersen, Bent, "The Danish Energy Conservation Programme in Housing", Ministry of Housing, 1984.

Petersen, H. Craig, *Survey Analysis of the Impact of Conservation and Solar Tax Credits, Final Report,* National Science Foundation, 15th July 1982.

Petersen, H.C., "Solar Versus Conservation Tax Credits", *The Energy Journal,* Vol. 6, No. 3, 1985, pp.129-135.

Pinto, Frank J.P., "The Economics of, and Potential for, Energy Conservation and Substitution", paper presented to Fifth Annual Meeting of International Association of Energy Economists, New Delhi, January 1984.

Randolph, John, "Energy conservation programmes: A review of state initiatives in the USA", *Energy Policy,* December 1984, pp.425-438.

Ray, G.F. and Morel, J., "Energy conservation in the U.K.", *Energy Economics,* April 1982, pp.83-97.

Ray, George F., "*Energy Management: Can we learn from others?*", Policy Studies Institute and Royal Institute of International Affairs, Energy Paper No. 16, United Kingdom.

Robinson, John B., Brooks, D.B. et al., "Determining the Long Term Potential for Energy Conservation and Renewable Energy in Canada", *Energy,* Vol. 10, No. 6, pp.689-705, June 1985, Pergamon Press.

Röglin, H.-C., *Evaluation of Consumer Information of AGV (Arbeitgemeinschaft der Verbraucher),* study ordered by the German Ministry for Economics, May 1985.

Rosenfeld-Hafemeister, "Energy Sources: Conservation and Renewables", *American Institute of Physics Conference Proceedings,* 1985.

Samouilidis, J.E., Berahas, S.A. and Psarras, J.E., "Energy conservation: centralized versus decentralized decisionmaking", *Energy Policy,* December 1983, pp.302-312.

Sant, Roger W. et al., *Eight Great Energy Myths: The Least-Cost Energy Strategy — 1978-2000,* Energy Productivity Report No. 4, Mellon Institute, Arlington, Virginia, 1981.

Sant, Roger W., et al., *Creating Abundance: America's Least Cost Energy Strategy,* McGraw-Hill, 1984, pp.90-92.

Sawhill, J.C., Cotton, R. (editors), *"Energy Conservation — Successes and Failures",* The Brookings Institution, Washington, 1986.

Schipper, L. and Ketoff, A., *Oil Conservation: Permanent or Reversible? The Example of Homes in the OECD,* Lawrence Berkeley Laboratory, University of California, October 1984.

Schipper, L., "Residential energy use and conservation in Denmark, 1965-1980" in *Energy Policy,* Vol. 11, No. 4, December 1983, pp.313-323.

Schipper, L., Ketoff, A. and Kahane, A., "Explaining Residential Energy Use by International Bottom-Up Comparisons", *Annual Review of Energy 1985,* 10: pp.431-405.

Schipper, L., Meyers, S. and Ketoff, A., "Energy Use in the Service Sector: An International Perspective", *Energy Policy* (forthcoming 1986).

Schipper, L., Meyers, S. and Kelly, H., *Coming in from the cold — Energy-wise Housing in Sweden,* American Council for an Energy Efficient Economy, Washington, 1985.

Schipper, L., et al., *Changing Patterns of Electricity Demand in Homes: An International Overview,* Lawrence Berkeley Laboratory, 1985.

Soderstrom, Jon, "Measuring Energy Savings — Problems", *Energy Policy, March 1984, pp.104-106.*

Solar Energy Research Institute, et al., A New Prosperity, Building a Sustainable Energy Future, the SERI Solar/Conservation Study, Brick House Publishing, 1981.

Southworth, Frank, *DOE National Rideshare Program Plan,* Oak Ridge National Laboratory, Martin Marietta Energy Systems, Inc., Oak Ridge, Tennessee, 1985.

Stern, Paul and Elliot Aronson, editors, *Energy Use: The Human Dimension,* Washington, National Research Council, 1984.

Stern, Paul et al., "The Effectiveness of Incentives for Residential Energy Conservation", in *Evaluation* Review, April 1986 (to be published).

Stern, Paul, editor, *Improving Energy Demand Analysis,* Washington, National Research Council, 1984.

Suding, Paul H., "Financial incentives to DH — Assessment of their benefits to the total economy", *Zeitschrift für Energiewirtschaft,* 4/82, pp.197-205.

Swedish Council for Building Research, *Energy '85: Energy Use in the Built Environment,* 1985.

Swedish Council for Building Research, *Local Energy Management Processes and Programs in Italy, Sweden, Germany and the United States,* IEA project within the Energy Conservation in Buildings and Community Systems Programme, Stockholm, December 1985.

Swedish Council for Building Research, *Energy Conservation Programs and their Impact on Rental Buildings in Italy, Sweden, Germany and the United States,* IEA project within the Energy Conservation in Buildings and Community Systems Programme, Stockholm, December 1985.

Swedish Industry and Commerce Energy Research and Development Foundation (NEFOS), *Energy Input Analysis in the Pulp and Paper Industry, Energy Audits in the Pulp and Paper Industry,* IEA projects, February 1984.

Thoreson, R., Rowberg, R., and Ryan, J.F., "Industrial Fuel Use: Structure and Trends", *Annual Review of Energy,* Volume 10, pp.165-199.

Tonn, B. et al., *The Bonneville Power Administration Conservation/Load/Resource Modeling Process: Review, Assessment, and Suggestion for Improvement,* ORNL-CON 190, Oak Ridge National Laboratories, January 1986.

UNIPEDE, European Community Committee, *Programmes and Prospects for the Electricity Sector, 1984-1990 and 1990-1995,* Paris, March 1986.

United Kingdom, *House of Commons, Eighth Report from the Energy Committee, Session 1984-85,* The Energy Efficiency Office (pages xx following).

United Kingdom, Department of Energy, *Energy Efficiency Office, Energy Use and Efficiency in UK Manufacturing Industry up to the Year 2000,* London, October 1984, p.60.

United Kingdom, Department of Energy, *Energy Conservation Research, Development and Demonstration,* Energy Paper No. 32, HMSO, 1978, p.60.

United Kingdom, Department of Energy, Chief Scientist's Group, Energy Technology Division, *The Pattern of Energy Use in the UK, 1980,* April 1984.

United Nations, Economic Commission for Europe, *An Efficient Energy Future, Prospects for Europe and North America,* 1983, Butterworths.

United Nations, Economic Commission for Europe, *Waste Energy Recovery in the Industry in the ECE Region,* 1985 (ECE/ENERGY/9).

United Nations, Economic Commission for Europe, *Study of regulations, codes and standards related to energy use in buildings,* New York, 1984.

United Nations, Economic Commission for Europe, *Efficient Use of Energy Sources in Meeting Heat Demand: The Potential for Energy Conservation and Fuel Substitution in the ECE Region,* New York, 1984.

United Nations, Economic Commission for Europe, *Waste energy recovery in the industry in the ECE region,* New York, 1985.

United States, Department of Housing and Urban Development, Division of Building Technologies, *An Analysis of the Feasibility of Energy Management Companies in Public Housing and Section 202 Housing,* 9th April 1982.

United States, Department of Energy, *Energy Conservation Financing for Public and Not-For-Profit Institutions,* May 1985.

United States, Department of Housing and Urban Development, *Residential Energy Efficiency Standards Study: Final Report,* Washington, 1980.

United States, Department of Energy, *Consumer Products Efficiency Standards Engineering Analysis Document,* Washington, March 1982.

United States, Department of Energy, *FY 1987 Energy Conservation Multi-Year Plan,* May 1985, Draft No. 3, p.53, p.136.

United States, Department of Energy, Energy Information Administration, *Commercial Buildings Consumption and Expenditures, 1983,* DOE/EIA-0318(83), 30th September 1986.

United States, Department of Energy, Energy Information Administration, *Energy Conservation Indicators 1983, Annual Report,* October 1984.

Victorian Department of Minerals and Energy, *HEAS Research and Evaluation Program: Key Findings and Progress in Domestic Energy Conservation Research in Victoria,* Melbourne, Australia, March 1985.

Victorian Department of Minerals and Energy, *Victoria's Energy Strategy and Policy Options,* Melbourne, Australia, November 1985.

Wickman, "Some Energy Policy Effects — A Documentation from Seven OECD Countries", *Scandinavian Energy,* No. 4, 1984.

Wilk, R.R. and Wilhite, H.L., "Why Don't People Weatherise their Homes? An Ethnographic Solution", *Energy,* May 1985, pp.621-629.

Williams, R.H., and Larson, E.D., "Steam-Injected Gas Turbines and Electric Utility Planning"; *IEEE Technology and Society Magazine,* March 1986.

Williams, Robert H. and Gautam S. Dutt, "Future Energy Savings in U.S. Housing" in Annual Review of Energy, Volume 8, 1983, pp.269-332.

OECD SALES AGENTS
DÉPOSITAIRES DES PUBLICATIONS DE L'OCDE

ARGENTINA - ARGENTINE
Carlos Hirsch S.R.L.,
Florida 165, 4° Piso,
(Galeria Guemes) 1333 Buenos Aires
Tel. 33.1787.2391 y 30.7122

AUSTRALIA-AUSTRALIE
D.A. Book (Aust.) Pty. Ltd.
11-13 Station Street (P.O. Box 163)
Mitcham, Vic. 3132 Tel. (03) 873 4411

AUSTRIA - AUTRICHE
OECD Publications and Information Centre,
4 Simrockstrasse,
5300 Bonn (Germany) Tel. (0228) 21.60.45
Local Agent:
Gerold & Co., Graben 31, Wien 1 Tel. 52.22.35

BELGIUM - BELGIQUE
Jean de Lannoy, Service Publications OCDE,
avenue du Roi 202
B-1060 Bruxelles Tel. (02) 538.51.69

CANADA
Renouf Publishing Company Ltd/
Éditions Renouf Ltée,
1294 Algoma Road, Ottawa, Ont. K1B 3W8
Tel: (613) 741-4333
Toll Free/Sans Frais:
Ontario, Quebec, Maritimes:
1-800-267-1805
Western Canada, Newfoundland:
1-800-267-1826
Stores/Magasins:
61 rue Sparks St., Ottawa, Ont. K1P 5A6
Tel: (613) 238-8985
211 rue Yonge St., Toronto, Ont. M5B 1M4
Tel: (416) 363-3171
Sales Office/Bureau des Ventes:
7575 Trans Canada Hwy, Suite 305,
St. Laurent, Quebec H4T 1V6
Tel: (514) 335-9274

DENMARK - DANEMARK
Munksgaard Export and Subscription Service
35, Nørre Søgade, DK-1370 København K
Tel. +45.1.12.85.70

FINLAND - FINLANDE
Akateeminen Kirjakauppa,
Keskuskatu 1, 00100 Helsinki 10 Tel. 0.12141

FRANCE
OCDE/OECD
Mail Orders/Commandes par correspondance :
2, rue André-Pascal,
75775 Paris Cedex 16
Tel. (1) 45.24.82.00
Bookshop/Librairie : 33, rue Octave-Feuillet
75016 Paris
Tel. (1) 45.24.81.67 or/ou (1) 45.24.81.81
Principal correspondant :
Librairie de l'Université,
12*a*, rue Nazareth,
13602 Aix-en-Provence Tel. 42.26.18.08

GERMANY - ALLEMAGNE
OECD Publications and Information Centre,
4 Simrockstrasse,
5300 Bonn Tel. (0228) 21.60.45

GREECE - GRÈCE
Librairie Kauffmann,
28, rue du Stade, 105 64 Athens Tel. 322.21.60

HONG KONG
Government Information Services,
Publications (Sales) Office,
Beaconsfield House, 4/F.,
Queen's Road Central

ICELAND - ISLANDE
Snæbjörn Jónsson & Co., h.f.,
Hafnarstræti 4 & 9,
P.O.B. 1131 – Reykjavik
Tel. 13133/14281/11936

INDIA - INDE
Oxford Book and Stationery Co.,
Scindia House, New Delhi I Tel. 45896
17 Park St., Calcutta 700016 Tel. 240832

INDONESIA - INDONÉSIE
Pdii-Lipi, P.O. Box 3065/JKT.Jakarta
Tel. 583467

IRELAND - IRLANDE
TDC Publishers - Library Suppliers,
12 North Frederick Street, Dublin 1.
Tel. 744835-749677

ITALY - ITALIE
Libreria Commissionaria Sansoni,
Via Lamarmora 45, 50121 Firenze
Tel. 579751/584468
Via Bartolini 29, 20155 Milano Tel. 365083
Sub-depositari :
Editrice e Libreria Herder,
Piazza Montecitorio 120, 00186 Roma
Tel. 6794628
Libreria Hœpli,
Via Hœpli 5, 20121 Milano Tel. 865446
Libreria Scientifica
Dott. Lucio de Biasio "Aeiou"
Via Meravigli 16, 20123 Milano Tel. 807679
Libreria Lattes,
Via Garibaldi 3, 10122 Torino Tel. 519274
La diffusione delle edizioni OCSE è inoltre assicurata dalle migliori librerie nelle città più importanti.

JAPAN - JAPON
OECD Publications and Information Centre,
Landic Akasaka Bldg., 2-3-4 Akasaka,
Minato-ku, Tokyo 107 Tel. 586.2016

KOREA - CORÉE
Kyobo Book Centre Co. Ltd.
P.O.Box: Kwang Hwa Moon 1658,
Seoul Tel. (REP) 730.78.91

LEBANON - LIBAN
Documenta Scientifica/Redico,
Edison Building, Bliss St.,
P.O.B. 5641, Beirut Tel. 354429-344425

MALAYSIA - MALAISIE
University of Malaya Co-operative Bookshop Ltd.,
P.O.Box 1127, Jalan Pantai Baru,
Kuala Lumpur Tel. 577701/577072

NETHERLANDS - PAYS-BAS
Staatsuitgeverij
Chr. Plantijnstraat, 2 Postbus 20014
2500 EA S-Gravenhage Tel. 070-789911
Voor bestellingen: Tel. 070-789880

NEW ZEALAND - NOUVELLE-ZÉLANDE
Government Printing Office Bookshops:
Auckland: Retail Bookshop, 25 Rutland Street,
Mail Orders, 85 Beach Road
Private Bag C.P.O.
Hamilton: Retail: Ward Street,
Mail Orders, P.O. Box 857
Wellington: Retail, Mulgrave Street, (Head Office)
Cubacade World Trade Centre,
Mail Orders, Private Bag
Christchurch: Retail, 159 Hereford Street,
Mail Orders, Private Bag
Dunedin: Retail, Princes Street,
Mail Orders, P.O. Box 1104

NORWAY - NORVÈGE
Tanum-Karl Johan
Karl Johans gate 43, Oslo 1
PB 1177 Sentrum, 0107 Oslo 1Tel. (02) 42.93.10

PAKISTAN
Mirza Book Agency
65 Shahrah Quaid-E-Azam, Lahore 3 Tel. 66839

PORTUGAL
Livraria Portugal,
Rua do Carmo 70-74, 1117 Lisboa Codex.
Tel. 360582/3

SINGAPORE - SINGAPOUR
Information Publications Pte Ltd
Pei-Fu Industrial Building,
24 New Industrial Road No. 02-06
Singapore 1953 Tel. 2831786, 2831798

SPAIN - ESPAGNE
Mundi-Prensa Libros, S.A.,
Castelló 37, Apartado 1223, Madrid-28001
Tel. 431.33.99
Libreria Bosch, Ronda Universidad 11,
Barcelona 7 Tel. 317.53.08/317.53.58

SWEDEN - SUÈDE
AB CE Fritzes Kungl. Hovbokhandel,
Box 16356, S 103 27 STH,
Regeringsgatan 12,
DS Stockholm Tel. (08) 23.89.00
Subscription Agency/Abonnements:
Wennergren-Williams AB,
Box 30004, S104 25 Stockholm.
Tel. (08)54.12.00

SWITZERLAND - SUISSE
OECD Publications and Information Centre,
4 Simrockstrasse,
5300 Bonn (Germany) Tel. (0228) 21.60.45
Local Agent:
Librairie Payot,
6 rue Grenus, 1211 Genève 11
Tel. (022) 31.89.50

TAIWAN - FORMOSE
Good Faith Worldwide Int'l Co., Ltd.
9th floor, No. 118, Sec.2
Chung Hsiao E. Road
Taipei Tel. 391.7396/391.7397

THAILAND - THAILANDE
Suksit Siam Co., Ltd.,
1715 Rama IV Rd.,
Samyam Bangkok 5 Tel. 2511630

TURKEY - TURQUIE
Kültur Yayinlari Is-Türk Ltd. Sti.
Atatürk Bulvari No: 191/Kat. 21
Kavaklidere/Ankara Tel. 25.07.60
Dolmabahce Cad. No: 29
Besiktas/Istanbul Tel. 160.71.88

UNITED KINGDOM - ROYAUME-UNI
H.M. Stationery Office,
Postal orders only:
P.O.B. 276, London SW8 5DT
Telephone orders: (01) 622.3316, or
Personal callers:
49 High Holborn, London WC1V 6HB
Branches at: Belfast, Birmingham,
Bristol, Edinburgh, Manchester

UNITED STATES - ÉTATS-UNIS
OECD Publications and Information Centre,
Suite 1207, 1750 Pennsylvania Ave., N.W.,
Washington, D.C. 20006 - 4582
Tel. (202) 724.1857

VENEZUELA
Libreria del Este,
Avda F. Miranda 52, Aptdo. 60337,
Edificio Galipan, Caracas 106
Tel. 32.23.01/33.26.04/31.58.38

YUGOSLAVIA - YOUGOSLAVIE
Jugoslovenska Knjiga, Knez Mihajlova 2,
P.O.B. 36, Beograd Tel. 621.992

Orders and inquiries from countries where Sales Agents have not yet been appointed should be sent to:
OECD, Publications Service, Sales and Distribution Division, 2, rue André-Pascal, 75775 PARIS CEDEX 16.

Les commandes provenant de pays où l'OCDE n'a pas encore désigné de dépositaire peuvent être adressées à :
OCDE, Service des Publications. Division des Ventes et Distribution. 2. rue André-Pascal. 75775 PARIS CEDEX 16.

70431-01-1987

OECD PUBLICATIONS, 2, rue André-Pascal, 75775 PARIS CEDEX 16 - No. 43853 1987
PRINTED IN FRANCE
(61 87 01 1) ISBN 92-64-12910-3

JAPAN
CRAFTS
SOURCEBOOK

JAPAN CRAFTS SOURCEBOOK

A Guide to Today's Traditional Handmade Objects

Japan Craft Forum

Introduction by Diane Durston

KODANSHA INTERNATIONAL
Tokyo • New York • London

Traditional Craft Centers

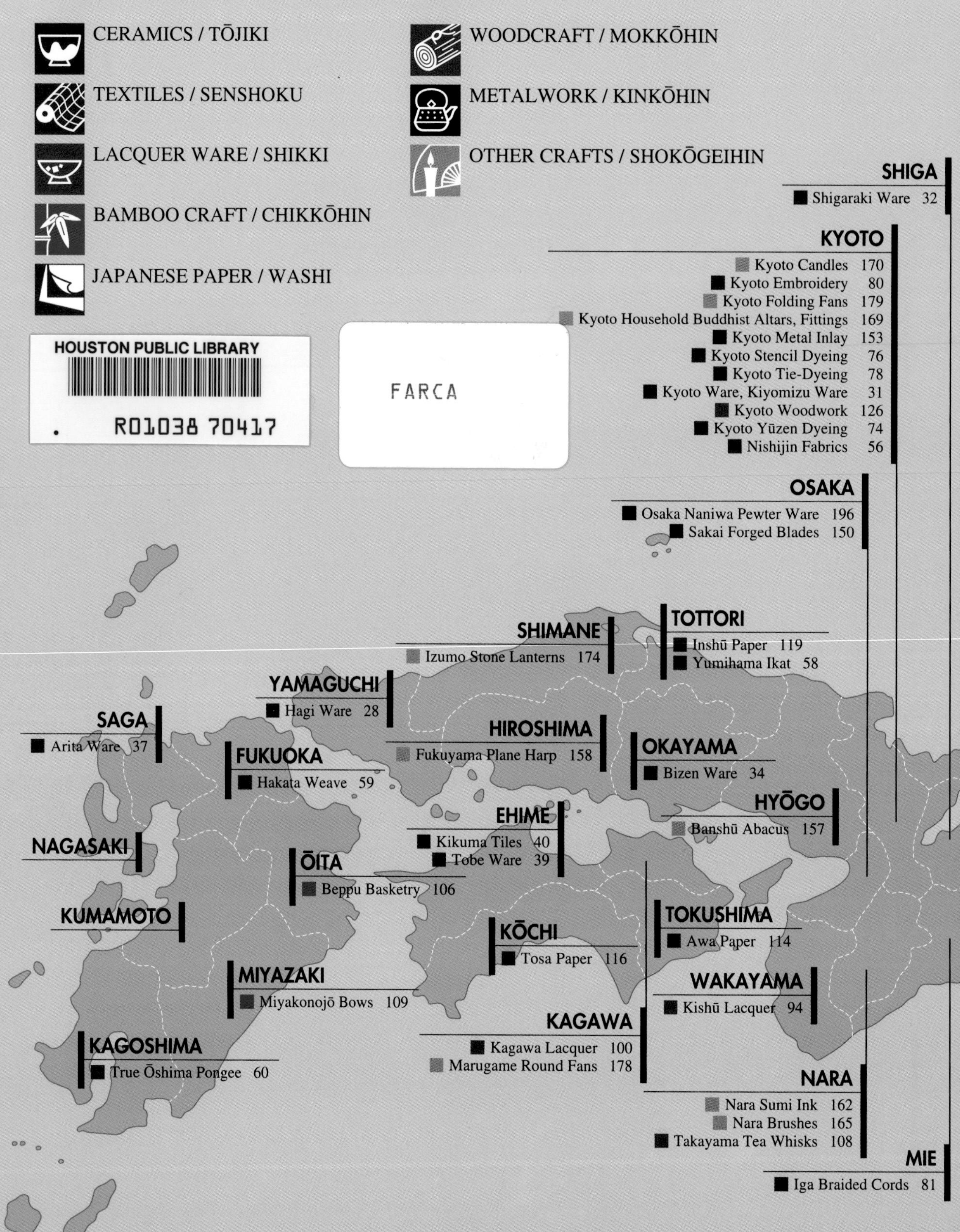

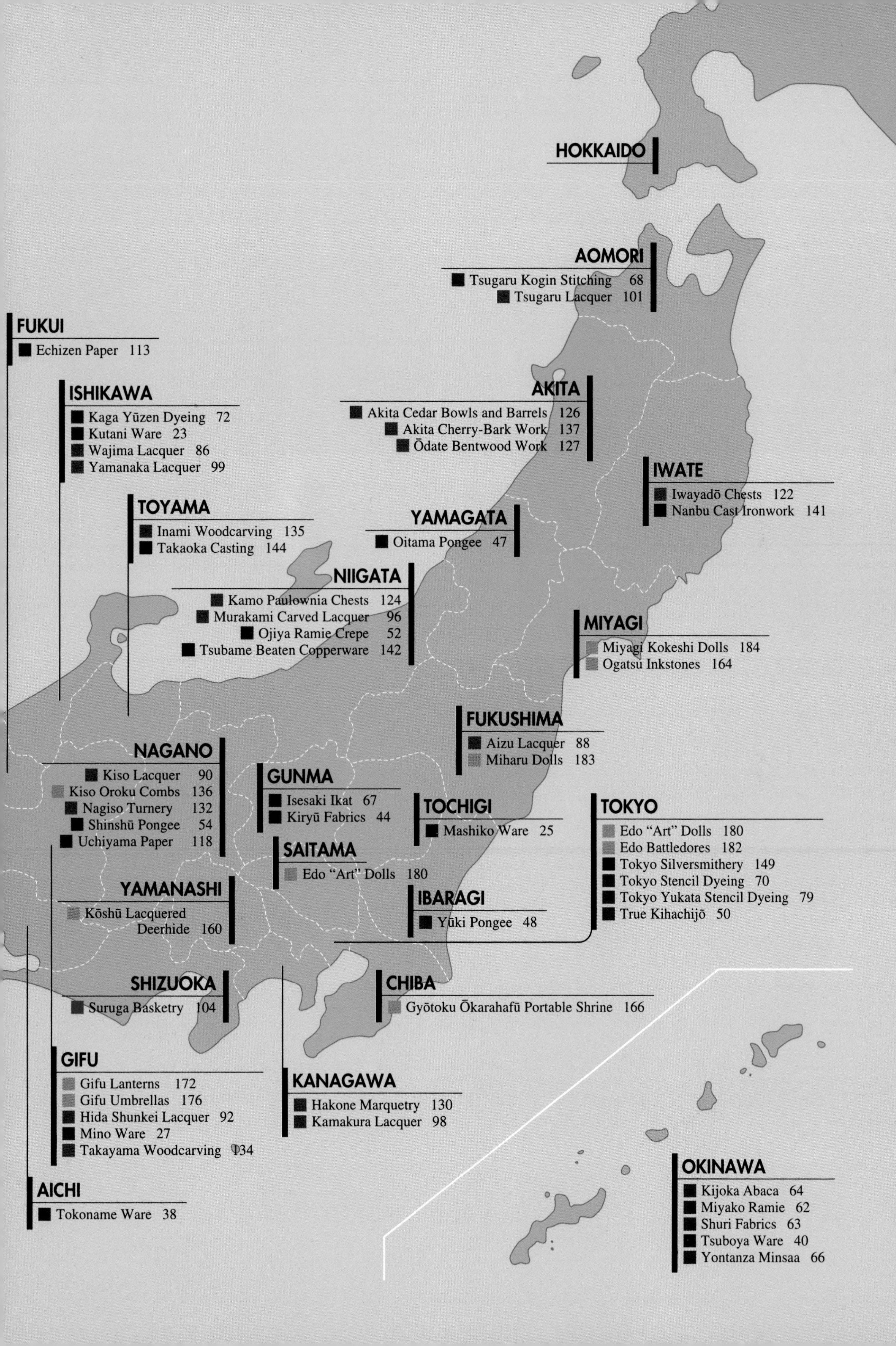

HOKKAIDO
AOMORI
Tsugaru Kogin Stitching 68
Tsugaru Lacquer 101
FUKUI
Echizen Paper 113
ISHIKAWA
Kaga Yūzen Dyeing 72
Kutani Ware 23
Wajima Lacquer 86
Yamanaka Lacquer 99
AKITA
Akita Cedar Bowls and Barrels 126
Akita Cherry-Bark Work 137
Ōdate Bentwood Work 127
IWATE
Iwayadō Chests 122
Nanbu Cast Ironwork 141
TOYAMA
Inami Woodcarving 135
Takaoka Casting 144
YAMAGATA
Oitama Pongee 47
NIIGATA
Kamo Paulownia Chests 124
Murakami Carved Lacquer 96
Ojiya Ramie Crepe 52
Tsubame Beaten Copperware 142
MIYAGI
Miyagi Kokeshi Dolls 184
Ogatsu Inkstones 164
FUKUSHIMA
Aizu Lacquer 88
Miharu Dolls 183
NAGANO
Kiso Lacquer 90
Kiso Oroku Combs 136
Nagiso Turnery 132
Shinshū Pongee 54
Uchiyama Paper 118
GUNMA
Isesaki Ikat 67
Kiryū Fabrics 44
TOCHIGI
Mashiko Ware 25
TOKYO
Edo "Art" Dolls 180
Edo Battledores 182
Tokyo Silversmithery 149
Tokyo Stencil Dyeing 70
Tokyo Yukata Stencil Dyeing 79
True Kihachijō 50
SAITAMA
Edo "Art" Dolls 180
YAMANASHI
Kōshū Lacquered Deerhide 160
IBARAGI
Yūki Pongee 48
SHIZUOKA
Suruga Basketry 104
CHIBA
Gyōtoku Ōkarahafū Portable Shrine 166
GIFU
Gifu Lanterns 172
Gifu Umbrellas 176
Hida Shunkei Lacquer 92
Mino Ware 27
Takayama Woodcarving 134
KANAGAWA
Hakone Marquetry 130
Kamakura Lacquer 98
AICHI
Tokoname Ware 38
OKINAWA
Kijoka Abaca 64
Miyako Ramie 62
Shuri Fabrics 63
Tsuboya Ware 40
Yontanza Minsaa 66

FRONTISPIECE: Painting a Miyagi Kokeshi Doll (page 184).
TITLE PAGE: Curing a hide, Kōshū Lacquered Deerhide (page 160).

Note: An asterisk following the name of a village or city in an entry heading indicates the craft is being produced exclusively at that location. For the majority of entries, the place name indicates the main or central area of production. • The names of Japanese historical figures follow the traditional manner, surname preceding given name. All other names take the Western order.

Distributed in the United States by Kodansha America, Inc., 114 Fifth Avenue, New York, New York 10011, and in the United Kingdom and continental Europe by Kodansha Europe Ltd., 95 Aldwych, London WC2B 4JF.

Published by Kodansha International Ltd., 17–14 Otowa 1–chome, Bunkyo-ku, Tokyo 112, and Kodansha America, Inc.

Printed in China.
First edition, 1996
1 2 3 4 5 6 7 8 9 10 99 98 97 96

LIBRARY OF CONGRESS CATALOGUING-IN-PUBLICATION DATA
Japan craft sourcebook: a guide to today's traditional handmade objects / Japan Craft Forum with an introduction by Diane Durston.
Includes bibliographical reference and index.
1. Decorative arts—Japan—History—1868–
I. Durston, Diane. 1950– II. Japan Craft Forum.
NK1071.J336 1996 745'.0952–dc20 95-51751 CIP

ISBN 4–7700–2073–2

CONTENTS

PREFACE

Centuries before there was any need to make a distinction between craft and mass-produced items, the work of Japanese artisans found favor outside the country among the fashionable and wealthy, especially in Europe. As is well known, the sheer quantity and quality of the lacquer ware which was exported from Japan encouraged people to tag this ware with the name of its country of origin. More recently, however, it is the giants of the electrical and automotive industries in Japan whose names have become household words throughout the developed world while the export of what are now known as Japan's traditional crafts is all but a trickle.

It could be argued that part of the reason for this decline is that insufficient efforts have been made to introduce these crafts to a non-Japanese public, which is now unaware of the wealth of artistic skill and technical genius embodied in Japan's traditional crafts. Admittedly, there are many people throughout the world who admire these crystallizations of Japanese life, culture, and nature. Nevertheless, there are few people who really know and understand these crafts and there are even fewer introductory publications to inform those who are willing or eager to learn more. Inevitably, this is due in great part to the difficulty in translating material related to the very core of Japan and Japanese life. Most Japanese people today find themselves as alienated from Japanese traditional crafts as Westerners are, and many of the terms and words associated with these crafts are beyond their comprehension. This is, however, no reason for not trying to inform.

At present, the consensus in most parts of the world is that it is important to preserve traditions of value, not simply for their inherent value but also as the possible roots from which the accouterments of contemporary life may sprout. In this belief, work on this book began despite the difficulties that were bound to be encountered. First of all, the legacy of each craft—its history, terms, materials, techniques, and present-day standing—had to be examined and verified. This required long hours of research through historical archives, as well as travel to every corner of Japan to interview and photograph craftsmen and craftswomen. Next, the information on each of the ninety-one unique crafts had to be translated as naturally as possible into English, and explanations for many of the traditional craft methodologies rendered in understandable terms. Consequently, seven years have now passed since the project's inception.

What has evolved into the *Japan Crafts Sourcebook* was one of the first major projects to be undertaken by the Japan Craft Forum. The Forum's present director, Yuko Yokoyama, who was also instrumental in its founding, has always believed in the value of Japan's traditional crafts and views them as a potential fountainhead of inspiration and technique for those involved in design, interior decorating, and contemporary craft both at home and abroad. It was undoubtedly her faith in the value of this project that brought it to fruition. The first stage of painstaking research was undertaken by Yujiro Sahara, a craft consultant to the Japan Traditional Craft Center. He compiled information on each traditional craft from throughout Japan. The arduous task of translating this material fell to Bill Tingey, who, as a craft researcher, designer, and photographer, has devoted much of his time in Japan to working for the Forum. More than simply translating, he significantly added to the original text and finally wrote what became the draft of this book. Thanks must also be extended to all those who contributed so much to this authoritative volume which, perhaps for the first time, provides basic information on so many of Japan's traditional crafts in an encyclopedic yet very readable form.

I hope, therefore, that this book will inspire its readers, giving them a better understanding of work that has been honed and refined over many generations by craftspeople who have sought to combine the alchemies of nature with the skills of hand and eye. The quality of the many crafts produced stands as a permanent testimony to Japanese creativity.

Shin Yagihashi
Professor
Kanazawa College of Art

ACKNOWLEDGMENTS

The sheer breadth of this project, which introduces crafts from all corners of Japan, required the willing participation of numerous people, some of whom are noted above by Professor Yagihashi. There are, however, a number of others who deserve special recognition, for without their diligence and dedication this book would not have come into existence. Considerable contributions were made in checking and clarifying the initial manuscript by Patricia Massy, Kim Schuefftan, and Philip Meredith, each of whom added his or her own specialist knowledge. Diane Durston provided an informative introduction. Takayuki Maruoka of the Japan Traditional Craft Center supplied updated information about local crafts and Kyoko Ishikawa of Jomon-sha assembled the lists of associations and craft centers. The fine photography of craftspeople at work was done by Soichiro Negishi, and Tetsuro Sato graciously allowed us to reproduce the color photographs from *Japanese Traditional Crafts*, published by Diamond Incorporated. Several supplementary photographs were provided by the craft center and the Haga Library/American Photo Library. Additional assistance was provided by Peter Dowling, Michiko Uchiyama, and Pat Fister, who also assembled the selected reading list. Finally, all of these loose ends were professionally and patiently edited and shaped into book form by Barry Lancet of Kodansha International. The handsome design of the sourcebook was conceived by Shigeo Katakura. Thank you all for unselfishly donating your time and energy toward the publication of this book.

Yuko Yokoyama
Director
Japan Craft Forum

JAPANESE HISTORICAL PERIODS

PREHISTORIC

Jōmon ca. 10,000 B.C.–ca. B.C. 300
Yayoi ca. 300 B.C.–ca. A.D. 300
Kofun ca. 300–710

ANCIENT

Nara 710–94
Heian 794–1185

MIDDLE AGES (MEDIEVAL)

Kamakura 1185–1333
Northern and Southern Courts 1333–92
Muromachi 1392–1573
Warring States 1482–1573

PREMODERN

Momoyama 1573–1600
Edo 1600–1868

EARLY MODERN / MODERN

Meiji 1868–1912
Taishō 1912–26
Shōwa 1926–1989
Heisei 1989 to present

INTRODUCTION

by Diane Durston

Down from the mountain forests that surround his tiny village runs an icy stream that flows through the life of the old paper maker. His hands are rough and gnarled. They have worked the cold waters of sixty winters, not yet numb to the pain. Swoosh—the fibers rush across the framed bamboo sieve, settling evenly into the hand-held frame, a technique that appears simple and is not. The craftsman has been making paper since he was fifteen years old and he doesn't make much money. He is not famous—not even known. He has no apprentice, but his work will live on beneath the masterful brushstrokes of a Zen priest and momentary insights of a haiku poet. Of this he is proud.

Drawn far from home by a yearning for something real, the young woman patiently struggles to make the tenacious bamboo strips obey her novice hands. She doesn't lose patience; she has chosen this path and she is no quitter. The master craftsman glances over at her with affectionate disgust. "Hopeless," he mutters. Praise doesn't come easily from a tenth-generation craftsman—lucky if it comes at all. With serious reservations (she was not from a bamboo maker's family), he relented and allowed her to work for a while in a corner of his workshop. He suspects she'll run home when winter sets in and the workshop is cold. Surprising, though, how long she has lasted and how well she's done, considering. When she first came to him, she said she "wanted to learn how to make beautiful things with her own hands." Well enough, he thought, perhaps she will *survive.*

The variety and bounty of fine decorative arts still produced by hand in modern Japan is astonishing. In the fields of ceramics, textiles, lacquer ware, bamboo craft, paper making, woodcraft, and metalwork, the Japanese practice what are among the greatest surviving traditions of fine craftsmanship in the world. Producing everything from wooden barrels to the luxurious brocades of wedding kimono, thousands of craftspeople across the country continue to follow the ways of their ancestors.

They are trained not in schools, but in workshops where, like their predecessors, they are apprenticed from a very young age. It is still a given among Japanese craftsmen that no skill of any worth can be learned in less than ten years. Apprenticeships are long and tedious; the results are impeccable.

Cutting corners is not a part of the Japanese craftsman's vocabulary. The real evidence of their ingenuity and meticulous workmanship is hidden beneath the silent layers of an immaculately finished surface. Wise customers know better than to press for arbitrary deadlines at the risk of a less-than-perfect product or an irritated master craftsman in whose hands the fate of their desire lies.

The quest for fame is not foremost in the minds of most traditional Japanese craftspeople. They ask only for a chance to demonstrate hard-won skills and for ways to survive in a society growing more and more dependent on mass-produced goods. Most of them work anonymously, as their grandfathers did, masters of one special step in a complicated process, expecting little more than a living and the respect of the patrons for whom they labor.

Supporting Traditional Crafts

The fact that so many craft forms have survived is due in large part to the survival of Japanese traditional arts such as the tea ceremony and ikebana. These practices require fine handwork as an integral part of their performance—the ceramic vases and tea bowls, the hanging scrolls and tatami mats, the fine kimono and white *tabi* socks. Masks for Noh actors, oiled paper parasols for geisha dancers, ink sticks for the serious calligrapher—all are supplied through the dedication of traditional craftsman.

Another factor in the survival of traditional Japanese crafts and craftsmanship is the continued encouragement of the Japanese government. Traditional crafts and the people who make them are recognized and promoted in many different ways, including the Living National Treasure program which has been honoring craftspeople and performing artists since 1950.

"Living National Treasure" is the popular term to describe those designated as *Jūyō Mukei Bunkazai Hojisha*, or Bearers of Intangible Cultural Assets. Craftsmen are selected by a government-appointed committee to receive this national honor. The purpose of the award is both to recognize and encourage exceptional craftspeople and to provide them with incentives to carry on their craft and hand it down to others. With an annual stipend and national acclaim comes the responsibility of teaching, lecturing, and appearing throughout the country at special exhibitions aimed at promoting an interest in Japan's artistic heritage.

The economic development and industrialization that took place in Japan following World War II left much traditional culture behind in the rush to embrace a new, more modern way. Traditional crafts were in danger of being lost altogether. But even as the new lifestyle took root, it was noted that "a certain feeling of dissatisfaction with the cold inhumanity of a technologically dominated and stereotyped lifestyle" had begun to pervade modern life in Japan.

In 1974, the Japanese government passed the Law for the Promotion of Traditional Craft Industries to protect and promote traditional crafts. They also undertook a nationwide study of the state of traditional craftsmanship and set up guidelines for ensuring its survival. The measures taken included the provision of subsidies for apprentices, the conservation of natural materials, and guidelines for healthier working environments.

The law also established criteria for the identification of craft objects to be protected. Craftspeople applying for official designation had to show that their products were used in everyday life, were made primarily by hand from natural materials, and followed techniques dating at least from the Edo period. They also had to be part of a craft tradition embracing at least thirty craftsmen in the same region. The designation was to be administered by local governments on behalf of the Minister of Trade and Industry in conjunction with a committee of craft experts, historians, and critics.

In the 1990s, a survey of regional crafts produced throughout Japan was conducted on the city and prefectural levels. Over 1060 distinctive crafts were identified, employing over 240,000 craftspeople, although not every craft met the full criteria for official designation. Today, over 184 objects bear the Mark of Tradition symbol identifying crafts that have qualified for government recognition. The mark is in the form of the stylized Chinese character for the word "*dentō*," or tradition, placed above a rising sun to form an official seal, a guarantee of quality and authenticity.

In 1975, a public foundation was created for the promotion of crafts bearing this seal. The Japan Traditional Craft Center in the Aoyama district of Tokyo opened an exhibition hall in 1979, providing craftsmen with an outlet for their products both in its permanent display and in rotating annual exhibitions that feature the crafts of a particular region.

In 1987, the Japan Craft Forum was established as a non-profit organization associated with the Japan Traditional Craft Center. The aim of the Forum is to develop a better understanding and awareness of Japan's traditional crafts, both at home and abroad, and to foster the development of new directions for craftsmen engaged in the making of traditional craft products.

One of the Japan Craft Forum's first projects was to compile information on one hundred traditional crafts from throughout Japan. The study was then translated into English with the aim of introducing a wide range of traditional Japanese crafts to an international audience. While a thorough treatment of the hundreds of crafts produced in Japan was beyond the scope of this project, the present volume introduces ninety-one traditional crafts and is intended as a reference for English-speaking readers with an interest in learning more about the subject.

The crafts presented in this book fall under the general category of *dentō kōgei*, or fine traditional craft, noted for refinement of technique and perfection of finish. This category does not generally include the roughly hewn objects referred to as *mingei*, or folk crafts for everyday use. The list does, however, include objects such as *kasuri* (ikat) fabrics and natural-ash glazed ceramics that have roots in a rustic tradition but are now created with a level of refinement that distinguishes them as fine rather than folk crafts.

History, Geography, and Craftsmanship

Although the beauty and integrity of the objects in this book speak for themselves, an understanding of the history behind individual crafts adds much to an appreciation of their importance as cultural survivors.

Each object has a story to tell—of the history and character of the Japanese people.

Perceived abroad as a homogenous island country, Japan has extreme regional differences that are often overlooked. The delicate, aristocratic porcelains of Kiyomizu in Kyoto belong to a different world from the earthy, blue-and-white merchant-class stoneware of Tobe, on the island of Shikoku. The handsomely ornate metalwork on the wooden chests of Iwayadō in Iwate Prefecture is far removed from the unvarnished simplicity and sophistication of the paulownia-wood Kamo cabinets of Niigata.

Historically, too, each object has its tale to tell. The small, leather *inden* purses decorated with intricately stenciled lacquer patterns come to life when you understand that this is the same craft that decorated the armor worn by samurai in the Middle Ages. The natural, wheat-colored texture of *bashōfu* textiles earns even more attention when it is noted that the "threads" from which it is woven are split by hand from banana leaves by a dwindling group of dedicated women on a remote island in Okinawa.

Razor-sharp *hamono* cutlery from Sakai seems even more formidable when you learn that these blades descended from the matchless swords of the samurai. The gorgeously dyed *yūzen* kimono fabrics gain social relevance if you understand that the technique for dyeing these elaborate patterns was the result of attempts by the low-ranked merchant class to rival the grandeur of the samurai's silk brocades, a cloth their class was forbidden to wear during the Edo period.

The Changing World of Traditional Crafts

Historically, the line between art and craft has always been ambiguous in Japan. No particular need for distinction was felt until the close of the nineteenth century, when the term *bijutsu* (art) was formulated to differentiate work done in a Western mode, such as oil painting. Until that time, the great names in Japanese art worked in many different media without differentiating greatly between painting, designing lacquer boxes, or decorating ceramic bowls.

In Japan, as in other industrialized nations, craftspeople have begun to create works that seek to expand traditional definitions of art and craft. Individual expression has become an important consideration. Some of the crafts presented in this book were made by young craftspeople who are searching for new means of personal expression, at times challenging traditional definitions of craft.

The concept of "tradition" itself is also being redefined today by these young craftsmen, eager to adapt the lessons and skills of the past successfully to the demands of the modern age. Crafts were formerly handed down from father to son, but in the contemporary setting some young Japanese are choosing a life of craftsmanship over the alternative of office work. In so doing, they bring a new eye and new ideas to the workshops of craftsmen carrying on centuries-old traditions.

Yet new ideas and dreams alone cannot halt the seemingly inevitable decline in the production of many of the traditional crafts described in this book. While the number of different objects recognized for official protection increases each year, the number of craftspeople who make them decreases. The median age of craftspeople in every category rises and the number of young people willing to carry on their family trade falls. The majority of people whose crafts bear the Mark of Tradition earn far less than the average wage, and prospects for increasing their incomes are meager.

Nonetheless, the role these people play within their society, both artistically and in terms of importance to community life, is vital. Their success is inextricably linked to neighboring craftsmen and merchants, who provide them with the finest materials and tools required to carry on their work at the present level of refinement. A master carpenter is lost without the best handmade chisel in his hand and the highest-quality natural sharpening stone on his workbench. Unless ample natural resources are preserved and fine craftsmanship among traditional toolmakers is carried on, the master craftsman must number his days and worry about his successor's future.

The Japan Craft Forum will continue to seek new ways of encouraging traditional craftspeople to carry on their honored crafts, of promoting an international exchange of ideas among craft cultures around the world, and of recognizing and rewarding lesser-known yet intrinsically valuable crafts.

There is much for a hardworking traditional craftsman to be proud of in the industrialized Japan of 1995. The survival of twenty centuries of unbroken artistic tradition is important in a culture that has otherwise lost patience with things that take time and care to create. For with that loss of patience—here as elsewhere in the modern world—has come a loss of appreciation for the finer expressions of the human spirit: the beauty of delicate painting on the edge of a porcelain cup, the natural softness of indigo fabric woven by hand, the sheen of a perfect lacquer finish on a serving tray offered respectfully to honored guests.

Beyond beauty and refinement, the Japanese craftsman's way of life has much to teach a modern world in which these simple, constructive human achievements are at risk of vanishing.

JAPAN CRAFTS SOURCEBOOK

TŌJIKI

Ceramics

Introduction

Japan is a mecca for ceramic enthusiasts. Many of the country's pottery traditions have been handed down from generation to generation and continue to prosper to this day, a rare circumstance in a highly industrialized nation. Several reasons exist for this paradox. Flourishing schools of tea ceremony (*chanoyu*) and flower arrangement (ikebana) instill an appreciation of and create a continuous demand for ceramic ware. Restaurants of all ranks, from the lowest noodle shop on up to the exclusive by-introduction-only Japanese haute cuisine establishments, rely on the availability of suitable ceramic vessels for the meals they prepare, their needs often changing with the seasons. But chief among the reasons for the ongoing popularity of handmade pottery is that a wide variety of dishes are still used in the average home. Because visual presentation remains as important as the requirements of the palate, an endless market exists for dishes, bowls, platters, cups, and plates.

Broadly speaking, ceramics in Japan can be divided into three categories: porcelains, glazed stonewares, and unglazed (or natural-ash glazed) stonewares. Porcelains from Kutani and Arita have long been popular both in and outside of Japan. Thousands of pieces were exported from as early as the mid-seventeenth century, often with designs particularly suited to the tastes of the targeted countries. Those familiar only with export porcelain will find the indigenous pieces quite different. Gone are the court gentlemen and ladies-in-waiting, the flamboyant polychrome dragons, and the clipper ships. In their place are what the Japanese themselves prefer: flowers, birds, perhaps a fallow field with Mount. Fuji in the background, and simple hand-painted patterns.

Stoneware is heavier than porcelain, with thicker walls and coarser clay, yet the better pieces are as striking as the most delicate porcelain. Among the glazed stoneware are Shino and Seto ware from Mino; Mashiko ware from the village of the same name, where the famed potter Shōji Hamada worked; and Hagi ware, with its creamy whites, pinks, and grays, and a web of crackle, or crazing.

Last, there comes the unglazed stoneware. Pieces in this category are not dipped in a vat of glazing liquid before being set in the kiln, but receive their final decoration during the firing of a wood-stoked kiln. Flames color the clay and in many cases the ash of the burning wood is allowed to settle on the pots, fusing at high temperatures with elements in the clay to form pools or drips or spatters of green, gray, brown, or gold, depending on the elements in the clay and the method of firing. The random ash deposits are often referred to as a "natural-ash glaze." Unglazed pots are quieter pieces—at their best serene, dignified, and unpretentious. Decorated by event rather than brush, they stand or fall by the kiln's whim and the potter's ability to play on the kiln's potential.

Instrumental in the rise of such unglazed stonewares as Shigaraki and Bizen was the development of the *noborigama*, or climbing kiln, a multichambered kiln built on a slope. The kiln is fed at the front with wood, usually pine, and then at side ports to reach the required temperatures and obtain the sought-after effects. There is a certain element of unpredictability with each firing since the fuel, stoking rhythm, weather, and interior conditions of the kiln will vary with each firing. This variation yields different effects and the skilled potter sets his pots in the kiln and then feeds the flame bearing in mind the potential of each firing.

In modern times, many of the glazed stoneware and porcelain makers have taken to using gas and electric kilns. Much of this interaction between fuel, flame, and potter has therefore been lost since the atmosphere of modern kilns is uniform, allowing a higher yield of salable pots but decreasing the variation of effects. Even potters of unglazed stoneware may rely on the less-expensive fuels in the early stages of firing their *noborigama* to conserve costs, a practice that often draws cries of protest from purists. Others argue that it is just such modern conveniences that allow cherished traditions to continue to thrive in an increasingly competitive environment. The tools may change, but in the end, as with any hand-crafted object, the skill and sensibility of the craftsperson determine the success of the outcome. Tools are the means, but the heart leads the way.

A Kiyomizu potter at the wheel.

KUTANI WARE • *Kutani Yaki*

Terai-machi, Ishikawa Prefecture

Porcelain has been produced in Japan for hundreds of years, but pottery making in Kutani only began in the mid-1600s when Maeda Toshiharu, the head of the Daishōji clan, discovered clay suitable for making porcelain in Kutani, which was within his domain. Maeda directed two retainers, Gotō Saijirō Sadatsugu and Tamura Gonzaemon, to set up kilns in the villages of Kutani and Suisaka to make tea-ceremony utensils. Soon after, Gotō's son Tadakiyo was sent to learn the techniques of porcelain making in Arita in Kyūshū. There he was fortunate to meet a potter exiled from Ming China. He returned with this exile, and the high-quality work produced by their joint effort is now termed *Ko Kutani*, or Old Kutani ware.

For reasons that are not completely clear, around 1694 the kilns in Kutani fell into disuse for more than a century. It was again with the help of an invited outsider, this time from Kyoto, that activities resumed. Aoki Mokubei (see Kyoto ware) came to Kutani in 1804 and fired a kiln in Kasugayama. His work eventually inspired the construction of other kilns in the surrounding areas. The pottery produced by this second wave of activity resulted in the reestablishment of Kutani ware, and was called *Saikō Kutani*, or what is loosely referred to as new Kutani ware. The general term "Kutani ware" came into use around 1803, but it was only the original, Old Kutani ware that was actually made in Kutani.

Red, green, yellow, purple, and Prussian blue make up the five-color palette of Kutani ware. This color scheme, emphasizing bold yellows and greens, remains the best known of the Kutani repertoire, which includes blue-and-white as well as red-and-gold schemes. The designs tend to be bold, in keeping with the purity of the overglaze enamel colors. The motifs are first outlined with cobalt blue, manganese black, and iron red. The enamels are then applied on top. After being fired, the colors exhibit a glasslike transparency, allowing the underdrawing to show through as part of the design.

While Kutani ware owes much to Chinese models, it has the distinction of being one of the first Japanese porcelain wares to find a place in the hearts of many people in Europe. A great deal of Kutani ware was exported there at the end of the nineteenth century, and now, inevitably, Japanese collectors and museums are busy buying it back. Meanwhile, potters are still actively producing utilitarian objects such as saké cups and bottles displaying traditional designs as well as modern innovations.

MASHIKO WARE • *Mashiko Yaki*

Mashiko-machi, Tochigi Prefecture

For many people the ceramic town of Mashiko is inexorably linked with the potters Shōji Hamada (1894–1978) and Bernard Leach (1887–1979). Leach, a British potter who was eventually to settle in England's St. Ives in Cornwall, and Hamada, later designated a Living National Treasure, worked closely with Sōetsu Yanagi (1889–1961) and others to promulgate the virtues of *mingei*, or folkcraft, of which Mashiko ware became a prime example.

While shards of *sueki* and *hajiki* have been unearthed from several ancient settlements and tumuli in the Mashiko area dating to the Nara period, the type of pottery being made there today has more recent origins. Mashiko ware is thought to have been initiated by Ōtsuka Keisaburō from the nearby village of Fukute (now Motegi-chō). After visiting the neighboring province to learn the techniques of Kasama ware, he established a kiln in Mashiko in 1853 and began producing household articles such as water jars, pouring bowls, and teapots. Ōtsuka was joined by others, and encouraged by the support of the local clan, the number of kilns gradually increased.

Although the production of Mashiko ware declined at the beginning of the twentieth century, it revived after the Great Kantō Earthquake in 1923, which devastated much of nearby Tokyo. Thereafter a large percentage of the tableware used in homes and restaurants in the capital came from Mashiko. Shōji Hamada settled in Mashiko the following year and began to make pottery drawing upon Mashiko traditions. It was in part through his efforts that others became interested in preserving the old folkcraft traditions and consequently the standard of Mashiko wares was raised.

The distinctive folkcraft appearance and the modest simplicity of Mashiko ware make it an appealing tableware. At present the kilns are producing primarily large platters, bowls, saké cups, and bottles as well as some cooking vessels. In addition to the conventional wheel-thrown articles, the slip trailing method of is used to create linear glaze designs that have become a trademark of Mashiko pottery. The old climbing kilns fired by wood were have given way to gas or electric kilns, particularly in the production of simple.

Mashiko ware was officially designated as a Traditional Craft Industry in 1979, and at present there are approximately eight hundred potters active in this pottery town. It remains, even today, as the best known of the folkcraft pottery traditions, and visitors still flock to Mashiko to purchase wares for their tables.

MINO WARE • *Mino Yaki*

Tajimi and Toki, Gifu Prefecture

Although primitive *sueki* dating from the seventh century have been discovered in the vicinity, the history of Mino ware itself dates back to the year 905, when it gained mention as one of the places that produced fine ceramics. At the time, ash-glazed stoneware was the main type of pottery. The potter's wheel came into use during the Kamakura and Momoyama periods, and the range of glazes and decoration expanded, leading to the creation of some relatively sophisticated pottery. During the Momoyama period, growing interest in the tea ceremony stimulated the production of tea wares. Many potters migrated to Mino, which had abundant supplies of clay and fuel, leading to the development of the distinctive stonewares discussed below. Porcelain was not produced in Mino until the end of the nineteenth century.

While Mino wares are colorful, they still retain a sense of serenity. One of the best known of Mino wares is Shino, made from the local clay called *mogusa tsuchi* and a feldspar glaze (*chōsekiyū*), which when fired turns a milky white through which the red body of the pot can be seen in places. There are many varieties of Shino ware, which can be left plain or decorated with simple motifs (*e-shino*). Shino wares that are predominantly gray in color are termed *nezumi shino*, and those that are a deep earthy red with creamy colored motifs are called *aka shino*.

Another traditional Mino ware is *ki seto*. Although the name suggests that it is yellow (*ki*), it is actually a bone-colored ware derived from a wood-ash glaze and fired at a high temperature. This causes the minerals in the clay and in the glaze to oxidize, and produces the particular color and texture of this ware. In *ki seto*, motifs are etched in the clay, then highlighted with judicious use of green glaze over the yellow.

Perhaps the most bold and colorful of Mino wares is Oribe ware, combining freely brushed designs of abstract shapes or semi-abstract scenes from nature. These are painted on while clay in iron pigment, and open areas are covered in an irregular manner with dark bottle green (the result of an ash glaze to which some copper has been added). While there are variations within this ware such as the salmon pink-colored *aka oribe*, it is the greens of *sō oribe* and *ao oribe* which are most admired. In addition to ash glazes, some Mino potters use a glaze containing iron to produce a black Oribe ware. Another Mino ware known as Ofuke ware is made from a special clay called *sensō tsuchi* containing iron. This ware is covered with a glaze containing some wood ash that turns a beautiful transparent pale creamy yellow when fired. *Seiji*, or celadon, ware is also produced in Mino.

In addition to coil-built and wheel-thrown pieces, Mino potters employ the technique of slab building (*tatara seikei*) to construct angular dishes and other asymmetrical forms. Among the common forms of decoration, aside from those mentioned above, are combing (*kushime*), impressing motifs into the damp clay with a stamp (*inka*), faceting the surfaces with a tool (*mentori*), and applying slip with a stiff brush (*hakeme*). One of the most distinctive decorative techniques is called *mishimade*, whereby impressions are made in the clay and then filled with a slip of a different clay before firing. Underglazes and overglazes are applied by dipping, ladling, or brushing.

After suffering a decline, interest in traditional Mino stonewares was revived in the mid-twentieth century by potters such as Arakawa Toyozō and Katō Hajime. At present there are approximately 150 potters working in the Mino district, producing primarily tea utensils, vases, household utensils, and religious paraphernalia and ornaments.

Shino ware.

Ki seto ware.

HAGI WARE • *Hagi Yaki*

Hagi, Yamaguchi Prefecture

The origins of Hagi ware can be traced back to Korea. When the shogun Toyotomi Hideyoshi dispatched troops to Korea in the last decade of the sixteenth century, among them was the leader of the Hagi clan, Mōri Terumoto. While in Korea, Mōri met two brothers who were potters, Lee Chak Kwang and Lee Kyung, whom he brought back to Japan. In 1604, a kiln was first fired in Matsumoto under their direction. In addition to this *goyōgama*, or "clan kiln," Lee Chak Kwang's son established another kiln in Fukagawa in 1657 that also received the full support and protection of the clan. In fact, the two types of pottery that were produced in these kilns are still being made today by the Saka and Miwa families, both of which have produced potters who became Living National Treasures.

Important to the development of Hagi ware was the discovery in 1717 of a type of clay called *daidō tsuchi*, which underwent a lot of subtle changes in color and texture during firing. If it was mixed with another more, fire-resistant clay called *mitake tsuchi*, the number of changes increased.

The muted qualities of Hagi ware are still among its special features. In addition, tea stays hotter longer in Hagi teacups because the clay actually absorbs some of the liquid and changes color the more they are used. This is called *chanare* (becoming accustomed to the tea) or *Hagi no nana bake* (the seven changes of Hagi). Flower vases and saké vessels are also common forms of Hagi ware.

Hagi ware is still made on a type of Korean kickwheel known as a *kerokuro*. After the forming of the foot, a wedge-shaped cut is made. This "split foot" (*wari kōdai*) is the formalized gesture of a gimmick devised to allow potters to sell their pottery to commoners during a time when Hagi ware was reserved exclusively for the use of the upper class. As "damaged goods," these "disfigured" pieces could be bought by commoners.

Hagi pots are almost never decorated with painted motifs. Instead they rely on wood-ash glazes (*dobaiyū*), an ash glaze called *isabaiyū* made from the wood of the *isu* tree (*Distylium racemosum*), or a straw-ash glaze (*shirohagi gusuri*). These glazes coupled with the changes that occur during the lengthy firing at a low temperature in a climbing kiln give them decorative interest and tone gradations that become more prominent with use. Kilns are usually fueled with pine and will probably continue to be as long as the qualities of Hagi ware are valued.

KYOTO WARE • *Kyō Yaki*
KIYOMIZU WARE • *Kiyomizu Yaki*

Kyoto

Kyoto is a treasure-trove of traditional crafts that have been perpetuated by generation after generation of craftsmen and craftswomen. Many of these crafts still thrive in the traditional atmosphere of the old capital, including a diversity of ceramics that are collectively referred to as Kyō *yaki*, or Kyoto ware. Of these, Kiyomizu ware is perhaps the most well known.

There is evidence that *sueki* ware was being made in the Kyoto area during the reign of Emperor Shōmu (724–49) and that tiles (*ryokuyū gawara*) were produced in the Enryaku era (782–806). Raku ware, favored for the tea ceremony, began to be made during the Eishō era (1504–21), but what is now called Old Kiyomizu (*Ko Kiyomizu*) is actually the pottery from kilns in Yasaka, Otowa, Kiyomizu, Seikanji, and Mizorogaike—all within the bounds of Kyoto—that emerged during the Kan'ei era (1624–44). It was not until the potter Nonomura Ninsei became active after 1647 that the overglaze enamel techniques and decoration associated with Kiyomizu ware were firmly established.

After Ninsei, several individuals played important roles in the development of Kyō *yaki*. Ogata Kenzan perfected several techniques for a lustrous pale yellow ware decorated with indigo designs modeled after pottery and porcelain imported from Holland. Between 1789 and 1801, Okuda Eisen succeeded in producing actual porcelain modeled upon Ming prototypes. Two of the techniques he was able to perfect were *sometsuke*, or the use of transparent glazes for decoration, and *akae*, a multicolored overglaze technique in which red predominated. Okuda also instructed a number of other potters, including Aoki Mokubei. Through the efforts of Okuda and those who followed in his footsteps, the range of porcelain produced in Kyoto increased significantly and included works modeled after Korean and Chinese examples. Apart from *seiji*, or celadon, ware, there was also *kōchi*, inspired by a three-color ware made during the Sung and Ming dynasties featuring green, yellow, and purple. This ware first reached Japan via Indochina and consisted mostly of small incense containers (*kōgō*) made from molds. Another ware emulating continental models was *gohon*, resembling the ceramics produced at a kiln at Pusan in Korea between 1661 and 1688. The clay was mostly white with a dull, reddish-yellow glaze, and it was decorated with white, indigo blue, or iron glazes.

The fact that much of what is produced under the heading of Kyō *yaki* is derived from foreign models is in itself one of the special features of this ware. With so much borrowing, it is not surprising that a wide range of techniques is used. Wheel-thrown pieces predominate, but some are hand-formed and a variety of mold methods are used. Designs are added by impressing with stamps (*inka*), combing (*kushime*), inlay (*zōgan*), and slip trailing (*itchin*). Glazes are applied in all of the common ways (dipping, dripping, trailing, painting) as well as by spraying (*fukigake*) and by allowing the glaze to dribble and collect. The glazes themselves are mainly mineral, iron, copper, or the Raku glazes associated with Raku ware. The methods of applying motifs with glazes are more limited; Dutch pigments are used on underglazes, and overglaze decoration is either transparent enamels or in alternating colors in the manner of three-color ware.

A large number of the articles produced in the kilns of Kyoto today are of a traditional nature. They include utensils for the tea ceremony, ikebana, and the less well-known *kōdō*—the pastime of enjoying the scent of incense. In addition there are the usual tablewares that can be found in homes, hotels, traditional inns, and restaurants throughout the nation.

SHIGARAKI WARE • *Shigaraki Yaki*

Shigaraki-chō, Shiga Prefecture

The origins of this rustic, rough-textured stoneware date back to very ancient times. During the reign of Emperor Suinin (29–71 B.C.), potters who had been making *sueki* in what was once called Kagami no Hazama (present-day Kagamiyama) apparently found a need for a different clay than the one they had been using. They thought they might find a clay to their liking in Shigaraki, where in fact, a good-quality clay was also being used to produce *sueki*. However, the true beginnings of Shigaraki ware date considerably later. While pottery tiles were made for Shigaraki no Miya, the capital established by Emperor Shōmu in 742, it was not until the end of the twelfth century that pottery production really commenced.

At first potters made primarily utilitarian wares such as small bowls for seeds (*tanetsubo*), mortars (*suribachi*), and large jars for storing liquids. Objects that were made before 1573 are referred to as Old Shigaraki, or *Ko Shigaraki*. The natural, unpretentious beauty of Shigaraki ceramics became recognized during the subsequent Momoyama period, especially by tea masters such as Sen no Rikyū and Kobori Enshū. The practice of drinking of tea actually helped to bring Shigaraki ware to the attention of the people at large. This was especially true during the years that Tokugawa Iemitsu was shogun (1622–51), when a Shigaraki tea jar (*chatsubo*) containing tea was transported from Uji, near Kyoto, to Edo (present-day Tokyo), the seat of the shogunate, in a formal procession known as the *chatsubo dōchū*.

Various kinds of household goods came to be made in large quantities in Shigaraki after climbing kilns were introduced during the Edo period. The reputation of Shigaraki ware was further heightened when an indigo blue glaze called *namakogusuri* was developed at the end of the nineteenth century. This glaze was used on hibachi (charcoal braziers), which were still a major form of domestic heating. Shigaraki hibachi were so popular that at one time eighty to ninety percent of the nation's braziers were being produced in Shigaraki kilns. Hibachi are still being made today, along with other household utensils, ornaments, and gardenware.

The Shigaraki "look" of the unglazed stoneware has a lot to do with the rugged natural elements of the finished pieces. Small pebbles of feldspar permeate the clay. When the ware is fired, the white feldspar "rises" to the surface. While no glaze is applied to the pieces, a natural-ash glaze forms on some pieces when ash from the wood fuel of the firing settles on the pieces and fuses with the clay, turning green or brown. The chunks of feldspar, the random ash-glaze effect, the warm reddish coloring, and the rough texture all combine to give the ware a natural quality that appeals to Japanese sensibilities for subdued but dignified work. It was the accidental qualities of Shigaraki ware that caught the eye of the tea masters, and more recently, have also gained the attention of people far from this corner of western Japan.

BIZEN WARE • *Bizen Yaki*

Inbe, Okayama Prefecture

The oldest of Japan's Six Old Kilns, Bizen ware is essentially a continuation of the stoneware tradition that started with *sueki*. During the ninth and tenth centuries, *sueki* was being produced in the village of Sue (now Nagafune-chō). However, at the end of the Heian period the potters of Sue moved to Inbe in search of good-quality clays and fuel for their kilns. During the Kamakura period more and more kilns were built, until potters were active all around Inbe. It was at this time that they began to produce pottery with the distinctive red to chocolate brown coloring for which Bizen ware is known today.

The ceramic industry became more centralized during the Muromachi period, and with the fashion for drinking tea during the subsequent Momoyama period, Bizen ware reached new heights. Inbe potters received the protection and encouragement of the local clan during the Edo period, when large quantities of practical pottery such as bottles for saké and other liquids, small water jars, mortar bowls, and seed jars were being made alongside the more artistically inclined articles offered to the court and shogunate. By the end of the nineteenth century, a new type of climbing kiln came into general use, and the potters of Bizen switched to making tea wares, saké cups and bottles, and flower vases. Since the end of World War II, the number of artists and potters in the area has risen to more than two thousand, some of whom have acquired the accolade of Living National Treasure. Selected works have been designated as Important Cultural Treasures by the prefecture, thus ensuring the position of Bizen ware in the contemporary world.

Bizen ware can be divided into six distinct types according to firing techniques and the accidental effects that occur during firing. Winding straw around a piece leaves red streaks on the pot when it is fired—a technique called *hidasuki*. Another effect known as *goma* (sesame), so named because it appears as if sesame seeds were sprinkled on the surface of the pot, is caused by ashes from the wood fuel spotting the surface and turning into a glaze in the intense heat. Those pieces having a dark gray color are referred to as *sangiri*. The gray occurs when a piece near a stoke hole is buried in ash during the firing so that the flames do not touch the surface directly. The fourth type is called *botamochi* and takes its name from a red rice cake. A small cup or pot is placed on another one so that the surface is partly covered. Those areas which are exposed will be completely fired, but because the covered parts are not directly exposed to the heat they turn red. Pieces which are fired with their openings facing down are called *fuseyaki*. This is a way of preventing the interior surface from becoming speckled with ash, which is less desirable for some dishes, platters, and other pieces. The last of these types is *ao bizen*, wares having a very dark blue color resulting from firing them in a kiln rich in carbon flames.

The natural beauty of Bizen ware stems largely from the unforeseen changes that occur during the firing, affecting the shape, color, and texture. Moreover, because glazes are not applied, the elemental feeling of the iron-rich clay is retained. Consequently, Bizen ware is often cited as the quintessence of Japanese ceramics.

ARITA WARE • *Arita Yaki*

Imari and Arita-machi, Saga Prefecture

Arita ware, popularly known as Imari ware, was first made at the end of the Momoyama period, around 1590. The earliest ware was actually a type of old Karatsu pottery made mainly in and around what was known then as Aritagō, located near the present-day Arita-machi. However, changes occurred as a result of shogun Toyotomi Hideyoshi's campaign in Korea (1592–98), when the Korean potter Lee Cham Pyung was brought back to Japan by the head of the local Nabeshima clan. Lee discovered the kaolin clay necessary for porcelain in 1616 on Mt. Izumi near Arita, and successfully produced some pieces of porcelain. This event marked the true beginnings of contemporary Arita ware.

At first the ceramics produced in Arita followed Korean pottery styles, but in 1643, Shodai Sakaida Kakiemon perfected the polychrome overglaze enamel technique called *akae* that he had learned from studying Chinese wares. The designs on what became known as Kakiemon ware were first outlined in blue on a white porcelain ground, then overglaze enamels, with red predominating, were applied on top. Around this time the Nabeshima clan built its own kiln, and with the clan's backing Shodai Imaemon created a ware known as Iro-Nabeshima. He was also successful in perfecting two much more individual styles of porcelain. One called *nishikide* displays red, blue, yellow, green, purple, gold, and other colors applied as overglaze pigments, resulting in some spectacular pieces. The second, *somenishiki*, combines the use of *nishikide* or polychromatic overglazes on a *sometsuke* or blue-and-white ground.

The porcelain from Arita kilns was shipped to various parts of Japan through the port of Imari, from which the popular name Imari derived. The ware was also exported to Europe, where it became known as Old Imari. Arita ceramics began to influence some of the major European porcelain wares such as Maissen in Germany, Delft in Holland, Chantilly in France, and Worcester and Bow in England. The influence of Arita porcelain was also felt at home, especially in the pottery-producing areas of Kyoto, Kutani, and Tobe.

A vast range of ceramics which are both practical and decorative continues to be produced in Arita today. What is still termed Old Imari is characterized by blue-and-white *sometsuke* ceramics and the more colorful *somenishiki*. In addition there is the detailed and refined Iro-Nabeshima ware and the brilliantly colored *akae* ware.

Among the glazes used are limestone, a wood-ash glaze made from the wood of the *isu* tree (*Distylium racemosum*), a celadon iron glaze (*seijiyū*), other iron glazes, and a cobalt oxide glaze (*rurigusuri*). Gold and silver colors are also added if both underglazes and overglazes are to be applied.

While mass-produced items are now made in molds and the glazes are applied by machine or sprayed on, the more exclusive pieces of Imari ware are still highly prized at home and abroad, and continue to carry the banner of Japan's first porcelain.

TOKONAME WARE • *Tokoname Yaki*

Tokoname, Aichi Prefecture

Tokoname is known primarily for the rugged, reddish brown stoneware produced there from the twelfth century onward. However, the history of ceramics in and around the city of Tokoname dates back even further. The remains of a large number of old kilns were discovered at the southwestern foot of Mt. Sanage in Mikawa, Aichi Prefecture, about 45 kilometers from Tokoname. This area appears to have had the largest concentration of kilns in Japan between the Kofun period and the Heian period. From the Heian through the Muromachi period, various kinds of stoneware, including teacups, plates, cooking pots, bowls, and large jars, were being fired in kilns scattered over the whole of the Chita Peninsula. The products of these kilns became known collectively as Tokoname ware, the origin of which is said to be around 1100 A.D. Large jars for household and ceremonial purposes in particular were made in great quantities. Characteristic of these dark stoneware vessels, formed by coiling ropes of clay and pinching, are the natural-ash glaze deposits and encrustation on the shoulders, which often were the result of droppings from the kiln roof.

The Tokoname ceramic industry suffered during the wars of the sixteenth century, but by the mid-eighteenth century, the demand for tea-ceremony ware had attracted a number of famous potters to the area. In the late nineteenth century, Tokoname potters also began producing unglazed tea ware made from a fine-grained clay containing a high percentage of iron. Pieces were fired to produce a range of shades from red to dark brown. The firing technique for this ware is similar to that used for the large jars. Both wares are fired in an oxidizing atmosphere with slow heating and cooling to avoid cracking caused by sudden changes in temperature.

In addition to being a popular domestic ware, by the 1920s, Tokoname ware was being exported to the United States and Italy. Today the pottery industry continues to flourish. A variety of items—both glazed and unglazed—are produced in Tokoname, including the traditional teacups, jars, flower vases, saké bottles, hibachi, and drainpipes for industrial use.

TOBE WARE • *Tobe Yaki*

Tobe-chō, Ehime Prefecture

Tobe ware had very plebeian beginnings. During the eighteenth century, boats called *kurowanka-bune* sold food and saké to people on the vessels that plied the waters of the Yodogawa River in the Hirakata area of Osaka. The simple bowls used by these venders were called *kurowanka chawan,* and they became very popular with the people at large. Present-day Tobe ware rice bowls with their thick bottoms are representative of this simple, unassuming ware, although the types of ceramics that take the name Tobe are diverse.

Stonewares were made in this area earlier, but the production of Tobe porcelain officially began when the Ōzu clan built a kiln in the village of Gohonmatsu (Tobe-chō). Porcelain was first fired there in 1777 with the help of some potters from the Ōmura clan in Hizen. Initially sponsored by the domain, Tobe kilns developed into private operations. Colorful porcelain began to be produced after 1825 using the *nishie* multicolored technique learned from Hizen potters, and by the end of the nineteenth century, Tobe ware was technically and economically on firm footing.

The ivory-colored Tobe porcelain found favor in the United States, and tableware decorated with stenciled underglaze blue designs (the so-called Iyo bowls) was also exported in large quantities to Southeast Asia. Although production declined during the period between the two world wars, attempts were made to revive the industry after World War II. Advice was sought from top designers, kilns were rebuilt, and an effort was made to modernize production by introducing machines. The outcome of this quiet revolution has been the production of a great variety of work. Now, traditional pieces are made alongside wares with a folkcraft flavor and others that are more contemporary in spirit.

Unlike some other porcelain, Tobe ware is thick with a slightly cloudy surface. While often decorated with simple designs painted in charcoal-black glaze, it can also be very colorful. The primary glazes used are a clear glaze, *seijiyū* (celadon), and a black iron glaze called *tenmokuyū.* Both the underglazes and overglazes are applied by hand and help to give Tobe ware the distinctive, slightly rustic flavor that still has the ability to delight the eye at the dining table.

TSUBOYA WARE • *Tsuboya Yaki*

Naha,* Okinawa Prefecture

Okinawa played a significant role as a trading center during the fourteenth and fifteenth centuries, when it was a nominal kingdom known as Ryūkyū. Ceramics, particularly celadon wares (*seiji*), were imported from China and then re-exported to Southeast Asia and Japan. Around this time potters in Wakita (part of Naha) were making a type of unglazed ware known as *arayaki* for the local brew.

In an effort to better organize the ceramic industry, the court authorities in Shuri appointed Kobashikawa Kōin to the post of official tile maker (*kawara bugyō*) during the period between 1573 and 1619. This position was later modified to cover pottery in general, but initially he was in charge of making roof tiles and porcelain. The ceramic industry was further influenced by the arrival of potters from China and Korea. Among them were a Chinese tile maker. These craftsmen set about instructing the local potters. One of the Korean potters in particular—Chang Hun Kong (who later adopted the name of Nakachi Reishin)—significantly influenced the ceramic art in Okinawa. Another potter called Hirata Tentsū, who was originally from China, was given a house in Wakitahara by the king of Ryūkyū. Hirata went back to China to study techniques, and upon his return to Okinawa, he started making glazed pottery. One of his apprentices, Nakasone Kigen, began making ordinary wares for the common people instead of the expensive Chinese-style porcelain tableware being made for the upper classes up until that time.

In 1682 the court authorities at Shuri consolidated the widely scattered kilns by establishing a center at Tsuboya. Pottery production continued there until World War II, by which time the potters were dividing their time between making pottery and farming. Mechanized mass production, coupled with the complete change of lifestyle after the war, resulted in a decline in the production of unglazed ware in favor of the glazed ceramics that now constitute the mainstream of Okinawan pottery.

Apart from conventional bowls, plates, tea ware, and the original kind of unglazed *nanban tsubo*, some of the products of the Tsuboya kilns are rather unusual due to the special nature of Okinawan customs. Examples are the *dachibin*, or curved hip flask; the mythical lion-dog (*shiisaa*) standing on the slope of the roofs that is thought to ward off evil spirits; and *zushigame*, highly decorated caskets for the bones of the deceased in the form of an urn or a boxlike shape resembling a miniature traditional building.

In keeping with the subtropical climate, Okinawan ceramics are decorated with bright, mostly primary colors. The designs are simple but powerful and are derived from the life and customs of Okinawan people. Both unglazed and glazed wares are formed on a wheel, in molds, or by hand. Glazes, which themselves are unique to Tsuboya, are applied by dipping (*hitashigake*), sprinkling (*furikake*), or using a cloth that has been dipped in the glaze (*nunokake*). Any application of motifs is done by hand. The unglazed ware is fired in a *nanbangama*, the type of kiln found in Southeast Asia. Thus, even today Tsuboya ware is still representative of the distinctive nature of Okinawa's culture and historical background.

KIKUMA TILES • *Kikuma Gawara*

Kikuma-chō,* Ehime Prefecture

Silver-gray roofs are one of the most distinctive features of traditional Japanese buildings. The tiles for such roofs, which are still being made on the island of Shikoku, have noble beginnings extending back to the Kōan era (1278–88), when Kikuma tile makers went to Iyo to work for the powerful Kōno family. Ming tile-making techniques were introduced around 1573, and with the support and protection of the local clan during the Edo period, the foundations of an industry were laid. While roofing tiles made in Kyoto have always been recognized for their high quality, those produced in Kikuma are equally fine and were used in the construction of the new Imperial Palace in the late nineteenth century.

Apart from roofs, tiles have traditionally been used in Japan for paving floors (*shikigawara*) and also for walls (*kabegawara*). However, the word *kawara* is more or less synonymous with roof tiles. There are several kinds of roof tiles, which differ in size and shape according to their function. *Hiragawara* are the simplest, being completely flat or gently curved. They are used in conjunction with *marugawara*, or channel tiles, to create one of the oldest styles of roofs, called *honkawarabuki*. Used

mostly at the corners of a roof and sometimes along the edges for decoration, *sumigawara* are appropriately called corner tiles. Also falling under the category of corner tiles are the two kinds of ridge caps; one is decorated with a demon mask (*onigawara*) and the other is in the form of a fish with its tail arched up over its head (*shibi*). Two *shibi* are usually placed at either end of a ridge inside the capping *onigawara*.

Two types of clay are combined for the making of Kikuma roof tiles. One is called *sanuki tsuchi* and comes from the neighboring Kagawa Prefecture. The other is a locally obtained clay called *gomi tsuchi* (clay with "five flavors"). Once mixed, the clay is pressed into molds. After removing the clay from the molds, the tiles are dried and then fired for a full twenty-four hours. Mica is applied to the surfaces and burnished before firing in order to achieve a silvery luster.

Just as tiles have more or less replaced thatch as roofing material, so, too, have other materials taken the place of roof tiles. But do the new roofs ever look as wonderful as a tile roof after a light snow or a rain shower?

S E N S H O K U

Textiles

Introduction

As the traditional textiles of Japan were made primarily for personal attire, what we know today as the kimono determined not only the construction of the weaves and the patterning of the fabric but also the width of the cloth itself. A single bolt, or *tan*, of cloth measures approximately 9 meters in length and 30 centimeters in width, and this usually suffices to make one kimono, whether for men or for women regardless of height and weight. Thus kimono fabrics as a rule are sold by the bolt and rarely by the meter. Essentially, the kimono consists of four strips of fabric, two forming the panels covering the body and two the sleeves, with additions for a narrow front panel and the collar.

Customarily, woven patterns and dyed repeat patterns are considered informal. Formal kimono have free-style designs dyed over the whole surface, or in the case of a married or older woman, along the hem and the front panel and perhaps also at the shoulder. Men wear woven fabrics, usually of blue or a subdued gray or brown, the only exceptions being for costumes worn at festivals or by entertainers. They will, however, take a hand-painted fabric for underclothes or a jacket lining.

Only fabrics of reeled silk are used for formal wear. These include silk crepes such as *chirimen* and satin weaves such as *rinzu*. The lightweight *habutae* is best suited for linings but also may be made into *obi* with a heavy backing. Pulled floss silk is called *tsumugi*. As it is made from waste cocoons, it is considered inferior in rank to reeled silk, although it is highly esteemed for its texture. A hand-pulled, hand-woven *tsumugi* may fetch a price many times above that of a reeled silk. *Kasuri*, the Japanese version of ikat, may also be very expensive, but it is not considered appropriate for formal occasions, nor are any of the bast fibers, such as ramie. The gauze weaves are worn at the height of summer, and if in silk, may be formal.

A woman's *obi* is usually 4 meters long. The 60-centimeter width is folded in half and the obi is wrapped twice around the waist, then tied in back, so ideally the fabric has some body and holds its shape well. *Obi* for casual wear may be as narrow as 10 centimeters. Formal *obi* are usually of a brocade or tapestry weave. The more pattern, the more formal is the basic rule. There are grandiose examples covered over their entirety with woven or embroidered designs. Such *obi* are now worn only by a bride. Casual *obi* can be made of *chirimen, tsumugi, habutae,* twill, satin, gauze weaves, cotton, or wool. Men's *obi* are either stiff (*kaku obi*), in which case they are about 5 centimeters wide, or soft (*heko obi*), in which case they are made of a thin, tie-dyed silk.

Today, despite its beauty and many advantages, the kimono has been superseded by Western clothes for everyday wear. This has greatly affected the producers of casual-type fabrics. At the same time, crafts like weaving, which once were lucrative skills, cannot match the attraction of more fashionable ways of earning an income. As the number of professional people engaged in weaving, spinning, tie-dyeing, and so forth dwindles, local craft industries are working hard to produce a younger generation of skilled labor, and the kimono industry is striving to capture the hearts of young men and women.

Kyoto *yūzen* dyeing.

KIRYŪ FABRICS • *Kiryū Ori*

Kiryū, Gunma Prefecture

Kiryū is situated due north of Tokyo on the edge of the Kantō Plain, in one of Japan's major areas for sericulture; other well-known cities in this connection are Isezaki and Ashikaga. In the Edo period, boats carried the cloth woven in Kiryū to Nihonbashi, the commercial center of Edo (present-day Tokyo), where the top-quality silks vied with those of Nishijin in Kyoto. The origins of weaving in the area are said to date back to the reign of Emperor Junnin (ruled 758–64), when Princess Shirataki went to Kiryū to teach people to raise silkworms and to instruct them in the art and techniques of weaving. At first the fabric was called Nitayama silk, Nitayama being the former name of Kiryū. Following the establishment of a silk market, business grew; trading between Edo and Kyoto began, and it is thought that the contacts thus made resulted in improvements in dyeing techniques. At some point a draw loom known as a *sorabikibata* was introduced from Kyoto, making the weaving of complex cloths such as gauze and twill more feasible. A woman or child was usually stationed high atop the loom to pull on the warp yarns in sequence in much the same way as a Jacquard loom works today. Later, *hatchōnenshiki* yarn-twisting machines driven by water wheels stimulated the production of crepe yarns, and after that yarn-dyed, figured-textile weaving techniques were also introduced. Jacquard looms have been employed in Kiryū since the end of the nineteenth century. These machines led to the specialization of production processes that began during the latter part of the 1920s.

Apart from raw silk, *tamamayu*, and *tsumugi*, such yarns as spun silk, cotton, linen, and gold or silver threads are also employed. Between four to eight thousand fine threads are used for the warp, and a thread that has been twisted more than two thousand times is employed for the weft. The actual method of weaving differs according to the type of cloth: weft brocade, warp brocade, double-weave, floating weave, warp *kasuri*, and twist weave. The representative *omeshi* crepe can be either a plain-weave yarn-dyed or pre-softened silk cloth, or a figured satin weave, or a *shusu* satin weave, or even a variation or combination of these. Rice paste is applied to the crepe yarn, which is used for the background weft after it has been lightly twisted by hand, then twisted further by machine.

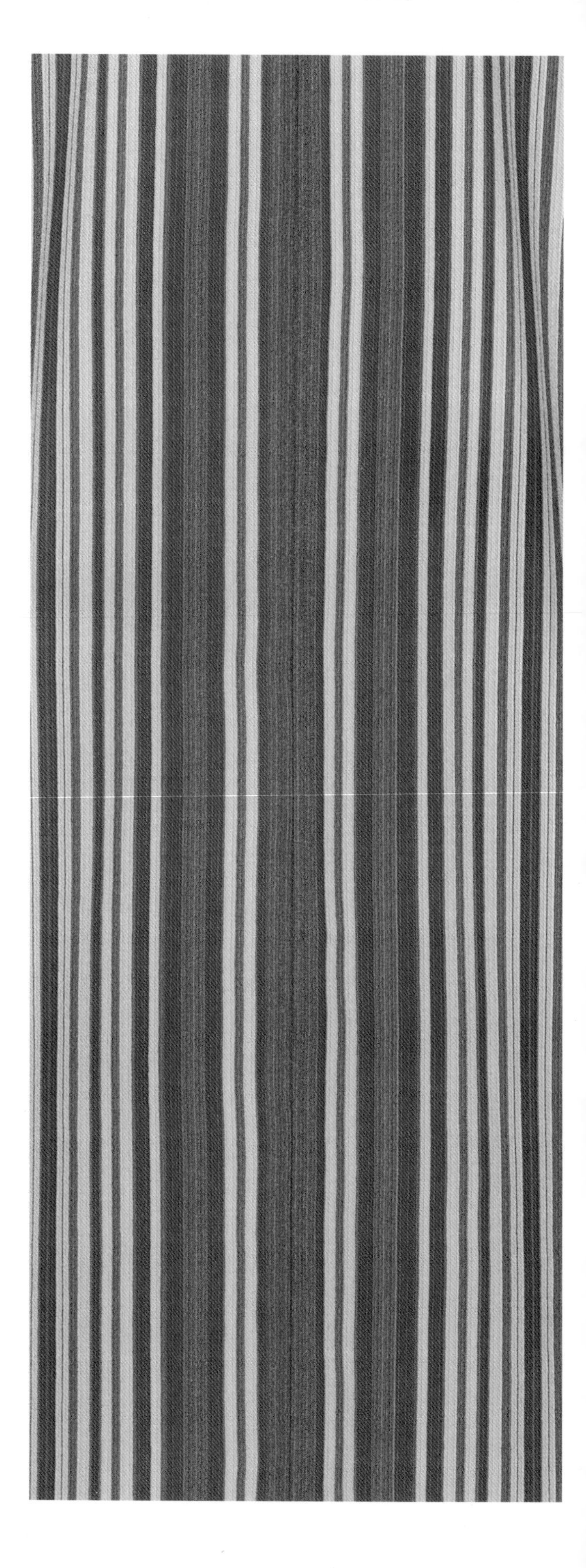

OITAMA PONGEE • *Oitama Tsumugi*

Okitama Region, Yamagata Prefecture

Oitama *tsumugi* is the collective name for the various types of silk *tsumugi* such as Nagai *tsumugi*, Shirataka crepe, and *Yoneryū* that are produced with natural dyes and motifs drawn from nature in the Okitama area of Yamagata Prefecture. This area has long been connected with sericulture and silk weaving: the Shiroko Shrine in the city of Nagai, which enshrines the deity of silk culture, seems to have been established in 712.

Basically, Oitama *tsumugi* is a yarn-dyed, plain-woven *kasuri* cloth that is occasionally striped. Slightly different combinations of yarn and *kasuri* techniques are used for the various types. For *Yoneryū* (to give it its full name, *Yoneryū itajime kogasuri*), the *kasuri* yarn is first dyed by the *itajime* technique. The warp is raw silk, and the weft is either raw silk, *tamamayu* silk, or a handspun *tsumugi* yarn. The yarn is dyed by the same technique used for Shirataka *itajime kogasuri* cloth, but both the warp and the weft are twisted raw silk. A yarn twisted three or four thousand times per meter is introduced in the areas having no pattern. After the cloth has been woven, a technique called *shibo dashi* is applied. This involves rubbing the cloth in hot water so as to crimp it. Raw silk, *tamamayu* silk, or *tsumugi* yarn is used for cloth in which the *kasuri* design is expressed either with just the weft, as in *yokosō gasuri*, or with both the warp and the weft as in *heiyō gasuri*. All the yarn for these is twisted, except the *tsumugi*, which is used for the warp of the *yokosō gasuri* cloth.

The natural dyes are obtained from logwood, a grass known variously as *kariyasu* or *kobunagusa*, and safflower. The latter is called *benibana* in Japanese, and the soft pink cloth it yields is known as *benibana zome*. Cloth dyed yellow with the grass is known as *kariyasu zome*. All the cloths woven with these techniques have an unpretentious, folk-cloth quality and are extremely durable. They are used for *obi* and kimono, and are usually worn by women over forty years of age.

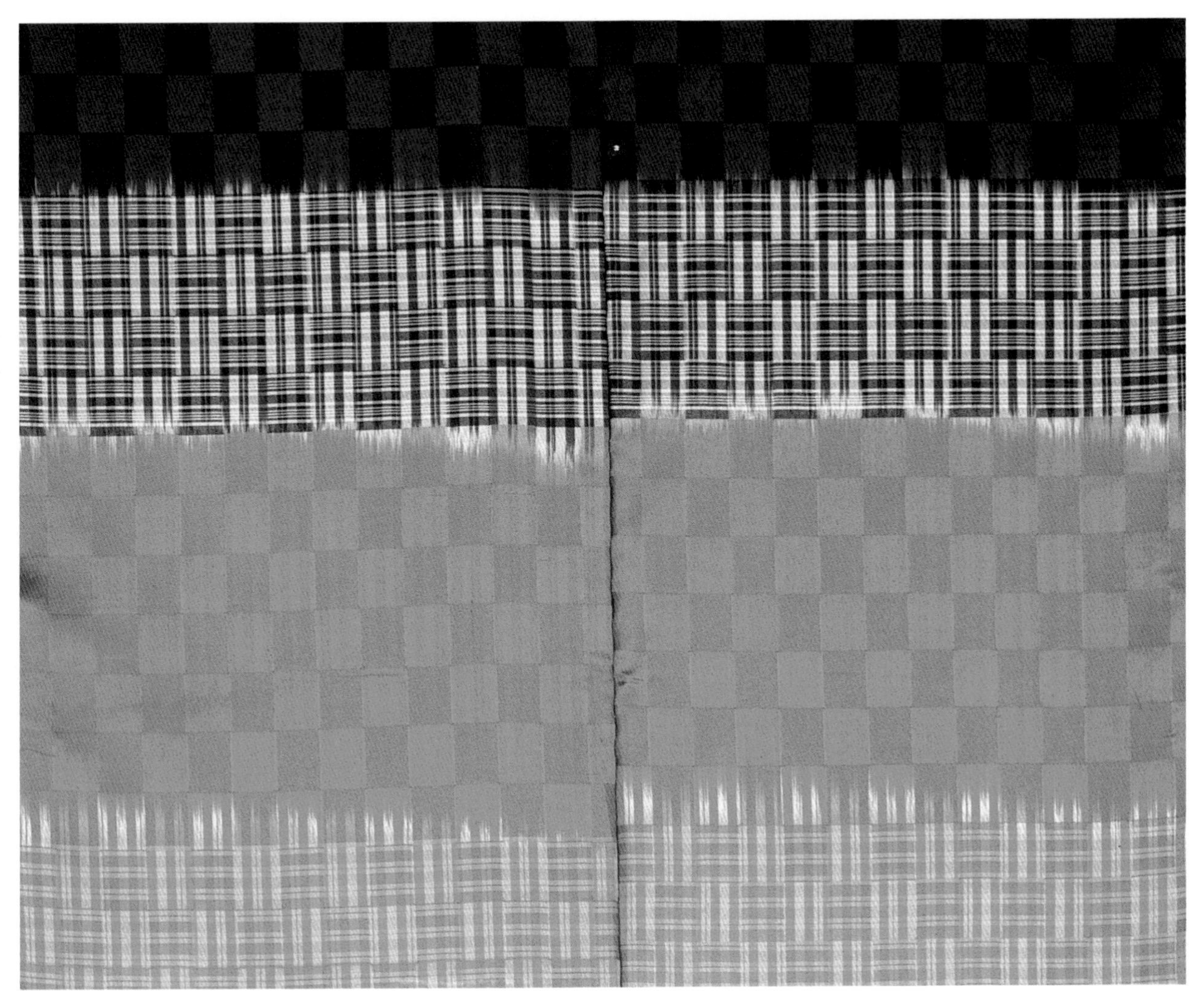

YŪKI PONGEE • *Yūki Tsumugi*

Yūki, Ibaragi Prefecture

Just as wine improves with age, so does this cloth. Its appearance and feel are reputed to become more interesting the older it gets, but the processes involved in its production make it very expensive.

First, the *tsumugi* is hand-spun from silk floss for both warp and weft, then the *kasuri* yarns are bound by hand or clamped in various places to resist the dye. The cost of the cloth is further augmented by the use of a back-strap loom known as *izaribata*, but there is no doubt that this also contributes to the final look of the cloth. Besides the plain weave cloth, Yūki *tsumugi* also includes a crepe known as *chijimi ori*. For this, a twisted yarn is taken for the warp, and the woven cloth is put in warm water and rubbed by hand to produce a crinkled crepe texture.

The finished cloth is expensive, but this is balanced out by its durability and tendency to improve in appearance as it is worn, and the overall restrained effect is enhanced by its peculiar dull luster.

TRUE KIHACHIJŌ • *Honba Kihachijō*

Hachijōjima Island, Tokyo

Situated some 334 kilometers south of Tokyo, Hachijōjima Island was under the direct control of the Tokugawa shogunate during the Edo period, and supplied salt and silk cloth to the state as annual taxes. Due to its rather poor soil, sericulture was encouraged in lieu of the usual form of taxation, rice. Although there are no detailed records, it seems that a type of silk *tsumugi* known as *ki* (yellow) *tsumugi* was presented to the authorities in 1498, and a 1665 document mentions a striped *tsumugi*. The type of cloth woven today first appeared sometime during the eighteenth century. It found favor at the seat of the shogun and was worn by maids and servants who worked at Edo Castle, as well as the feudal lords and lower-ranking officers. The island, in fact, takes it name from the cloth, which was woven in lengths of *hachi-jō* (eight *jō*, one *jō* being equal to about 3 meters). After the end of the feudal era, the cloth became popular throughout the country. Production has actually increased in recent years through the efforts of local intellectuals and the combining of new equipment with traditional techniques.

Vegetable dyes, often mordanted with mud, are used to color the raw silk, *tamamayu,* or *tsumugi* yarn. The yarn is then either plain woven or woven up into a striped or plaid twill. Brilliant and distinctive colors are obtained from the various plants used. A bright yellow (*ki*), for example, is obtained from *kariyasu,* a grass similar to pampas grass. The main cloth produced on the island is known as *kihachijō,* and is primarily of this yellow. A deeper, reddish yellow is derived from the *tabunoki* (*Machilus thunbergii*), and cloth in which this color predominates is known as *tobihachijō.* Another of the island's outstanding cloths—*kurohachijō* (black Hachijo)—obtains its intense yet dull black from a type of chinquapin (*Castanopsis cuspidata*).

Though primarily made up into kimono and *obi,* Hachijō cloth is also used for ties and other small articles such as purses. As a silk cloth it has a sober, refined appearance that improves with age. It must be one of the few craft products that have had a place named after them.

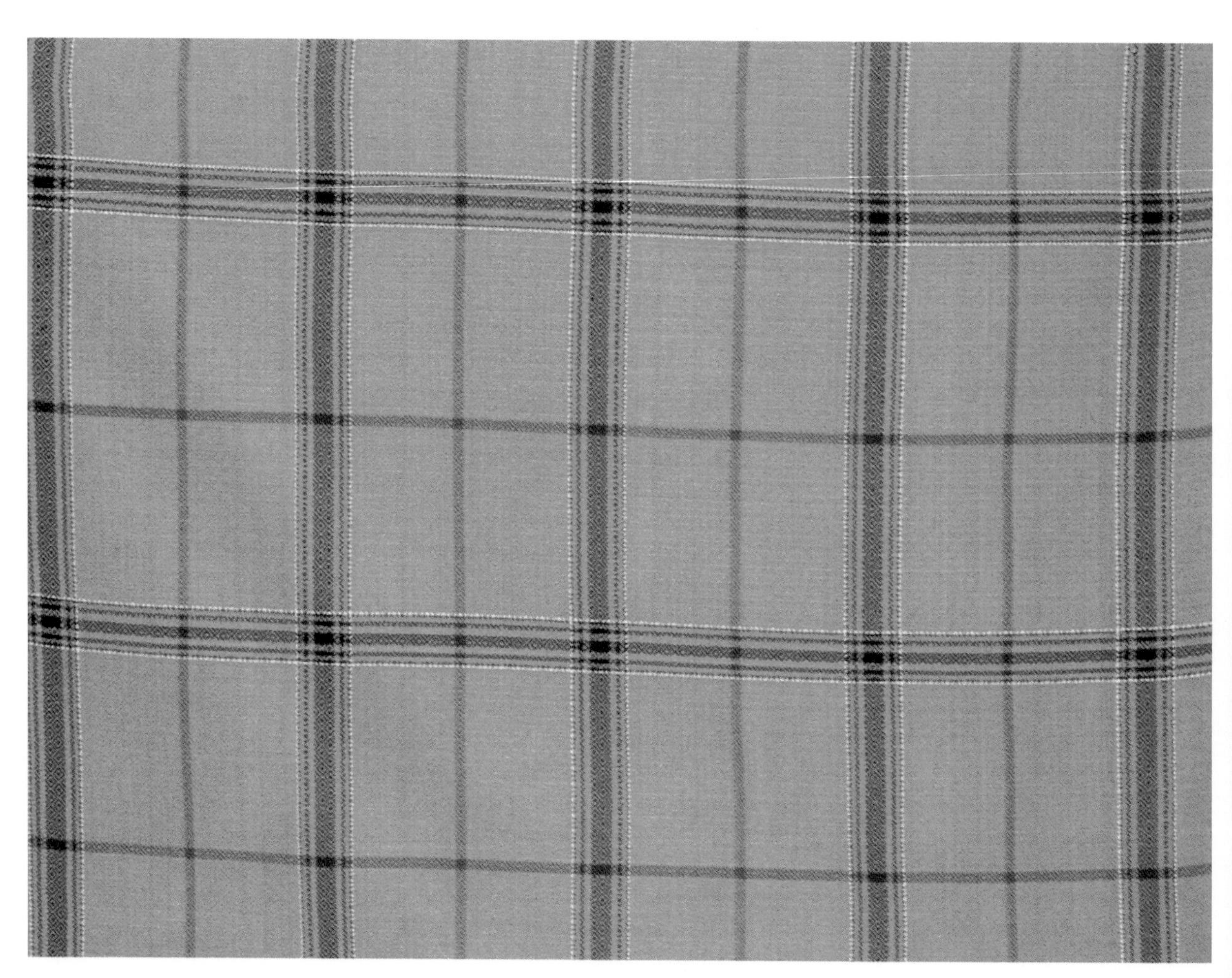

OJIYA RAMIE CREPE • *Ojiya Chijimi*

Ojiya, Niigata Prefecture

Environmental conditions in the region where this cloth is produced are the primary reason it is still being made. The Chinese silk plant or ramie (*Boehmeria nivea*), from which the yarn is produced, grows exceptionally well in Niigata Prefecture, and the high humidity in this area facing the Japan Sea makes the job of handling the yarn a good deal easier.

During its history, Ojiya ramie crepe has gone under several different names. It has frequently been lumped together with a cloth known as Echigo *jōfu*, both cloths in the past being variously called *Echigofu, Eppu, Hakuetsu,* or *Hakufu*, and then Echigo *chijimi*. During the Kanbun era (1661–72), Hori Jiro Masatoshi, a retainer of the Akashi clan, moved to the Ojiya area and passed on the method of making crepe cloth using a tightly twisted crepe yarn that is known as *chijimi*—a technique also used for another cloth called Akashi *honchijimi*. The production of *chijimi* and the number of locations where it was woven gradually increased, with the areas around Uonuma and Kubiki being most noteworthy. Ojiya, along with places such as Tōkamachi and Horinouchi, developed into distribution centers for the cloth. By the Meiji period, machine-spun yarn was being used along with hand-spun ramie. Apart from the very basic kind of back-strap *izaribata* looms, more conventional types of hand looms were also being used, with power looms coming into operation during the early part of the twentieth century. The work of weaving this cloth has, in other words, changed from a home industry to one organized along the lines of a mill, albeit one that still relies on some handwork.

Essentially, Ojiya *chijimi* is a yarn-dyed, plain-weave crepe *kasuri* cloth of ramie. The *kasuri* pattern is either in both warp and weft, or in the weft alone. The warp yarn is usually bound up by hand, and dye rubbed into it, again by hand, for a multicolored design. The weft is tightly twisted in order to produce the crepe effect when the woven cloth is washed in hot water. In the case of the plain *hakumuji* and slightly figured *jihaku* versions of this cloth, the lengths of woven cloth are stretched out on the snow in the early spring to be bleached by the rays of the sun, the melting snow, the oxygen in the air, and ozone. It is a splendid sight characteristic of this region of Japan.

The finished cloth is extremely fine and light. Kimono made of it are particularly comfortable in the summer, and it is also used for such items as cushion covers. Production of this cloth and of its sister cloth, Ojiya *tsumugi*, keeps some 435 people employed. While their primary interest is clearly to earn a living, their work nonetheless enables the long tradition of Ojiya *chijimi* to survive.

SHINSHŪ PONGEE • *Shinshū Tsumugi*

Nagano Prefecture

The attics in the great farmhouses of cold areas such as Nagano Prefecture provided ample, predator-free space for rearing silkworms as well as the requisite warmth in winter from the open fires below. Although technology has advanced, in Nagano Prefecture today the silk still comes mainly from the cocoons of silkworms reared indoors under controlled conditions. Some "wild" tussah silk is also included in the finished yarns. The tussah silkworms are farmed outside and allowed to free-range over their oak trees—which, by contrast with their indoor cousins, means a less than 25 percent harvest of cocoons.

Production of Shinshū *tsumugi* is centered in the city of Matsumoto, but includes a number of *tsumugi* cloths from Ueda, Iida, and Ina as well as Matsumoto. Historical references from as early as 1665 speak of a striped *tsumugi* being produced in Ueda. By the middle of the eighteenth century, *tsumugi* cloths from the area were well known throughout Japan, largely because the local clans vied with each other in their pursuit of sericulture as a source of income in places where rice did not grow well.

The yarns used are raw silk—including, as we have seen, some wild silk—and hand-spun *tamamayu*. The cloth itself is usually a yarn-dyed, plain-weave *tsumugi*, either *kasuri* or striped. Raw silk on its own is not used for the weft. Instead the thread used is produced on the *hatchō-nenshiki* yarn-twister by combining two or more raw silk and *tamamayu* yarns. Once a yarn of the required thickness is obtained, it is lightly twisted. *Kasuri* threads are bound up by hand, then dyed.

When wild silk is included, it lends its refined sheen to give this cloth a distinctive character. Shinshū *tsumugi* is also imbued with a sober, restrained quality resulting from the use of vegetable dyes, and it makes up well into kimono and *obi*.

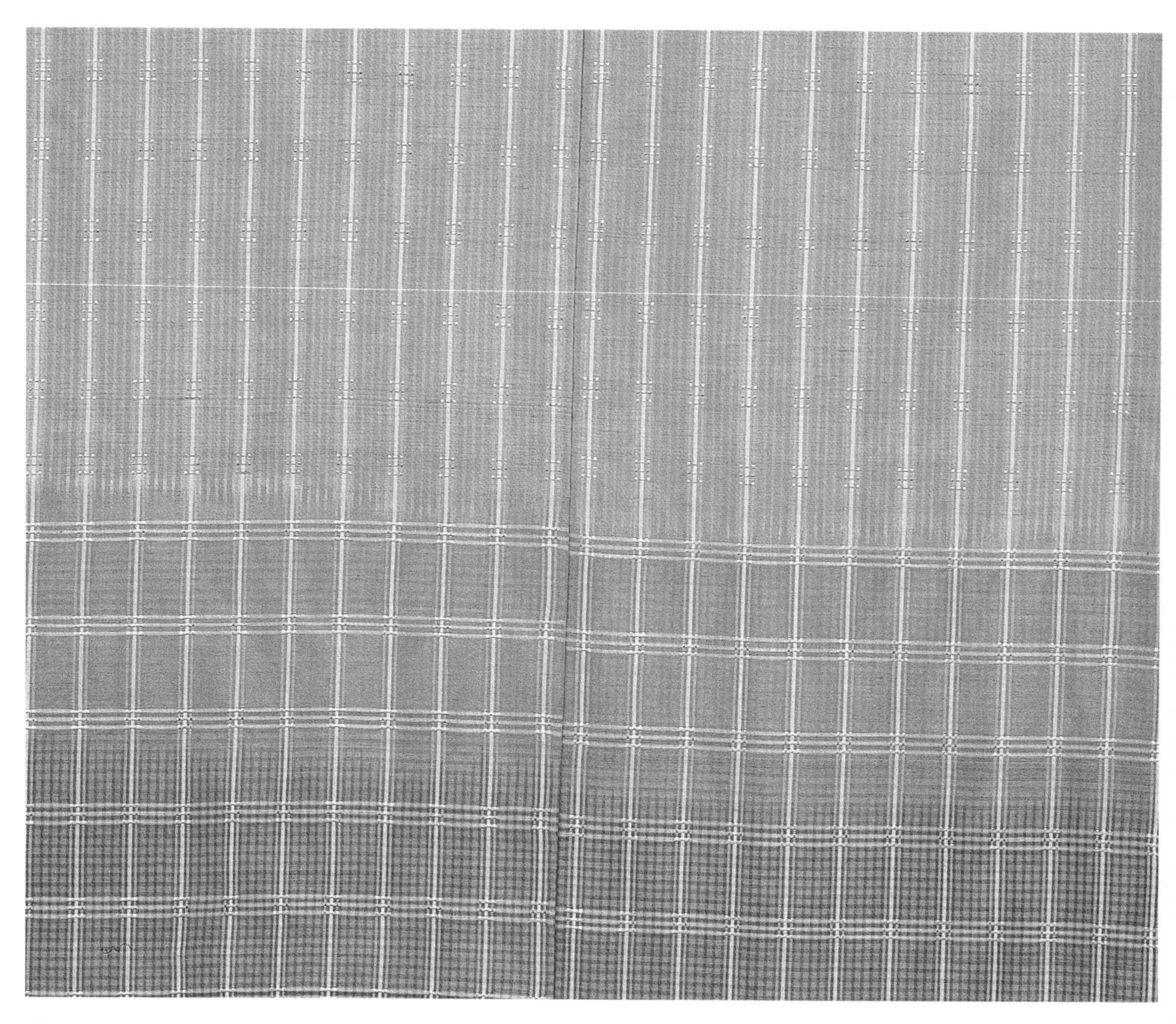

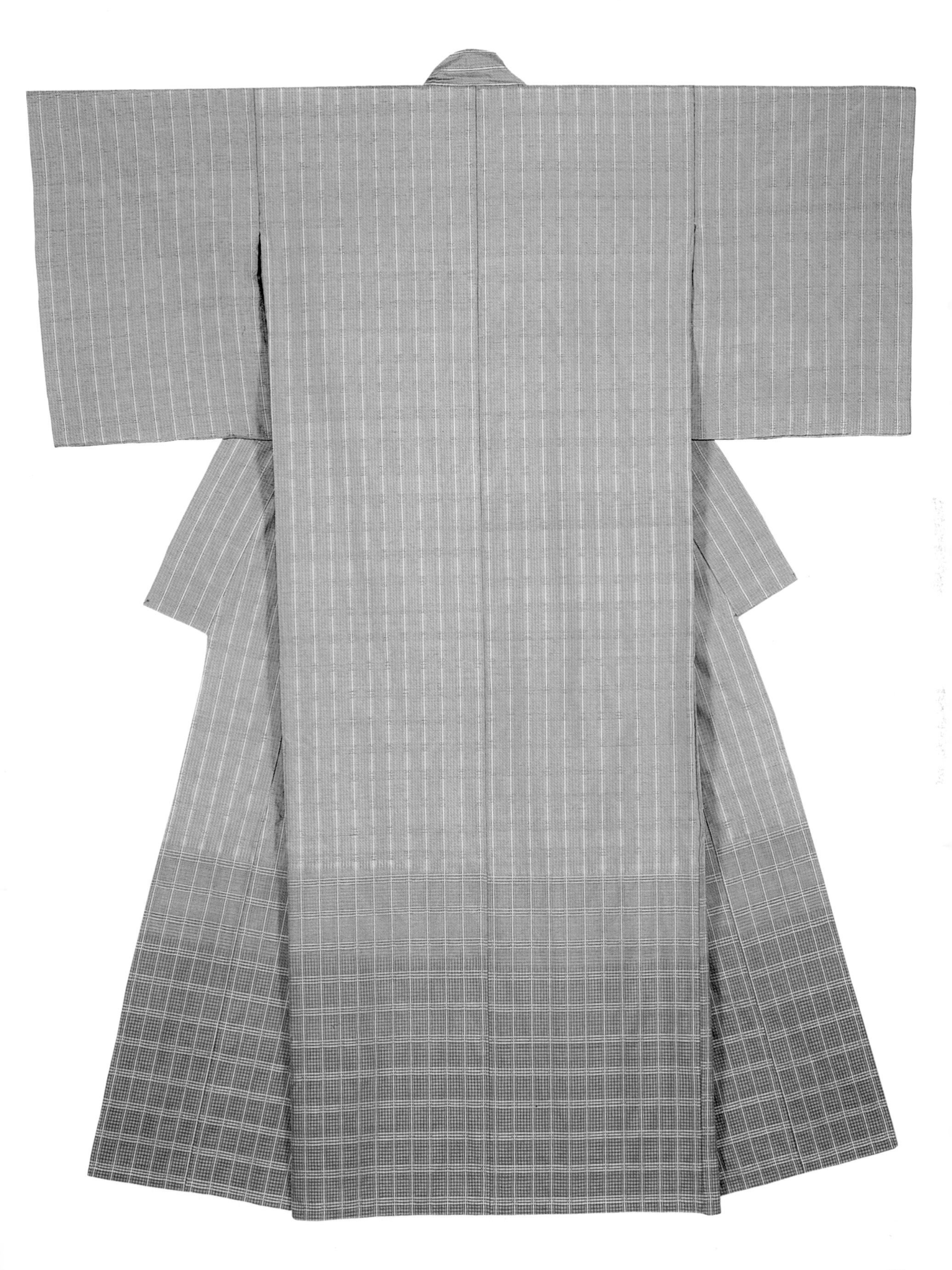

NISHIJIN FABRICS • *Nishijin Ori*

Kyoto

Having been spared from bombing during World War II, Kyoto is in many respects Japan's most traditional city, as well as being the seat of many traditional crafts. Among these, Nishijin *ori* is widely considered to be one of the nation's foremost achievements in any craft.

The history of Nishijin *ori*—actually, a collective title for silk cloth produced in the Nishijin district of Kyoto—goes back to the sixth century, when sericulture and silk weaving were started in the western part of Kyoto by the powerful Hata family, who may have originated from Korea or China. The Heian period witnessed the formation of an official weaver's guild, but civil strife during the eleven years of the Ōnin Wars (1467–77) laid Kyoto waste. Some weavers took refuge in the port town of Sakai, where they studied Ming weaving techniques. When calm was restored to the city they went back to the remains of the eastern camp of Kyoto, Tōjin, and began weaving *habutae*, a lightweight plain-weave cloth of untwisted raw silk, and *nerinuki*, a cloth using raw silk for the warp and glossed-silk thread for the weft. Meanwhile, the weavers who had settled in the western camp, or Nishijin, were busy bringing twilled cloths back to life, and it was they who established the foundation of today's Nishijin *ori*. At the end of the nineteenth century, the weavers sent a delegation to Europe to learn European methods and introduced the Jacquard loom to Japan. Development was further encouraged by the perfection of home-produced power looms.

Perhaps the most representative of Nishijin fabrics is the tightly woven tapestry cloth known as *tsuzure*. It is a yarn-dyed plain-weave cloth, and because the design is woven in the weft the density of these threads is

between three to five times that of the warp around which they are woven, so that the warp cannot be seen in the finished cloth. The small weft shuttles, moreover, do not go the full width of the cloth except where it is part of the ground of the design, but travel back and forth solely within their areas of color. Therefore a slight gap between the different colors sometimes appears. The fingernails of the weaver are used in place of a beater.

The cloths known as *nishiki* are a Japanese version of brocade woven on Jacquard machines using numerous colored yarns. One of these is *tatenishiki*, a yarn-dyed plain weave or twill with the design woven in the warp. *Nukinishiki* has a weft design. It is either a plain weave or a twill of pre-glossed silk, or sometimes a plain-woven cloth with variations in the warp and weft (*henka ori*). Figured satin weaves are represented by *donsu* and *shuchin*. Yet another variant, *shōha*, is a heavy twill of highly twisted yarn with a minute V patterned surface. And finally there is a figured double cloth called *fūtsū*.

The cloths produced in Nishijin have earned international renown, reflecting an endless wealth of designs and techniques. *Obi* and kimono fabrics are still being made as they were in the past, but the Nishijin looms these days also turn out fashion fabrics and cloth for interior decoration, bags, and ties. The variety is as great as the artistry and technique are brilliant. Today, the sounds of the looms in this quarter of Kyoto have been joined by a new one—the digital clicking and whirring of the computers now being used to control some of the looms, carrying the Nishijin tradition on irresistibly into the twenty-first century.

YUMIHAMA IKAT
Yumihama Gasuri

Sakaiminato, Tottori Prefecture

If local documents are to be believed, the cultivation of cotton in this area facing the Japan Sea started during the Enpō era (1673–81). By the time the production of indigo began around 1750, it seems that *kasuri* cloth was already being produced. In those days, *kasuri* weaving was a cottage industry, with the finished cloth being distributed through wholesalers. The amount of cloth increased year by year, supported by a stable supply of good cotton fertilized with sardine and seaweed from the Japan Sea and by equally good supplies of indigo. By the end of the Edo period it was a major source of income for the local clan, and the industry came under the direct control and protection of the fief. An official document on the production of *kasuri* cloth in 1874 ranked the area third after Ehime (Iyo *gasuri*) and Hiroshima (Bingo *gasuri*), but production has since declined. Yet today, the traditions of the cloth are supported by local patrons.

Yumihama *gasuri* is a yarn-dyed, plain-weave cotton weft *kasuri*. The yarn is bound by hand with a coarse type of ramie thread called *araso*, which is obtained by steaming the stems of the plant and drying the parings. Vegetable or synthetic dyes are used as well as indigo. The cloth is hand woven; adjustments are made to the pattern as the cloth is woven.

The designs are pictorial, using birds, flowers, scenery, and farm life as motifs. The cloth is made into kimono, *obi*, cushion covers, and the like. Though not so common in modern times, Yumihama *gasuri* once held such an important place in the local culture that, at the age of fourteen or fifteen, local girls were already accomplished weavers and would have started weaving cloth for their trousseau. Today, it is prized in a different way, by all those who admire traditional Japanese folk weaving.

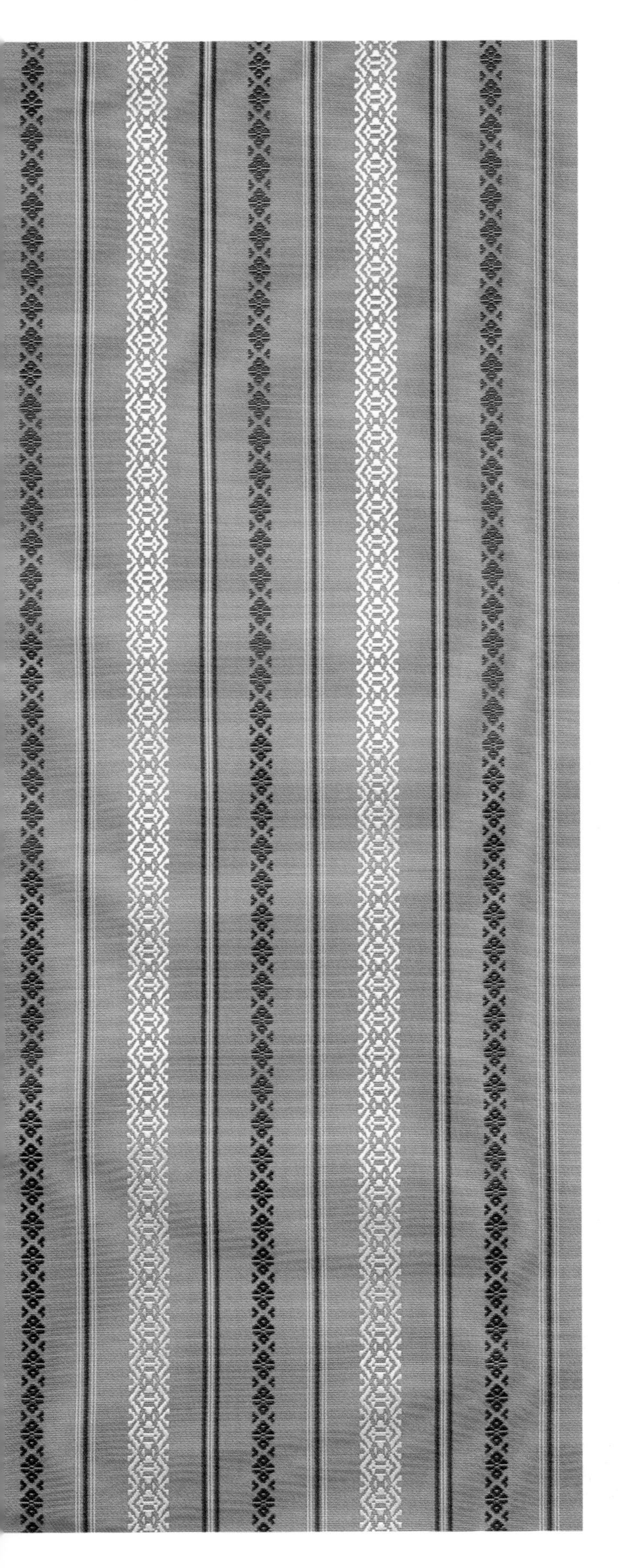

HAKATA WEAVE • *Hakata Ori*

Fukuoka, Fukuoka Prefecture

As a bouquet is to a bride, so the *obi* is to a kimono. The *obi* should complement and set off the color and pattern of the main garment; without an *obi*—it was traditionally said—the kimono is nothing. In this sense, Hakata *ori*, which is mainly an *obi* fabric, adds the finishing touch to kimono weaves.

The generally accepted theory of the fabric's origin is that one Mitsuda Yazaemon went to Sung China in 1235 to learn the techniques of weaving, and on his return initiated what has since become the craft's long tradition in his family. Later, descendants and other weavers to whom they passed on the trade improved and developed the techniques. In 1599, the head of the local clan began presenting some of this sturdy cloth to the shogunate every year; consequently it became known as *Kenjō* (Tribute) Hakata, and its fame spread. The technology of the Hakata weave greatly influenced other fabrics made in Nishijin, Kiryū, Ashikaga, Kōshū, Yonezawa, and Edo (present-day Tokyo).

Kenjō is a *tateune*, or ribbed weave. The density of the warp is extremely high; around 4,800 threads are used at the lower end of the scale, but most have a thread count of between seven to eight thousand, and in some cases more than ten thousand threads are contained in the approximately 30-centimeter width of the cloth. There are more than eight threads per *hane* or reed dent, and in less than 4 centimeters there are 72 *hane*. The yarn—a yarn-dyed glossed silk—is sometimes combined with raw silk, spun silk, or gold and silver thread, as well as a lacquered thread called *urushi ito*. The latter is made of cotton yarn wrapped with narrow strips of a Japanese paper called *torinome* and then coated with colored lacquer. The raised (floated) warp threads are the principal medium for expressing the designs. Stripes figure strongly, along with a motif called *dokko*, a metal object used in Buddhist rituals. *Hanazara*, another design derived from a Buddhist object, is expressed as a striped motif.

This stiff, lustrous cloth is nowadays made up into ties and small bags in addition to *obi*. The *obi* made of it do not slip once tied, but it is somehow indicative of the times that, although the techniques of the Hakata weave have managed to survive, the number of women who can actually tie such an *obi* on their own decreases year by year.

TRUE ŌSHIMA PONGEE
Honba Ōshima Tsumugi

Naze and Kagoshima, Kagoshima Prefecture

The Amami Islands, located between Kyūshū and Okinawa, enjoy a subtropical climate particularly suited to the cultivation of mulberry bushes. This, together with the warm, moist conditions, means that three or four batches of silkworms can be raised in one year. Production of what is known as *Honba*—"true"—*ōshima tsumugi* did not really get started until the beginning of the seventeenth century, however, and then only with the help and protection of the Shimazu clan, which collected the cloth as tax payment. Although originally produced solely on Amami-ōshima, the largest island of the group, by the Meiji period it had become more widely distributed. At the beginning of the twentieth century, the invention of the *shimebata*, a loom used for weaving the *kasuri* pattern into the warp and weft yarns before dyeing, made the production of the cloth feasible in other locations, principally in Kagoshima itself. Around the same time, raw *tamamayu* silk began to replace the original hand-spun silk yarn. In 1922, combined production in these two locations was over 6.6 million meters, and in 1977 it reached 8.7 million meters. With the trend away from the kimono as an everyday garment, however, it has since declined.

True Ōshima *tsumugi* is a plain-weave yarn-dyed silk *kasuri* cloth, with the *kasuri* design either in both warp and weft, or in the weft alone. Once the *kasuri* yarn has been woven on an *orijime* loom, the material is dyed twenty times. The resulting red-brown is achieved by using a dyestuff obtained from *sharinbai* (*Rhaphiolepis umbellata*), a member of the rose family, then mordanting it with mud, a procedure that deepens the color and softens the cloth. The *kasuri* yarn is unraveled and woven up again, this time into the most delicate of patterns, individual threads being adjusted at regular intervals during the weaving to make the pattern match properly. The precision of the patterns, the natural coloring, and the snappy feel of this light but warm cloth make it greatly admired for everyday kimono.

MIYAKO RAMIE • *Miyako Jōfu*

Hirara and Miyako Island, Okinawa Prefecture

Tradition attributes the origin of Miyako ramie (*jōfu*), to sometime during the second half of the sixteenth century, when Miyako Island was part of the nominally independent kingdom of Okinawa, or the Ryūkyūs. A high-quality cloth, it was at one time used as a means of paying taxes.

Jōfu in fact means "superior cloth," and the term basically indicates lightweight, finely woven cloth hand woven from a fine, hand-twisted ramie yarn. Compared to hemp, ramie produces a thinner filament and can be woven much more finely, and it was accordingly favored by the upper classes for summer kimono.

In its traditional form, Miyako *jōfu* is a plain-woven *kasuri* cloth of yarn-dyed ramie. The warp is twisted eight times on a spinning wheel and the weft seven times. The *kasuri* yarn is prepared either by binding it up by hand to create the resist, or by weaving it tightly (*orijime*) and repeatedly dyeing it with indigo before unraveling it. Either Ryūkyū indigo (a type of *indigofera*) or the indigo found on the Japanese mainland is used. The actual setting up of the warp yarns is unconventional. The eyes of a heddle called *musō sōkō* are arranged at different heights, a pair of heddles making a set. One warp thread is passed through above the eye of a low heddle, then the next warp is passed through below the eye of a high heddle. This setup provides extra play to the warp yarn and produces a compact weave. The warp is then passed through the reed, and weaving can begin. The weft is hand beaten while the shuttle is passed back and forth in the traditional manner by hand. The woven cloth is washed in warm water to remove the starch applied to the yarns and then, after being dried, is finished by being laid out on a wooden block and beaten with a mallet.

Used almost exclusively for kimono, the resulting cloth is distinguished by its fine *kasuri* patterns and has the quiet luster befitting a top-quality, expensive cloth.

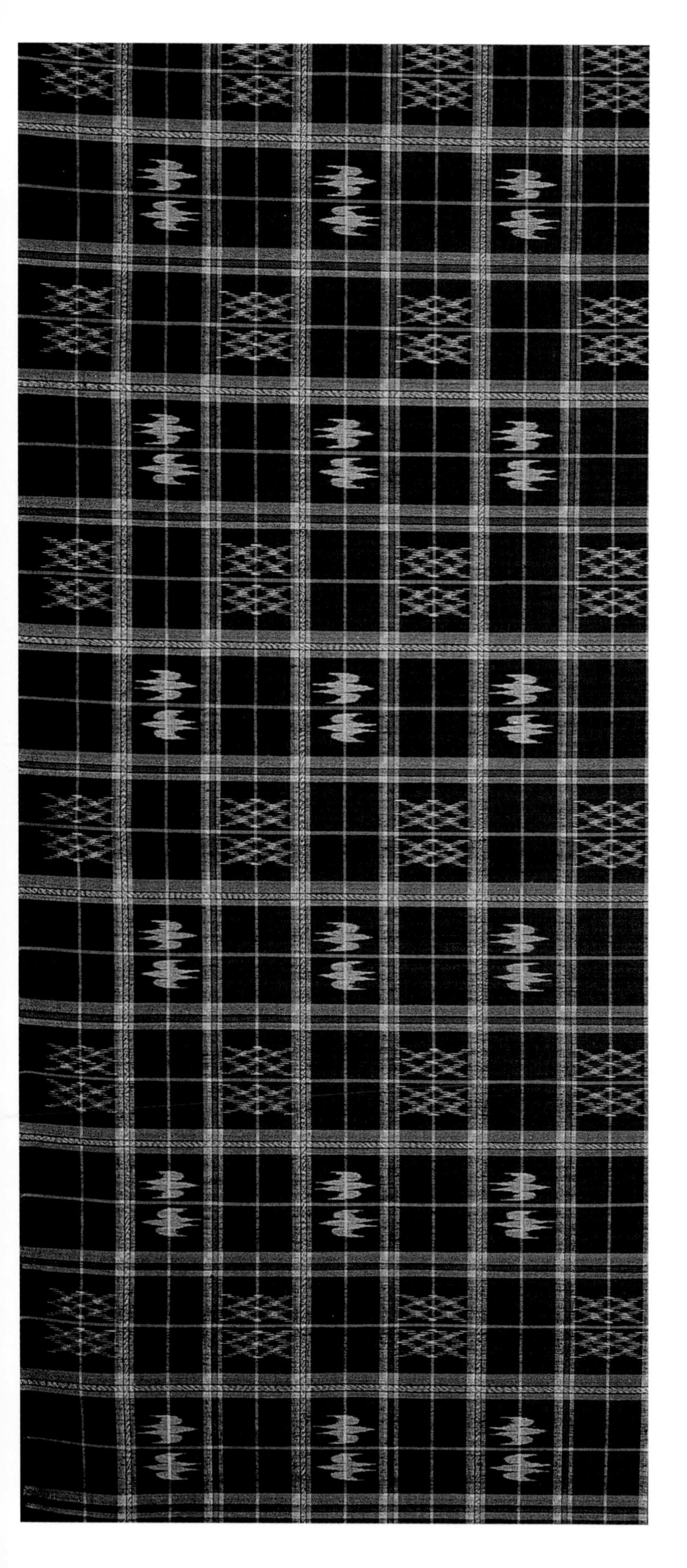

SHURI FABRICS • *Shuri Ori*

Naha, Okinawa Prefecture

Ikat techniques that originated in India around the eighth century are thought to have reached the kingdom of Ryūkyū sometime in the fourteenth century through trading links with Southeast Asia. Taken up and developed further on the islands, this complex way of producing patterned cloth gave rise over time to what might be called an individual *kasuri* culture. *Kasuri* cloth was particularly favored by the Ryūkyūan court, and the system of delivering cloth as a form of taxation that was established in the seventeenth century marked the real beginning of *kasuri* in Japan.

Apart from *kasuri*, other figured cloths with distinctive techniques that were first introduced from China and Southeast Asia include *Hana ori*; *Dōton ori*, which is basically a Chinese cloth woven in a locally developed manner; *Hanakura ori*; and *Minsaa*. The yarns used include raw silk, *tamamayu*, hand-spun *tsumugi*, cotton, ramie, and a yarn obtained from the Japanese banana plant (*Musa basjoo*). The dyestuffs are Ryūkyū indigo, *fukugi* (*Garcinia subelliptica*), smilax (*Smilax china*), and *sharinbai* (rose family; *Rhaphiolepis umbellata*). The generic name Shuri *ori* signifies the plain-woven yarn-dyed *kasuri* cloths, the *kasuri* yarn being either woven up on an *orijime* loom, bound by hand, or worked from memory. *Hana ori* and *Dōton ori* are yarn-dyed plain-woven figured cloths. The design of the former is inserted with an embroidery shuttle, whereas the design of the latter is expressed by the use of more than four heddles. *Hanakura ori* is a yarn-dyed, figured plain weave, so light and fine that its use was at one time restricted to the court. *Minsaa* is a yarn-dyed ribbed and figured cloth, produced for use in *obi*.

The designs look abstract at first sight, but are in fact derived from natural objects and even household articles. Textures and colorings are equally distinctive, yellow having been particularly favored at the Ryūkyūan court.

KIJOKA ABACA • *Kijoka no Bashōfu*

Village of Ōgimi,* Kunigami-gun, Okinawa Prefecture

An abaca cloth, *bashōfu* has been worn by the people of this chain of subtropical islands for many hundreds of years. It was certainly in existence before cotton was introduced to the islands, and its origins appear to date back to at least the thirteenth century when there were banana plantations under the direct management of the Ryūkyū court. Despite such royal patronage it knew no social barriers, and was indispensable for clothing in the long, hot summer months. In the past, people not only wove *bashōfu* for their own clothing but also planted the banana trees and prepared the yarn, thereby fulfilling the roles of both producer and consumer.

Fiber from the plant is first pared away in a process called *obiki*. It is then hand-twisted into a usable yarn. The cloth itself is yarn-dyed and either plain woven or figured, the *kasuri* threads being bound by hand. Normally only two colors are used: a brown obtained from *sharinbai* (rose family; *Rhaphiolepis umbellata*), and an indigo blue produced with Ryūkyū indigo. The weft is soaked in water before it is woven, and the woven cloth is mordanted with wood ash, exposed to the sun, and softened by being immersed in *yunaji*, a fermented mixture of soft rice, rice flour, and water. The final finish is given by rubbing the cloth with a teacup or rice bowl and pulling it both diagonally and across its width.

The *kasuri* designs take the form of wavy lines, vaguely described squares spaced regularly across the cloth, or steady grids in dark brown on a cream-tinted ground. The cloth has a cool look to it and feels light and comfortable against the skin. In use, it takes on a refined, dull luster, regardless of whether it is made into kimono, cushion covers, the short doorway curtains called *noren*, or tablecloths.

Kijoka is the birthplace of *bashōfu*, and although the fabric completely disappeared for a time before World War II, it has made a recovery and today holds a position among the best of the traditionally crafted cloths of Okinawa.

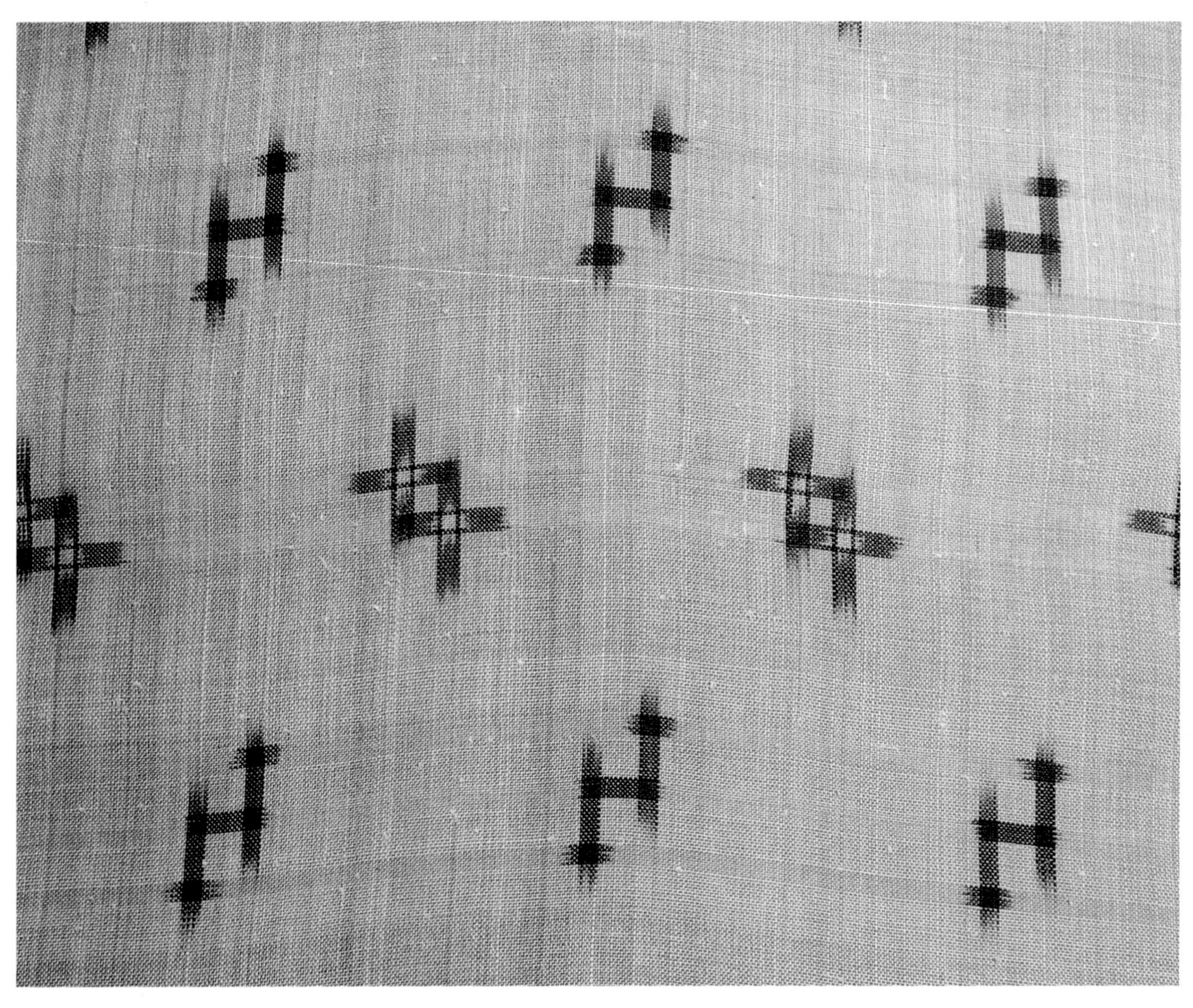

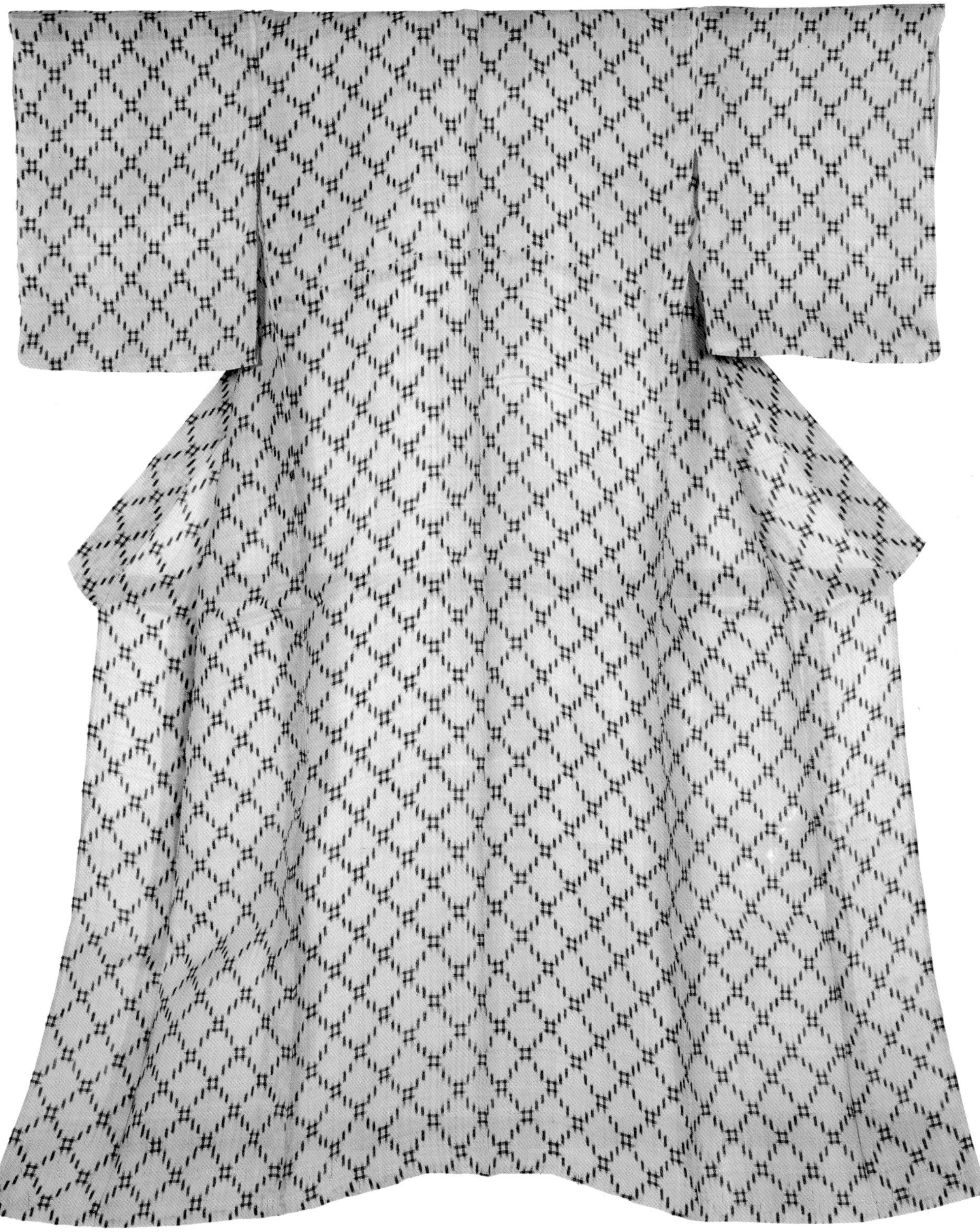

YONTANZA MINSAA • *Yomitanzan Minsaa*

Village of Yontan,* Nakagami-gun, Okinawa Prefecture

It was once the custom for a girl to give one of the narrow *obi* known as a *minsaa* to the man of her choice as a sign of her love, especially before he set off on a journey. Used exclusively by men, these *obi* were first woven around the same time that a figured *Hana ori* cloth, based on a type of weaving introduced from Southeast Asia, was being woven in the village of Yontan. Production of the cloth ceased for a time, but was started up again with the assistance of the village authorities in 1964.

Yontanza *minsaa* is a figured cloth with geometric flower designs in color on dark grounds, and is yarn-dyed and ribbed; *kasuri* designs are also used on occasion. The dyestuffs are obtained from local flora such as Ryūkyū indigo, *sharinbai* (rose family; *Rhaphiolepis umbellata*), *fukugi* (*Garcinia subelliptica*), and myrica (*Myrica rubra*). In some cases *kasuri* threads run through the warp, while in others the cloth is a simple striped one. Other designs are of abstracted flowers created with heddles or, more simply, by lifting the warp yarns by hand in order to pass the weft under them.

The *minsaa*'s continued existence today is partly due to the fact that its techniques have been successfully handed down from an older generation, that may still have believed in its value as a talisman, to a younger generation able to recognize its value and beauty as a craft worth preserving and sustaining.

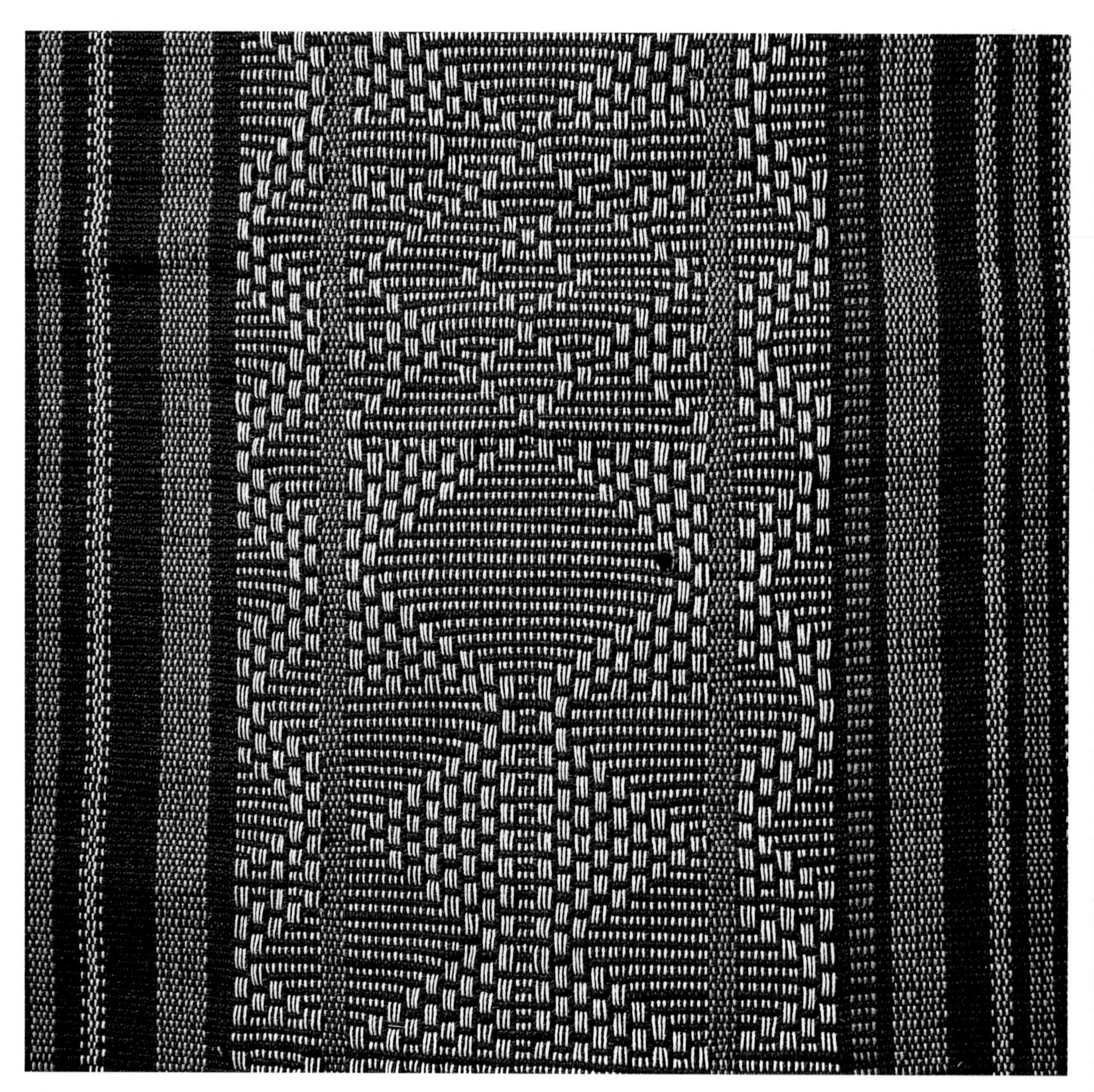

ISESAKI IKAT
Isesaki Gasuri

Isesaki, Gunma Prefecture

The silk cloth woven in the environs of Isesaki first gained notice between 1521 and 1527, when a market opened in the city for the sale of raw silk and cloth. In the late seventeenth and early eighteenth centuries, the cloth was further promoted when the city itself became an important distribution center for silk and silk goods. At that time there were two main types of cloth: a striped fabric called Isesaki *shima*, and Isesaki *futo-ori*, a sturdy fabric for working clothes originally woven by farmers during the winter using leftover silk floss and *tamamayu* yarn. By the Meiji period, however, a plain-weave silk *kasuri* called Isesaki *meisen* had appeared and rivaled in quality similar cloths from Kiryū, Ashikaga, and Hachiōji. Nowadays, *meisen* refers to the cloths from Isesaki, almost all of which are large-patterned *chin gasuri*, or *heiyō gasuri* (a warp and weft *kasuri*).

The *kasuri* is made by one of three methods: hand-binding, *itajime*, or stenciling. In the latter method, a resist may be applied through the paper stencil or the dye may be brushed through the stencil, a technique known as *katagami nassen*. For *heiyō gasuri* (warp and weft *kasuri*) and *yokosō gasuri* (weft *kasuri*), the warp is either raw silk or *tamamayu*, and the weft is raw silk, *tamamayu*, or *tsumugi* yarn.

Utilizing a number of different motifs in its patterns, Isesaki *gasuri* has a real country feeling and is ideal for casual wear.

TSUGARU KOGIN STITCHING
Tsugaru Kogin

Hirosaki,* Aomori Prefecture

Situated at the northwestern tip of Honshū, Aomori Prefecture experiences some of the coldest winters in Japan. Despite the need for warm clothing, work clothes during the feudal period continued to be made of hemp and other native bast fibers that provide little comfort. Cotton cloth was not employed even after it had become widespread in other areas because local law virtually banned its use. The problem was partially overcome by stitching a thick white cotton thread along the weft of the hemp or similar fabric, thus making the cloth thicker and softer to the touch. Gradually, counted-stitch embroidery weft patterns developed. Because the embroidery yarn follows the weft yarn precisely over and under the warp yarns, it resembles a woven pattern.

Precisely when this technique began is unclear, but there are records of patterns that date from the Tenmei era (1781–89). As the number of patterns increased, the cloth also became a required costume for festive wear as well as part of a bride's trousseau among the farming villages north of Hirosaki. In the early part of the twentieth century *kogin* declined, but the crafts movement brought about a revival in the 1950s. Although colored yarns at one time had a run of popularity, the original white yarn on an indigo ground is now more common, and instead of hemp, a coarse wool or linen cloth is generally used. *Kogin* is now applied to wall hangings, table runners, handbags, business-card holders, and other items of daily use, as well as to *obi*, jackets, ties, and shawls. There are six or so basic patterns from which numerous variations can be made.

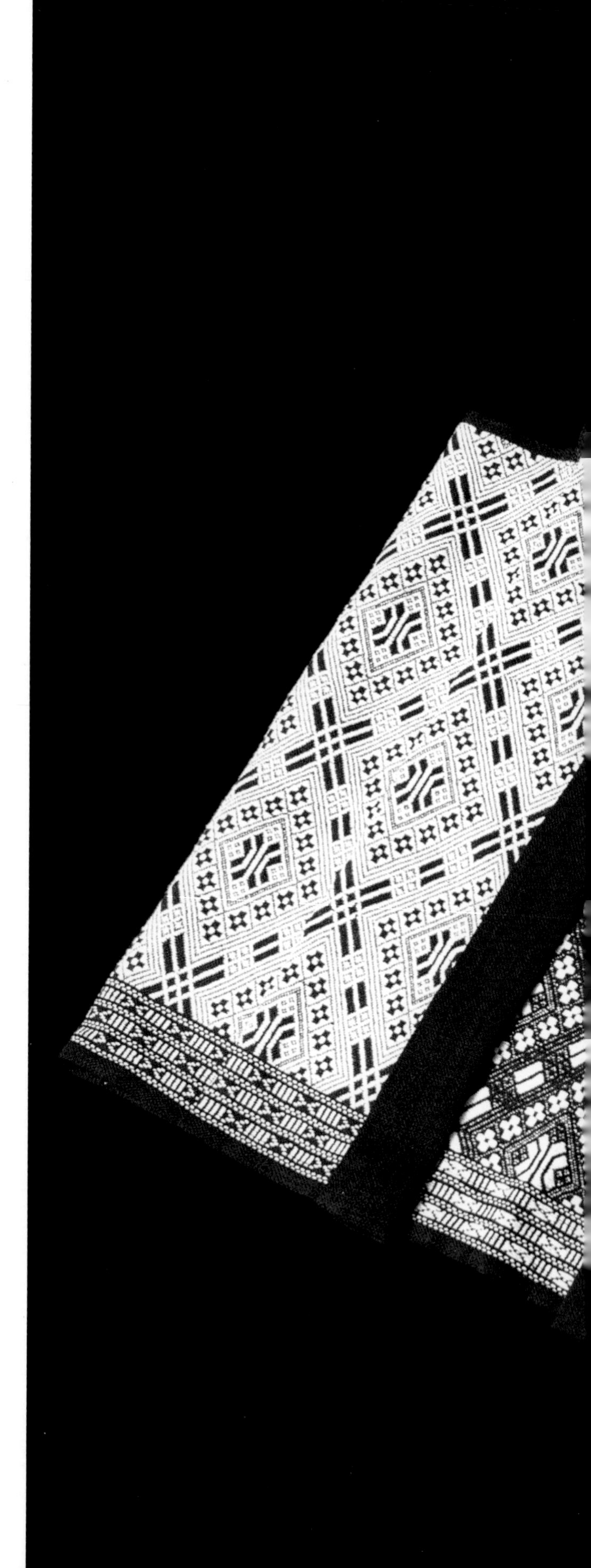

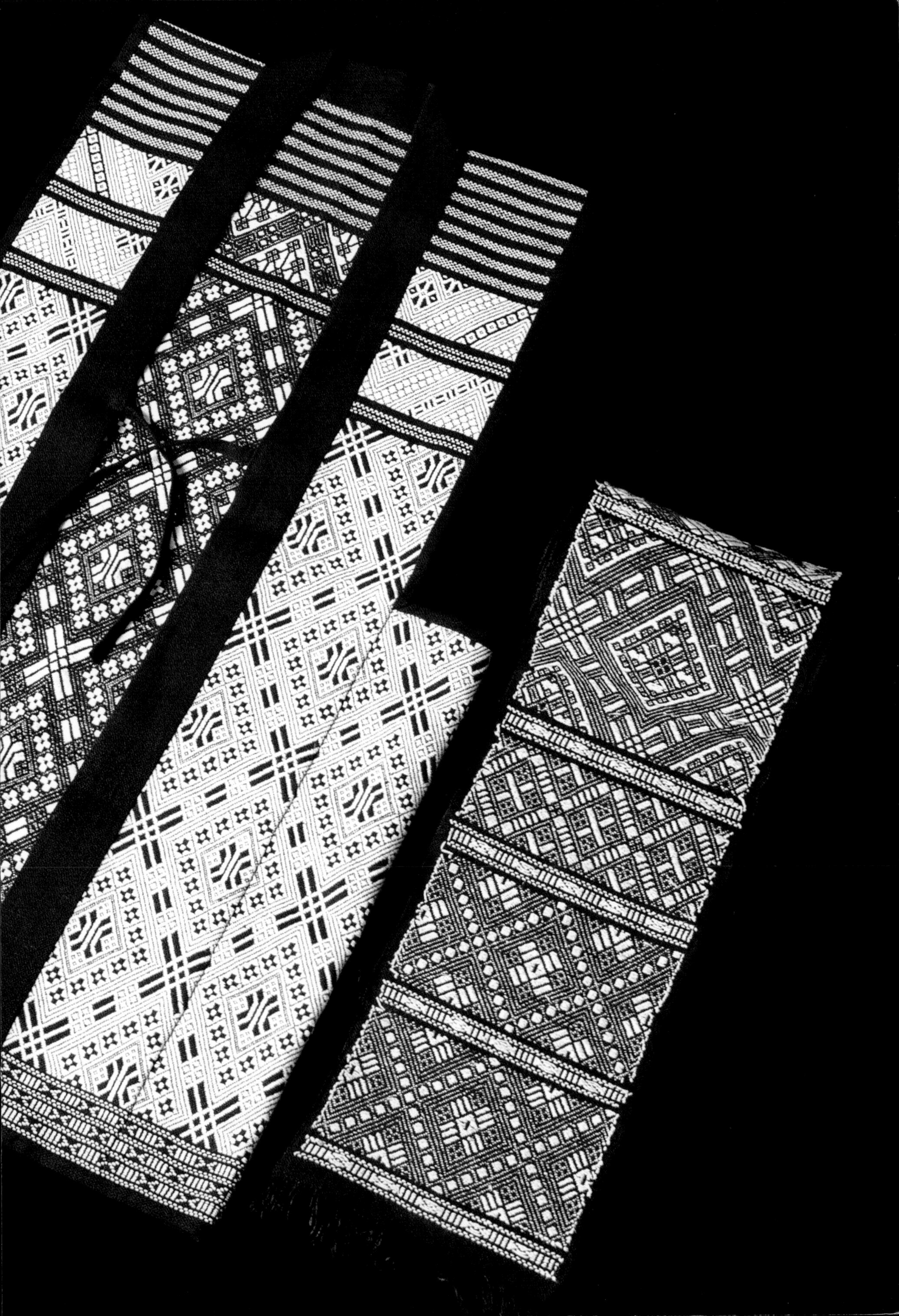

TOKYO STENCIL DYEING
Tokyo Somekomon

Tokyo

This type of stencil dyeing, which is also known as Edo *komon* ("Edo" was an early name for Tokyo), developed for use by samurai men during the Edo period. Originally, the small repetitive patterns (*komon*) were employed primarily on *kamishimo*, a samurai's ceremonial outfit consisting of culotte-type trousers worn over kimono and a separate vestlike top with exaggerated shoulders. Around the beginning of the seventeenth century the patterns were often of medium size, but by midway through the following century clans were competing to produce ever finer and more delicate patterns, and the dyers and stencil cutters were also drawn into this spirit of competition. From this were born patterns of extraordinary minuteness, often produced by multitudes of minuscule dots, that looked from a distance like a texture rather than a pattern.

With the ascendancy of the merchant class during the second half of the Edo period, this exquisite monochromatic cloth, previously worn exclusively by the warrior class, became fashionable among the townspeople, men and women alike. Since the end of the nineteenth century, *komon* has been regarded as more suitable for women. Because it refrains from ostentatious display, it is the favored fabric for the tea ceremony and other semiformal occasions that require a dignified but quiet presence.

Today the patterns are dyed on silk and applied with paper stencils that sometimes are so delicate that they must be reinforced with silk threads or a gauze backing. In *shigoki zome* the dye is mixed with a rice paste resist that is spread through the stencil; in *hiki zome* the dye is brushed directly onto the fabric through the stencil. Steaming sets the dye, and afterward the cloth is washed to remove the resist.

KAGA YŪZEN DYEING • *Kaga Yūzen*

Kanazawa, Ishikawa Prefecture

Situated close to the Japan Sea and backed by high mountains, Kanazawa, which was the seat of the Kaga clan, has long had a rich local culture. With the encouragement of its administrators, the province of Kaga toward the end of the seventeenth century became a center of art, industry, and learning rivaling Kyoto, where the *yūzen* process of dyeing seems to have originated.

The origins of today's Kaga *yūzen* date back to the Edo period. The dyeing techniques of Kaga, which had been perfected before the beginning of the eighteenth century, seem to indicate that *yūzen* dyeing was introduced by Miyazaki Yūzensai, who had moved to Kanazawa from Kyoto, with valuable assistance being given by Tarodaya, the official dyer to the local clan. Later on, stenciled *yūzen* was employed to dye the fine overall repeat patterns for *kamishimo* and *haori*, although the freehand *yūzen* is perhaps what characterizes Kaga *yūzen* most.

Nowadays, Kaga *yūzen* dyed cloth is used for various types of patterned goods including *furoshiki* (wrapping cloths), *noren* (a short, split curtain for hanging over a doorway), and *futon* (quilt) covers, as well as for *obi* and the more formal or dressy types of kimono, for which the cloth is usually silk. The designs are mostly of flowers and plants depicted realistically (to the extent, even, of showing the occasional insect-nibbled leaf). There is some shading of the color from the edges to the center of shapes, but in general, the designs are very colorful, with indigo, carmine, yellow ocher, grass green, and a warm grayish-purple figuring prominently on a cloth that is a remarkable tribute to the dyer's art.

KYŌTO YŪZEN DYEING
Kyō Yūzen
Kyoto

Perhaps more than anyone else it was Miyazaki Yūzensai who contributed to the spectacular advances in the art of dyeing in Japan. In 1687 this fan painter turned his attention to developing contemporary dyeing techniques into a method for producing freehand designs. Not only was his style distinctive, he also made revolutionary improvements in dyes and methods of dyeing, as well as in the coloring and complexity of designs.

In freehand (*tegaki*) *yūzen,* the design is first traced with *aobana,* a fugitive blue made from the dayflower. Afterward a fine line of rice-paste resist is applied on the traced design in order to prevent dyes from running. Next, dye is rubbed into the fabric within the confines of the resist lines with stubby brushes, then the dyed areas are covered with resist (a step known as *fusenori oki*), and the background color is brushed over the entire fabric. When the resist is washed away, a thin white outline remains around the designs. This process, which is called *sashifuse yūzen,* has two variations. One is *sekidashi yūzen.* In this case, rubber resist is used, and around it another resist paste is applied in such a way that no white line remains after the resist has been removed. In the other variation, only the parts of the kimono that will carry a family crest are treated with resist, and the design itself is painted directly onto the cloth. This is called *musen yūzen.*

Kata yūzen (stenciled), which was first made in 1881, calls for the dye to be brushed through paper stencils. Numerous stencils are employed, each for a specific color. In 1892 it was found that the cloth could also be dyed by combining the dyestuff with rice-paste resist. This new technique was called *hikiotoshi yūzen.* Before the dye is set by steaming, the cloth is stretched and pulled diagonally in order to crack off the resist.

Kyoto *yūzen* designs generally feature birds, flowers, landscapes, and court scenes that are portrayed in a stylized manner. The colors tend to be brighter than those used for Kaga *yūzen,* and the fabric is often embellished with gold and silver leaf and embroidery, making it the aristocrat of dyeing and design.

KYOTO STENCIL DYEING • *Kyō Komon*

Kyoto

This type of patterned cloth, like the Tokyo *komon* stencil dyeing, takes natural and geometric motifs and reduces them to the point where they assume a textual appearance. The word *komon* literally means "small crest," and the rich yet restrained character of this type of pattern was always favored by the warrior classes.

Although its origins are unclear, such examples of Kyoto *komon* stencil dyeing as the armor of Tokugawa Ieyasu, the first Tokugawa shogun, and the thin ramie kimono coat of Uesugi Kenshin, a warlord of the late sixteenth century, prove that the technique was already fairly well established at the end of that century.

In the early seventeenth century, *komon* was used for *kamishimo*, the ceremonial garment that was a symbol of the warrior class, each important family having its own exclusive design. The Tokugawa family, for instance, used *matsuba komon*, an elegant design based on pine needles, and another, *jūji komon*, made up of small "plus signs." The great family of Shimazu used a design known as *daishō arare komon* that utilized the family crest in two sizes to make a random overall dotted design. The *gokushō same* (sharkskin) *komon* used by the Kishū family is perhaps the finest and most characteristic of all *komon* designs.

This overall textural type of design soon found followers among other sections of society in the Edo period. At the end of the nineteenth century, the time-saving technique of *utsushinori zome* was developed to dye these fine patterned cloths, which by then were being used, as today, for kimono. Instead of dyeing the cloth after it had been stenciled with rice-paste resist, the dye was mixed directly into the paste. Nowadays both techniques are emphycal. The fabric is white silk. After the dye has been fixed by steaming and the cloth thoroughly washed, the positive white part of the design is then dyed faintly in the same color as the background so as to give the cloth a more subdued appearance. If, however, color is to be used in the resist-covered areas of the design, the cloth is first piece-dyed

Differing technically very little from the *komon* stencil dyeing found in Tokyo, Kyoto *komon* stenciling is characterized by its brilliant coloring.

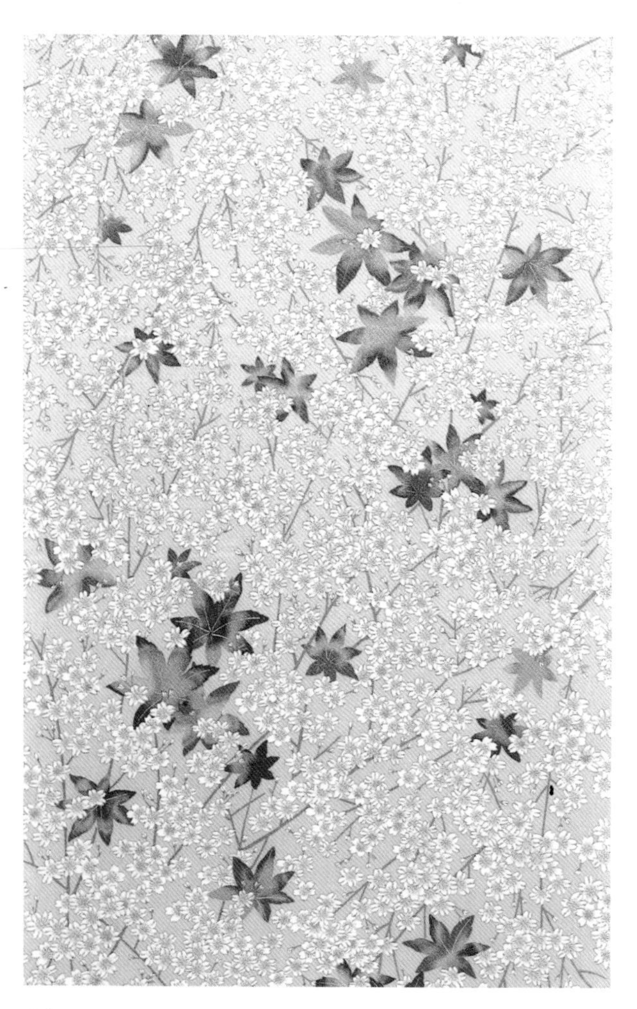

KYOTO TIE-DYEING • *Kyō Kanoko Shibori*

Kyoto

The well-known manual dexterity of the Japanese, coupled with an almost saintly patience and an ability in the abstraction of images, is surely what makes the elaborate tie-dyeing techniques of this island nation so special.

The origins of tie-dyeing in Japan, as in the rest of the world, are ancient. It would seem that shaped resist dyeing was already widespread by the six and seventh centuries. Tie-dyeing found its way into the Heian period in various forms, one of which, *mokkō kōkechi,* was the one most like the *kanoko* (fawn spot) tie-dyeing in use today.

The *kanoko* tie-dyeing associated with Kyoto earned numerous mentions in literature during the early part of the seventeenth century, and its praises were even extolled in a song around the same time. From the late 1680s on into the early part of the next century (Genroku and Kyōhō eras), when it was very fashionable to wear a tie-dyed kimono, the work of making this type of cloth had become highly organized, each process being carried out by different artisans, and the whole business of production and wholesaling was strictly controlled.

Numerous variations have been developed over the years, and they are still being added to. *Hitta shibori* most closely resembles the basic form of fawn spot tie-dyeing. A small portion of cloth is pinched and bound tightly with a silk thread—a relatively simple process, yet even someone highly skilled at the technique will still need six months to complete enough cloth to make one kimono. The minute dots of the technique known as *hitome shibori* have a diagonal slant, whereas other methods, tying up larger amounts of fabric, produce ringed effects.

Besides *kanoko* tie-dyeing, there are also techniques requiring folding and sewing. The result is a ribbed effect, or one resembling netting, depending on the way the cloth is tied. Sometimes techniques are combined: banded areas to be left undyed are sandwiched between boards while other areas are minutely tied. In the technique known as *oke shibori,* even larger areas can be kept out of the dye by putting them into a bucket and strapping a lid on it before the whole is immersed in the dye vat. But common to all to a greater or lesser degree is the blurred effect characteristic of tie-dyeing.

TOKYO YUKATA STENCIL DYEING • *Tokyo Honzome Yukata*

Tokyo

While many people know what a kimono is, very few people outside Japan have actually worn one. The *yukata*, on the other hand, has gained a large following during the last twenty years, especially in America and Europe. A simple cotton kimono that requires minimal care and can be worn as a bathrobe or dressing gown, its origins can be traced to a robe known as *yukatabira* that was worn in the bath. During the Heian period, when nakedness was frowned upon even in the bathroom, well-mannered people customarily wore an unlined linen garment known as *hitoe*. Later, the *yukata* became a garment to put on after taking a bath; in the Edo period, it became fashionable to wear cotton *yukata* dyed with bold designs after taking a bath at the *sentō*, or public bath, which had become a social center for the townsfolk. By the end of the nineteenth century *yukata* were being worn as a summer garment, becoming an essential part of the summer scene at festivals, firework displays, and other events during the hot, humid months.

The craft of Tokyo *yukata* stencil dyeing became established around the middle of the eighteenth century, using Edo *komon* stencil techniques on cotton. By the early part of the twentieth century such yukata were universally popular, and in the 1920s mass production started, using the *chūsen* technique of dyeing first developed at the end of the nineteenth century and still in use today. With this technique, the dye is poured through the cloth; first, plain white cotton is stretched out and lightly pasted on a long board and the design in resist is applied, then the back is treated the same way so that the design matches precisely on front and back. If the design is complex, separate stencils are used for the front and the back. The dye is then poured on by hand and soaked up from behind. Folding the cloth at each repeat and vacuuming the dye through the cloth is a modern labor-saving method.

Small hand towels called *tenugui* are also produced by this dyeing process, but the fresh-looking designs and the feel of the cotton are particularly suited to the *yukata*, which by now has—somewhat to the amusement of the Japanese themselves—become a kind of ambassador of Japan abroad.

KYOTO EMBROIDERY • *Kyō Nui*

Uji and Kyoto, Kyoto Prefecture

When Buddhism was introduced into Japan, the crafts and architecture associated with it came too. Carpenters, metal workers, and even artisans skilled in the techniques of embroidery were brought in to train the local people in skills developed on the continent.

Embroidery (*shishū*) in Japan really started at that time, a fact that is recorded in the chronicle of early times, the *Nihon Shoki*. It was soon being used for the ceremonial robes of the nobility, and in 603, some one hundred years after the arrival of Buddhism, Prince Shōtoku wore embroidered robes at a court ceremony. The oldest existing piece of ancient embroidery, executed in 622, is the *Tenjukoku Mandara*, or "Heavenly Paradise Mandala," at the temple of Chuguji in Nara.

In 794, when the capital was moved to Kyoto, an official embroidery department called the *nuibe no tsukasa* was established at the court of the emperor. By the Kamakura period embroidery was being employed for sword-hilt covers, sashes, and other decorative elements of the warrior's wardrobe. New stitches were added, and the more elaborate techniques become popular among the people at large. During the Muromachi period, *nuihaku*, embroidery combined with gold and silver leaf, produced Noh costumes of outstanding brilliance. Embroidery was also combined with tie-dying in the *tsujigahana* robes of the sixteenth century. During the early part of the Edo period, embroidery was applied to tie-dyed *kanoko* cloths to make richly embellished kimono. Since the end of the nineteenth century, embroidery has continued to be used on kimono and *obi* as well as on religious articles, and has also been applied, in a more pictorial manner, to wall hangings and screens.

The types of stitches correspond more or less to those used in Western needlework; those more commonly used in Kyō *nui* include *nuikiri*, outline satin stitch; *matsui nui*, backstitch outline; *sagara nui*, French knotting; *sashi nui*, long and short stitching; and *komatsui*, couching. The commonest threads are yarn-dyed silk, a thread coated with *urushi* (lacquer), gold and silver threads, and also thread that has been coated with gold or silver leaf. After the design has been transferred onto the cloth, the cloth is spread out over an embroidery stand or frame to facilitate easy working.

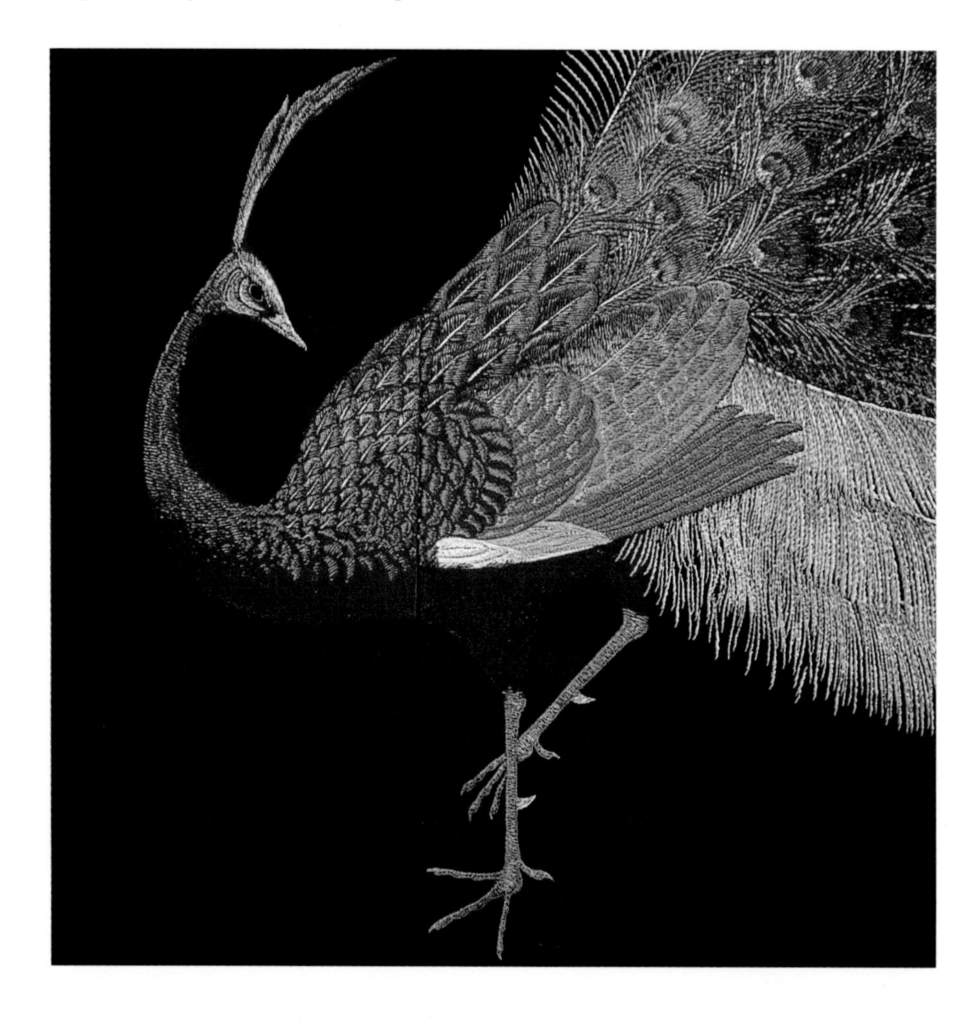

IGA BRAIDED CORDS • *Iga Kumihimo*

Ueno and Nabari, Mie Prefecture

Decorative cords in Japan were originally used to tie sutra scrolls and on the robes of Buddhist monks. Later their primary use was for linking the components of a suit of armor and on sword hilts, but they also were an important part of court costumes as well as of various bags, such as those in which tea-ceremony caddies are kept.

Iga Ueno—known as the birthplace of the haiku poet, Matsuo Bashō and for its associations with ninja and intrigue—is also famous as one of the centers of braided cord making. Late in the sixteenth century, during the Momoyama period, specialist craftsmen were competing with each other technically in what seems to have been a golden age for braided cord. With the break-down of feudal society, there was a decline in the use of such cord, but fortunately the traditions of Iga *kumihimo* were carried on by the armor makers to the Tōdō clan.

The foundations of Iga *kumihimo* as it is known today were laid in 1892 when Hirosawa Tokusaburō established a business for producing decorative cord for tieing such things as *obi* and *haori* after having mastered the techniques in Tokyo. Although machine braiding has come into extensive use since then, Iga is particularly known for hand-braided cords and accounts for 80 percent of the country's total production, with Kyoto, Ōtsu, Tokyo and Sakura in Chiba Prefecture providing most of the remainder.

The techniques have now grown so sophisticated that almost any pattern can be produced. The essential equipment consists of a wooden stand called *marudai*, which looks like a small, round stool. The top surface is usually round with a hole in the middle. Pre-dyed raw silk, spun silk, and gold and silver thread are the usual materials. The thread is wound onto small, weighted spools that hang over the edge of the stand. The head of the braid is passed through the hole in the top plate and kept in place with a small counterweight. Braiding can now begin by crossing opposite threads or by passing one thread over or under another in a set sequence, both hands usually being used at once. There is also the *takadai*, which resembles a loom, with simple frames to the left and right of the operator over which the braiding spools are hung. The principle of braiding is basically the same as the *marudai*, except that the completed braid is reeled up on a bar located in front of the operator. A similar way of taking up the braid is used with the *ayatakedai*, but this has a series of long, projecting "teeth" directly in front of the person doing the work. These are used to separate the individual threads as they are being braided, and it produces a flat braid. The semi-mechanized *naikidai* was developed in the Edo period and caused something of a revolution in braid making at the time.

GLOSSARY OF TEXTILE TERMS

AOBANA: An extract from the spiderwort plant. The flower from which this fugitive liquid is made only lasts a day, hence its other name, dayflower. It is used to draw out designs on fabric.

BENIBANA: (*Carthamus tinctoric*) Safflower; produces a pink or yellow dye.

CHIJIMI: Crepe produced by the use of highly twisted wefts and/or warps in any fiber. The twist is secured by starching the yarn. When the woven cloth is washed and rubbed in hot water, the yarns pull together, producing a crinkled texture.

CHIRIMEN: Silk crepe produced by alternating weft yarns having opposite twist directions; favored for *yūzen* and stencil dyeing.

HAORI: The jacket worn over a kimono.

ITAJIME: As a dyeing process, it is used to create *kasuri* threads or patterns on cloth by tightly sandwiching lengths of yarn or cloth between wooden boards that have been appropriately gouged out. The flat parts of the boards which are retained act as a resist when the dyeing takes place. This method of dyeing is mostly used for geometric designs that repeat throughout the cloth.

KAMISHIMO: A man's formal attire consisting of culotte-type trousers and a separate vestlike top with exaggerated shoulders.

KANOKO SHIBORI: A method of tie-dyeing in miniature. Small bits of the cloth are pinched and then a thread is wound around them to create the resist. This results in what looks like an overall texture or an image expressed by dots resembling the dappling on the back of a fawn. Hence the name *kanoko* (literally, "fawn spot tie-dyeing"). Apart from this basic technique, there are many other ways in which the cloth can be sewn in order to create a variety of effects.

KARIYASU, also known as *kobunagusa*: (*Arthraxon hispidus*) A tall grass producing a vivid yellow dye.

KATAZOME: A form of stencil dyeing. Employed to a greater or lesser degree in the production of Tokyo *somekomon*, Tokyo *honzome yukata*, Kyō *komon*, and in *yūzen* techniques; the durability of the stencil and the water solubility of the rice-paste resist are of particular importance. The paper for the stencil is a kind of traditional *washi* made from the *kōfōzo*, or paper mulberry (*Broussonetia kazinoki*). Several sheets are glued together before it is coated with persimmon juice and then subjected to smoking for several days. While the paper is strengthened by the persimmon tannin and the smoke, it is still possible to cut it with razor-sharp knives and specially designed tools, to leave clean edges of the bold and sometimes highly intricate designs. The rice-paste resist is made from a mixture of rice flour and rice bran that is steamed and then kneaded into the right consistency so that it can be spread evenly. The stencil is cut and, if the design is very delicate, reinforced with silk threads or silk gauze. Before the stencil can be used it is soaked in water to make it pliable so that it will adhere more closely to the fabric. The rice-paste resist can then be applied through the stencil with a wooden spatula. After the resist has dried, the cloth is sized with an extract of soybean liquid, or *gojiru*, and then the dye can be applied. After being left for several days for the dye to cure, the cloth is soaked in water to remove the resist.

OMESHI CREPE: A crepe of pre-dyed, highly twisted yarn in both warp and weft.

ORIJIME: A way of pattern-dyeing *kasuri* threads by weaving them tightly on a loom. This technique of creating a resist is also sometimes called *shimebata* or *kariori*. Reputedly developed by Nagae Iemon in 1907 on Amami-ōshima, the technique involves the use of a special handloom to weave *kasuri* pattern-yarn as the weft, with a cotton warp to "bind" it tightly before dyeing. The number and interval of the warp threads depend on the pattern that is to be applied to the *kasuri* yarn. The cloth must be tightly beaten with the beater to achieve the resist, so this work is usually done by a man. It is possible to produce very detailed designs using this technique.

TAMAMAYU: Two cocoons that have been spun together, dupion, producing a nubby yarn when reeled.

TSUJIGAHANA: A combination of tie-dyed designs and flowers hand-painted in ink popular in Kyoto during the sixteenth century.

TSUMUGI: A rough hand-spun silk. Cloth going under this name is woven from a hand-spun silk yarn produced from imperfect cocoons. Originally this was done mostly by farmers to produce working clothes for themselves and their families. The floss from the cocoons is first degummed by immersing the cocoons in hot water containing baking soda and sulfurous acid. The floss from five or six cocoons is then combined to make it easier to work with, and after rinsing it in water, the process is repeated. The small bags that result, known as *fukuro mawata,* are then dried before being soaked in a mixture of powdered sesame seed and water in order to make the thread easier to spin. The *fukuro mawata* are then attached to a *tsukushi,* a device made of bamboo and corn or cane stalks. Several filaments are drawn from the floss at the same time and twisted into a single thread, the spinner wetting the thread with saliva. A middle-aged woman is considered best for this job, as her saliva contains just the right balance of hormones. The resulting thread is strong, elastic, and glossy and gives the cloth from which it is woven a rustic, home-spun quality, similar to that of pongee.

YŪZEN: This type of textile dyeing can be divided into two main types. The one known as *tegaki yūzen* can be called freehand *yūzen,* as both the resist paste and the dyestuff are applied to the cloth by hand. First, the design is drawn onto the fabric with *aobana.* Next, the full length of cloth to be worked on is stretched out, and tensors called *shinshi* are inserted at regular intervals under the width of the piece. The *aobana* lines are then covered with a fine line of paste resist, using what looks like an icing bag called *tsutsu.* These are made of waterproofed paper with a metal tip and come in various sizes. The juice from the persimmon is used to render the paper waterproof and to strengthen it. Having clearly defined the areas of the design to be dyed, a thin, liquid resist called *gojiru* that is made from a soybean extract is spread over the first resist and the remainder of the cloth. When dry, this will aid the absorption of the dyes and prevent unwanted running. Water is first brushed over the area to be dyed and then the dyestuff is applied with a small, flat brush. The dye is then steam-fixed and the paste rinsed off. What remains is the fine white outline of the design. A resist paste is used again if other dyeing such as the background is to be done before the final steaming and washing. The cloth is then stretched on the bamboo *shinshi* to dry in an airy place. The other type of *yūzen* dyeing is called *kata yūzen.* This is a type of stencil resist dyeing in which the design is transferred onto the cloth using the dye directly or using a dye-infused paste.

Yumihama ikat.

SHIKKI

Lacquer Ware

Introduction

Oriental lacquer—the term evokes images of folding screens inlaid with semiprecious stones, or gold-decorated panels incorporated into baroque furniture, or those cunning little Japanese pillboxes (*inrō*) with a proliferation of intricate detail. In short, "Oriental lacquer" is usually equated in the West with decoration, particularly gold on a shining black ground. Although not strictly inaccurate, this popular image is an anachronism and does not do the lacquer craft justice.

The word "lacquer" as it is used today is itself misleading. Any glossy, hard paint can now be referred to as "lacquer"—even one made in a chemical plant and bought at the hardware store to spray-paint model airplanes or bicycle fenders. As such, common parlance creates an additional, serious stumbling block to the understanding and appreciation of Oriental lacquer.

True Oriental lacquer is an organic substance: it is tree sap that has been aged and refined. It is a paint, to be sure, but a paint with special properties. Used widely in Southeast and East Asia, and in some places in India, lacquer is derived from the sap of a variety of trees and shrubs, most of which belong to the same family. In Japan, the sap of the *urushi* tree (*Rhus verniciflua*) is used; lacquer as a substance is called *urushi*, and lacquer ware, *shikki* ("lacquer ware") or *nurimono* ("painted things").

Japan has yielded the oldest known examples of lacquer in the world—red-and-black-lacquered earthenware pots dated to around 4500 B.C., in the neolithic Jomon period (ca. 10,000–300 B.C.). These lacquered pots pose more questions than they answer. One question arises because the lacquer used on these pieces is already highly developed—it has been refined, and pigments have been added. This would argue that the origins of lacquer are to be found at a much earlier date still. Again, it has always been assumed that lacquer craft came to Japan from the Asian mainland. So far, no lacquer of comparable antiquity has been found outside Japan. The possibility of a Japanese origin of lacquer is therefore worthy of serious consideration. Furthermore, although there are scattered examples of lacquer applied to late-neolithic earthenware pots, there is no evidence of a continuity of lacquer craft into the protohistoric and historic periods. In short, there is still much to discover in this field.

The chemical processes and properties of *urushi* were not fully understood until relatively recently. This highly complex organic substance hardens (it does not dry) when applied thinly as a coating, and this hardening occurs best in conditions of high humidity and high temperature. *Urushi* contains urushiol (the same stuff as is found in poison oak and ivy), which is responsible for lacquer's material properties as well as giving any susceptible person who enters into contact with it a month or so of severe itching. (Lacquer craftsmen are generally immune to these effects.)

Hardened lacquer forms a highly protective coating that repels water and prevents rotting in addition to resisting the action of acid and alkali, salt and alcohol. It is even a good insulator of heat and electricity. A dramatic revelation of *urushi*'s preservative qualities was the discovery in the 1970s of Han-dynasty tomb furniture in China, still in perfect condition after two thousand years of burial.

Lacquer has to be applied on something—a "core"—and more than one coating is necessary if the object is to last at all. The ways in which coatings of *urushi* are applied differ among the various local lacquer traditions in Japan.

Wood is the most common material for cores, but basketry, leather, and paper are also traditional, and plastic too has come into use, especially for drop-it-and-watch-it-bounce eatery ware. Ceramics and metal may also be coated with *urushi*. Wooden cores themselves are divided into three different types, depending on how they are made, and each method is a mature, fully formed craft in itself. The types are turnery (*hikimono* or *kurimono*; lathework), bentwood (*magemono*), and joinery or fine cabinetry (*sashimono*).

There are three types of lacquer coats: undercoats (*shitaji*), middle coats (*naka nuri*), and the final coat (*uwa nuri*). Some styles do not require a middle coat. An undercoat and middle coat may be single or multiple, but the final coat is always a single coating of the finest, most highly refined lacquer. After the final coating, the piece is decorated. Again, local lacquer traditions in Japan often have their own techniques of decoration. With the gold decoration known as *makie*, a final coat of high-gloss, transparent lacquer (*roiro*) is often applied last.

Turning the wooden base for a Yamanaka lacquer bowl.

WAJIMA LACQUER
Wajima Nuri

Wajima,* Ishikawa Prefecture

The famed durability of Wajima lacquer ware is the result of both the high quality of the lacquer refining process and the composition of undercoatings unique to Wajima. To summarize, a wooden core receives: 1) a primer and hole filler; 2) lacquer-saturated fabric strips applied to edges and seams for reinforcement; 3) the first undercoat (a mixture of burned earth, rice paste, and lacquer); 4) the second undercoat (finer burned earth, rice paste, and lacquer); 5) the third undercoat (the finest burned earth, paste, and a greater proportion of lacquer in the mixture); 6) the first middle coat; 7) the second middle coat; 8) the final (top) coat. Between all these coatings, the lacquer is allowed to harden, after which it is burnished and smoothed in various ways. (For example, the final burnishing before the top coat is applied is done only by elderly women because they have little or no oil in their hands.)

Albeit an oversimplification, this outline gives some idea as to why lacquer work remains a community craft. No single individual could possibly do all the stages of the process, from making core forms to finished surface decoration, and still earn a living.

The burned earth for the undercoatings unique to Wajima is diatomaceous earth dug from a hill in the town. This is reduction fired and sieved into three grades. This earth–rice paste–lacquer mixture is used nowhere else in Japan. Both *makie* and *chinkin* decoration are used in Wajima. *Chinkin* (see Aizu Lacquer), in particular, is traditional here and has been developed into very sophisticated decorative forms.

The origins of Wajima lacquer are still in dispute. Different authorities have also put forward conflicting theories of when lacquer first became an industry in Wajima. That folk lacquer has a long history in the Noto Peninsula is not in doubt, but it seems that the lacquer industry in Wajima is probably no more than about two centuries old.

AIZU LACQUER • *Aizu Nuri*

Kitagata and Aizu Wakamatsu, Fukushima Prefecture

The production of lacquer ware in the Aizu area of northern Japan got off to a solid start in the latter part of the sixteenth century. There is extensive documentation of these first flourishings. After 1599, for example, efforts were made to develop a system for cultivating lacquer trees, accompanied by efforts to improve lacquering techniques. In the eighteenth century, the use of colored lacquers and *makie* became widespread, and by the latter half of the nineteenth century Aizu lacquer ware enjoyed a veritable vogue. The year 1878 represented a peak in output. Thereafter the industry made efforts to reduce prices, but this had an unwelcome impact on quality. As the buying public lost confidence in the quality of Aizu ware, production in the area entered a long period of decline.

Despite such long-term scaling back of this once formidable industry, some 4,000 people are still engaged in the making of Aizu lacquer ware. The contemporary craft is well known for its colorful decorations. Typical products are small plates, bowls, round trays, and traditional tray-tables (*ozen*). Distinctive kinds of small boxes are made, as are the tiered boxes (*jūbako*) used for special foods during New Year's celebrations.

Two types of priming are employed. The first is known as *shibu shitaji*. Lamp black or powdered charcoal is mixed with persimmon tannin (the juice of unripe persimmons). Several layers of this mixture are applied as a primer, which is allowed to dry and then burnished. Finally, a coat of the persimmon tannin alone is applied, and after this dries the object is again burnished. The item can then be given its finishing layers of lacquer. In the other type of preparation, *sabi shitaji*, a claylike primer (*sabi*) is applied, allowed to harden, and then burnished. A lacquer undercoat follows the *sabi*, and after burnishing intermediate and final coats are applied.

Decorative techniques are particularly numerous. There are as many variations in techniques of applying colored lacquer as there are in *makie* techniques of metallic decoration (particularly those involving gold). Aizu has also developed the art of *chinkin*. This involves incising a design into a lacquered surface, applying a thin layer of lacquer into the incised lines, then applying either gold dust or gold foil over the tacky lacquer. As the lacquer dries, the gold is cemented into the incised design.

KISO LACQUER
Kiso Shikki

Village of Narakawa, Kiso-gun, Nagano Prefecture

Kiso *shikki* is an attractive and resilient lacquer ware that uses a clay undercoating found in Kiso County, in the mountains of southern Nagano Prefecture. A thriving industry based in the county produces a range of Kiso wares to this day.

Lacquer ware from Kiso was originally of two different types, Hirawasa *shikki* and Fukushima *shikki*. These date from the Keicho era (1596–1615) and Ōei era (1394–1428) respectively, and were of plainly finished *shunkei nuri* and *hana nuri* styles. At the end of the nineteenth century, however, craftsmen from Wajima introduced the use of clay undercoatings to Kiso. In place of the *jinoko* powder used in Wajima, Kiso lacquerers turned to *sabitsuchi*, a fine clay from the mountains near the village of Narakawa.

When combined with lacquer, *sabitsuchi* forms a tough ground that is easily absorbed by wood. Applied as an undercoat, it provides a sealing and protective layer that guarantees really durable lacquer ware.

In addition to household goods like trays and bowls, Kiso *shikki* also includes fishing rods and other angling gear. Some small pieces of furniture are also made using *hegime* grained wood. The main woods worked are *katsura* (*Cercidiphyllum japonicum*), Japanese cypress, and Japanese horse chestnut.

Kiso lacquering is not confined to any one technique. Bentwood goods or thin boards with a *hegime* grain are produced using a technique similar to Hida *shunkei*. Kiso *kawari nuri* employs a tough ground on which several layers of colored lacquer are built up. These are then rubbed down to produce a pattern. *Nuriwaki roiro nuri* also features a tough ground and applications of different colored lacquers.

In general, Kiso *shikki* is a happy medium that retains the beauty of the wood while making an article more functional through the application of lacquer.

HIDA SHUNKEI LACQUER
Hida Shunkei

Takayama, Gifu Prefecture

Takayama, in the Hida region of central Honshū, is one of Japan's most attractive traditional towns, and the home of much woodcraft as well as a type of lacquer ware called *shunkei*. The latter involves a relatively simple lacquering technique in which wood is first lightly stained and then coated with a clear yellow lacquer to produce a resistant coating and draw out the grain of the wood. The term *shunkei* is used to describe both the ware and the technique.

At some stage in its development, *shunkei* ware became associated with the tea ceremony. During the Edo period, when the tea ceremony steadily gained ground among the general population, *shunkei* became correspondingly popular. Production in Takayama increased under the patronage and protection of the local lord, who fostered the development of *shunkei* lacquer as a local industry. Toward the end of the nineteenth century, what is now known as Hida *shunkei* started to be distributed beyond the boundaries of the province, and the craft flourished as it became more widely known.

In spite of the simple, unpretentious air of *shunkei* wares, their appeal never flags, largely because the beautiful grain and texture of the wood is visible through the lacquer. *Shunkei* tea ceremony utensils are still being produced today, along with other traditional items such as tiered food boxes (*jūbako*), lunch boxes (*bentō bako*), confectionery bowls with lids, stationery boxes, flower vases, trays, and pieces of furniture.

Wood used to make bentwood goods and objects assembled from boards or strips of wood is processed in one of three ways. In the first, wood is cut and smoothed in the conventional manner. In the second, the wood is split or cleft, thereby retaining its rough, textured grain. In the third, the wood retains some of the tool marks of the plane used to finish it. The wood for bentwood pieces is steamed, and may be scored on the inside of corners to facilitate bending. The ends are fastened by cherry-bark strips. Turned goods (*hikimono*) are produced on a lathe in the conventional way.

The wood is colored with one of two different stains. Yellow stain is obtained from gardenia seeds or from the resin of a plant that grows in Thailand. Red stain is traditionally obtained from red ferric oxide, but a synthetic substitute is now available. After staining, raw lacquer is rubbed into the wood. The finishing coat of clear *shunkei urushi* is then applied.

Being one of the simpler lacquering techniques, Hida *shunkei* is not so expensive to produce and, as a result, is not so hard on the pocketbook. Also, compared to some of the more ornate lacquer wares produced in Japan, it has a simplicity that puts it in step with modern aesthetics.

KISHŪ LACQUER • *Kishū Shikki*

Misato-chō, Wakayama and Kainan, Wakayama Prefecture

The Kainan region is located at the end of the Kii Peninsula, jutting out into the Pacific Ocean and washed by the warm Kuroshio current. Since as long ago as the Muromachi period, the region has been producing Kishū *shikki* (also known as Kuroe *nuri*, Kainan *shikki*, and Kuroe *shikki*). Its commonly accepted point of origin is the Negoroji temple, established in 1288 in Wakayama city.

Monks at Negoroji made all their daily utensils themselves and did their own lacquering in a style known as Negoro *nuri*. Bowls and other objects were simply lacquered with a vermilion top coat on a black penultimate coat, a style known as *aka negoro*. Over time the top coat would be worn away in places to reveal the black lacquer beneath, an attribute that was particularly admired. Later, as the tea ceremony came into vogue, lacquer ware with a black finish (*kuro negoro*) was also made.

In 1585, a dispute with the abbot of Negoroji led many monks to escape to Kuro, where they turned to lacquering full-time. Thus established, the lacquer ware industry became known for producing functional items with distinctive lacquers. From the Kan'ei era (1624–44) onward, bowls were given a priming of persimmon juice. Later, a thick, tough ground called *kataji atsunuri* came to be applied.

In modern times, the area is still a major producer of everyday lacquer ware at a reasonable price. Most common are household goods and furniture, which may be bentwood, turned, or assembled (*sashimono*). The main decorative treatments are various kinds of *makie*, *chinkin*, *aogai zaiku* using mother-of-pearl, and illustrative work in colored lacquers.

At the same time as maintaining traditional techniques, Kishū *shikki* makers are adapting to modern demands. Wood composites are being used along with plastics for the base material, and the element of craft is giving way to product specialization. Yet whether traditional or modern, Kishū *shikki* remains a lacquer ware that can be used for many years until it takes on the worn appearance of its forebears.

MURAKAMI CARVED LACQUER
Murakami Kibori Tsuishu

Murakami, Niigata Prefecture

Known in China as *ti hong,* the Japanese *tsuishu* is the technique of building up layer after layer of lacquer and then carving it. A related technique, *kibori tsuishu,* uses a piece of wood carving as a base on which to layer lacquer, and is associated with the castle town of Murakami, north of Kyoto.

Although the origins of Murakami *kibori tsuishu* are not clear, production is believed to have been carried out since the Ōei era (1394–1428), when lacquerers from Kyoto settled in the area to work on the building of temples. Much later on, during the Bunsei era (1818–1830), the foundations of modern Murakami carved lacquer were laid. Clan leaders traveled to Edo (present-day Tokyo) to study *tsuishu.* Distinctive carving techniques were developed by the master craftsman Ariso Shusai, and, in the Tenpō era (1830–44), Yabe Kakubei developed a method of applying lacquer with the fingertips.

The elaborate nature of the carving is one of the special features of Murakami *kibori tsuishu.* After the wood is shaped by a specialist woodworker, a carver employs small knives to depict traditional landscapes, flower-and-bird arrangements, or contemporary scenes. The lacquerer then takes over. Numerous layers of lacquer are applied, allowed to dry, and polished, prior to the final polishing of the top coat with a special oily clay.

A large range of carved lacquer articles is being produced in Murakami today. Some, such as trays, tea caddies, and coasters, are turned before carving. Chopsticks and boxes for inkstones show Murakami carving at its delicate best, while shelves, vases, and *tansu* chests gain a unique glow under the brilliance of this elaborate craft.

KAMAKURA LACQUER • *Kamakura Bori*

Kamakura, Kanagawa Prefecture

Kamakura was Japan's capital city during the period of the same name from the late twelfth to the mid-fourteenth centuries. Situated about 40 kilometers south of Tokyo, Kamakura is a great repository of history, famous for its many temples and shrines and its enormous *daibutsu*. Less conspicuously but of equal cultural importance, the town is also the home of Kamakura *bori* lacquer ware.

Kamakura *bori* is essentially a highly developed type of wood carving, the quality of which is enhanced by a coating of lacquer. Traditional work of this kind is found not only in Kamakura. Nikkō *bori* comes from the famous town of Nikkō in Tochigi Prefecture, Murakami carved lacquer ware from Niigata Prefecture, Takaoka carved lacquer ware from Toyama Prefecture, and Kagawa carved lacquer ware from Kagawa Prefecture in Shikoku.

Kamakura *bori* appears to have originated in religious carving done by specialist woodcarvers during the Kamakura period. (Original examples of this type of woodcarving can still be seen at two temples in Kamakura today—on a desk at Enkakuji temple and on a Buddhist image dais at Kenchōji temple.) Descendants of Buddhist sculptors who had come from Nara formed a group of skilled craftsmen in Kamakura, where they set about producing things needed at the great temples, including small pieces of furniture. Such work was to become prized for gifts among the upper classes in the late fifteenth century. The techniques used then have been handed down to the present day in two families of craftsmen engaged in religious work—the Gotō family and the Mihashi family.

Originally, Kamakura *bori* was something that specialist woodcarvers engaged in during their spare time. But during the latter part of the nineteenth century and the beginning of the twentieth, it became a full-time job, and this laid the foundations of the industry. While each family of lacquerers has its own particular way of carving and applying lacquer, there are many people today who are attempting to use modern techniques—particularly since Kamakura *bori* is currently a popular hobby practiced throughout the country.

Although small pieces of furniture may be embellished by this technique of carving and lacquering, it is more commonly used for still smaller items. Trays, plates, coasters, hand mirrors, writing boxes, and inkstone boxes are the mainstays of production.

Several kinds of wood are used, including *katsura* tree (*Cercidiphyllum japonicun*), *hōnoki* (*Magnolia obovata*), Japanese cypress, and ginkgo. Carving is done with chisels and sculpting knives. Before undercoats are applied, the carved wood is primed with either raw lacquer alone or a rice paste. Two different kinds of undercoat are used—one a mixture of *urushi* and powdered charcoal, the other of *urushi* and clay (*tonoko*).

There are eight different kinds of final coat. The technique most often used today is called *hikuchi nuri*, whereby lamp black is sprinkled over wet vermilion lacquer to give an antique effect.

YAMANAKA LACQUER • *Yamanaka Shikki*

Kaga and Yamanaka-machi, Ishikawa Prefecture

All the principal types of lacquer ware in Japan have wood as their base material, and Yamanaka lacquer is no exception. In fact, Yamanaka ware began as uncoated woodcraft; only later did it develop into lacquer ware.

Woodcarvers settled in this part of Ishikawa Prefecture during the Tenshō era (1573–92), and began selling unlacquered ware to visitors to Yamanaka spa and Ioji temple. The technique of *sujibiki*—turned goods with uniform parallel lines—emerged in the mid-seventeenth century. In the Hōreki era (1751–64), lacquering techniques began to be introduced from other parts of Japan, and by the beginning of the twentieth century, Yamanaka *shikki* was well established as lacquer ware.

Since the end of World War II, competition from plastic goods has caused a decrease in production of most lacquer ware. In an attempt to compete, Yamanaka's lacquerers have pioneered the use of synthetic lacquers and fiberboards. But it is still the traditional techniques that attract the greatest following.

Most Yamanaka *shikki* are turned goods in the *sujibiki* style. There are four main types: *sensuji*, produced with a two-edged chisel; *inahosuji*, a series of lozenge shapes produced with a special plane; *tobisuji*, a series of nicks also shaped by a plane; and *hirasuji*, a line of projecting or receding wave forms.

Using various combinations of colored lacquers and undercoats accentuates the effects of turning. Common lacquering techniques include *mokumi tamenuri*, which enhances a wood's grain, and the colored applications *kuro nuri*, *shu nuri*, and *bengara tamenuri*. Woods that are suitable for turning are favored: zelkova, pine, birch, and Japanese horse chestnut.

Some *makie* is used for Yamanaka *shikki*. Yet in general most of this lacquer ware is simple in its feeling and enhances the natural beauty of the woods used.

KAGAWA LACQUER
Kagawa Shikki

Takamatsu, Kagawa Prefecture

This lacquer ware from the island of Shikoku is also known as Sanuki lacquer. Whatever the name used, it refers to a number of types of lacquer ware produced in and around the city of Takamatsu—all of them somewhat different from the other types of lacquer art to be found in Japan.

The origins of this foreign-influenced lacquer ware go back to the year 1830. The lacquerer Tamakaji Zōkoku, who was then in the employ of the Takamatsu domain, completed a study of the lacquer ware of Thailand and China. Thereafter, he developed the decorative techniques that came to be known in Japan as *kinma*, *zonsei*, and *zōkoku nuri*. However, it was his brother, Fujikawa Kokusai, who succeeded in putting these wares into production during the latter half of the nineteenth century. At about the same time, Gotō Tahei developed the Gotō lacquer style.

Both *kinma* and *zonsei* wares are made with cores of woven bamboo—basketry, as it were. A variety of goods are produced in this fashion, including low tables, shelving, trays, various containers both for foods and flowers, and some decorative items. Wares made in the conventional manner are also produced.

Kinma is a kind of colored lacquer inlay common in Thailand and Myanmar. The undercoating process is complex, since many coats are needed to provide durable means of smoothing over the irregularities of a woven bamboo core. Undercoats of paste, *urushi*, and rough sawdust are applied. The final coat is covered with paper after it has hardened. To this paper covering is applied a layer of paste, lacquer, and fine sawdust. Fabric is spread over joints and edges for strengthening in the conventional way, and conventional undercoats of lacquer and whetstone (*tonoko*) are applied. These are burnished and then covered with a layer of raw lacquer, followed by a layer of black and various layers of colored lacquer. Only at this point are designs incised into the lacquered surface; these incised lines are later painted with colored lacquers. When it has hardened, the surface is burnished. Commonly, a *kinma* design is expressed in vermilion and perhaps yellow on a black ground.

The basic lacquering of *zonsei* is the same as for *kinma*, since it uses a woven bamboo base. The design is painted on the colored lacquer surface in black. After burnishing, "shadows" are painted into the design with colored lacquers. Clear lacquer is then applied. Parts of the design, such as outlines and veins of leaves, are highlighted by shallow incisions with a sharp tool. Gold is often used for emphasis. A final coat of clear lacquer is standard.

In the case of *zōkoku nuri*, the wood is carved and a primer of black lacquer is applied; this is then sprinkled with *makomo* (rice powder). After this is dry, lacquer is rubbed into the surface.

For the colorful *chōshitsu* technique, several layers of different colored lacquer are applied. The design is then carved into the lacquer to expose the different layers of color.

The Gotō lacquer effect is achieved by patting the surface of tacky vermilion lacquer with such objects as a sponge, a brush, or a tiny soft bag, or just the tips of the fingers. This textures the surface in an interesting manner. Another coat of lacquer is applied with a cloth.

Takamatsu itself, both when it was under the feudal system and since, has always afforded the lacquer ware industry much support, encouragement, and understanding. There is a publicly funded crafts school as well as a research and testing facility in the city. In 1950, the Takamatsu region received official designation as an important lacquer producing area, along with Aizu Wakamatsu (Fukushima Prefecture) and Wajima (Ishikawa Prefecture). Today, the Takamatsu region is producing some of the nation's very best lacquer ware, on a par with the excellent products of Wajima.

TSUGARU LACQUER
Tsugaru Nuri

Hirosaki, Aomori Prefecture

Layers of lacquer that create polychrome patterns are the signature of lacquer ware from the Tsugaru district of Aomori Prefecture, in northern Honshū. The style, which dates from the Edo period, is most frequently seen today in the spotted *kara-nuri* style.

The origins of Tsugaru lacquer ware can be traced to the second half of the sixteenth century and are attributed to the lacquerer Seikai Genbei. Sometime between 1661 and 1704, Genbei made a chance discovery. While burnishing down a work table on which various colored lacquers had accumulated, he noticed that a mottled pattern appeared. Thereafter Genbei worked to create the effect intentionally; objects displaying this polychrome mottling became known as Tsugaru *nuri*.

Various kinds of Tsugaru ware have developed over the years, yet the process remains the same as that discovered by Genbei. Several layers of different-colored lacquer, to which egg white has been added, are applied to the chosen surface using a perforated spatula. Rather than spreading the lacquer on as is the convention, Tsugaru craftsmen pat the lacquer on with the spatula, thereby creating different patterns and textures. The layers of lacquer are then burnished down to produce the characteristic mottled patterning.

Alternatives to this basic *kara nuri* process—still the most popular Tsugaru style—make use of millet, rape, or hemp seeds. In *nanako nuri*, the seeds are sprinkled over the still-tacky lacquer, and removed after one day. A different-colored lacquer is then applied to produce a pattern of small rings. Polychrome "brocade" patterns can be formed by painting frets or arabesques over these rings, in the technique known as *nishiki nuri*.

When it originated, Tsugaru ware was confined to such luxury items as scabbards and gifts for the shogun. Some lower-ranking samurai even took up the craft themselves during the late Edo period. Following industrialization during the Meiji period, Tsugaru techniques are now applied to a broad range of everyday items, including trays, bowls, chopsticks, and small pieces of furniture.

CHIKKŌHIN

Bamboo Craft

Introduction

Bamboo could almost be called the perfect material. It is light, strong, extremely flexible, and has a hard, smooth outer skin. It is easy to work, and grows in abundance in many parts of Southeast Asia. Japan is particularly fortunate in having more than six hundred varieties, the many of which are well suited to basketry. The particular characteristics of bamboo make it difficult to craft and weave by mechanical means, so most of the work is still done by hand with simple tools. Although the work as such is relatively simple, it takes skill and experience to produce objects of beauty.

There are various kinds of bamboo crafts. Some depend on a particular type of bamboo; others depend on a particular way of processing. Generally speaking, however, basketry is the first thing that comes to people's minds. Baskets are woven from flat, narrow strips or constructed from round splints of bamboo. The skin is usually left intact, but material obtained from the thicker varieties can be used without the skin. In its familiar forms as cages for birds, stag beetles, and other insects, bamboo craft is relatively simple. However, the potential to form beautiful shapes and graceful curves is enormous, and innumerable things apart from baskets and cages can be made from it. "The question is, rather," as Basil Hall Chamberlain said, "what does it *not* do?"

Splitting the bamboo.

計量米びつ
花冠

SURUGA BASKETRY
Suruga Take Sensuji Zaiku
Shimizu, Shizuoka Prefecture

In contrast with much of Japan's basketry, which is made from flat, narrow strips of bamboo (*hira higo*), Suruga basketry employs round splints without the skin shaved off (*maru higo*). The origins of bamboo craft in Shizuoka Prefecture can be traced back to ancient times, but what is known today as Suruga basketry got its start in the Edo period when samurai took up the craft to supplement their incomes. They made insect cages and confectionery containers to sell as souvenirs to the travelers who passed through Shizuoka on the Tōkaidō, the main route between the capital, Edo (now Tokyo), and the imperial seat in Kyoto. During the succeeding Meiji period the production of basketry developed into a major industry. Suruga bamboo ware was exhibited and highly praised at the 1873 World's Fair in Vienna; the attention it attracted led to the export of basketry to Europe. Today, modern articles such as lamp shades and other light fittings have assumed places alongside traditional items such as household basketry and bird and insect cages.

The main species of bamboo used in basketry are *madake, mōsōchiku, hachiku,* and *kurochiku.* The technical process comprises five principal steps: the making of the round splints, the forming of the framing rings, the bending of the splints, the assembling and gluing, and the finishing. To make the round splints characteristic of Suruga basketry, thin strips of shaved bamboo are pulled through a steel plate with a hole in the middle. After they are boiled and dried, the round splints are coerced into graceful curves by applying a heated iron—a technique that requires years of expertise. The bent round splints are then connected to bamboo framing rings in bird-cage fashion, but with a great deal more artistry. Sometimes the design calls for rows of delicate splints that overlap one another, but there is always rhythm and movement in the final form.

Bases and lids for Suruga basketry, if needed, are usually woven from flat strips of bamboo. The weave patterns most commonly used are a plain "twill" weave (*ajiro ami*) and one based on the hexagon (*nankin ami*). Lacquer or Japanese tallow is used for the finish.

BEPPU BASKETRY • *Beppu Take Zaiku*

Beppu, Ōita Prefecture

Situated on the island of Kyūshū, Beppu has an abundance of raw material at hand. Basket-making in Beppu existed in ancient times, but it was the patronage of visitors to the local hot springs that stimulated craftsmen and provided an outlet for their work, which included household basketry as well as more artistic pieces. Over time, Beppu became widely known for its basketry, and what had started as a side business for farmers gradually grew into a full-fledged industry. This in turn led to the creation of some exquisite work by artists and highly skilled technicians. Great efforts were made to raise standards of design as well as to develop new products, and presently Beppu is the country's leading producer of bamboo goods.

Beppu bamboo ware can be generally classified into two groups: natural ware, which highlights the colors of the skin and black ware, which is dyed and lacquered. A tremendous variety of bamboo objects are being made in Beppu today, the largest proportion consisting of basketry that uses flat splints. The processing of the bamboo differs according to the article to be made, but with most baskets, the oils are removed by boiling the bamboo in a caustic soda solution, then the bamboo culms are sun-bleached, seasoned, and cut to appropriate lengths. The most prominent parts of the nodes are removed and, in some cases, the skin is scraped off. The culms are then split lengthwise, sometimes with the aid of another piece of bamboo, but usually with a hatchetlike tool having a long, narrow blade. The internal membranes at the nodes are then removed. Flat strips of the required thickness are obtained by repeated splitting, and adjustments in width are made by drawing a strip between two blades stuck into a block of wood, or by using a special tool. In some cases the initial steps of splitting the bamboo can be done by machine, but the final splits can be done only by experienced craftsmen.

Apart from a plain weave, there are many other styles of weaving: hexagonal, octagonal, mat, chrysanthemum, wickerwork, and numerous combinations and variations of these. When the weaving is complete, the ends of the woven strips are finished off in one of three different ways: splitting the ends and forming an edge (*tomo fuchi*); fixing the ends from both sides (*ate fuchi*); or concealing the ends of the woven strips of bamboo by wrapping split rattan around them (*maki fuchi*). Feet or handles are attached as necessary, and baskets made of bamboo without its skin are often coated with natural lacquer after priming. The finest basketry from Beppu exemplifies the skill with which material of little intrinsic value can be converted into objects of extreme beauty.

TAKAYAMA TEA WHISKS • *Takayama Chasen*

Ikoma,* Nara Prefecture

The adaptability of Japan's bamboo combines with the dexterity of its craftsmen to produce in the *chasen*—the whisk used in the tea ceremony to whip powdered green tea into a frothy, hot brew—one of the most exquisite types of bamboo ware in existence. It is not only beautiful to look at but superbly functional as well. A good tea whisk, for example, should produce a good froth on the surface of the tea; it should not splash the precious liquid; and it should by no means damage prized tea bowls.

Each school of tea has its own traditions and procedures calling for slight variations in utensils. In the case of *chasen*, differences are found in the length of the handles, the length and number of the splines, and the shape of the heads.

Chasen are made in Kyoto and Nagoya, but those made in Ikoma fill nine-tenths of the country's demand. There are four basic types of *chasen*, each for a different type of tea. The number of the splines ranges between 60 and 120, and the bamboo is either the light cream, sun-bleached type (*shiratake*) or the "smoked" *susudake* variety, which is usually of *madake* bamboo that has also been bleached. After sun-bleaching and the removal of oils, the well-seasoned bamboo in its untreated state is cut to a particular size according to the type of *chasen* to be made. The skin of the portion that is to form the head is scraped off, then split into between 12 and 24 splines. These are then split again into alternating thick and thin splines, forming an outer ring (*sotoho*) and an inner ring (*uchiho*), respectively. The numbers of outer splines and inner splines are usually in a ratio of 6:4. Next, the splines are carved so that they are thinner toward their ends—a delicate process, peculiar to the *chasen*, known as *aji kezuri* (literally, "flavor-whittling"). A thread is then tied around the base of the outer splines, and the inner splines are arranged together in the center of the whisk. Finally, the ends of the splines are made to curl inward and adjustments are made to produce a perfect shape.

The *chasen* illustrates well the dilemma that a fine object of Japanese craftwork often presents—whether to use it, or just to admire it as a work of art.

MIYAKONOJŌ BOWS • *Miyakonojō Daikyū*

Miyakonojō, Miyazaki Prefecture

Bows have been used in Japan for many centuries, indispensable as weapons and symbols of power. The making of bows in Miyakonojō began when this old castle town came under the rule of the Shimazu daimyo family, which governed Satsuma from the end of the twelfth century. Local artisans began making bows to supply the constant demand from warriors who enjoyed *kyūdō*, the art of archery that is still practiced today. These craftsmen developed a tradition that subsequently became known throughout the country. After World War II, when the teaching of martial arts was banned at schools, the making of bows ceased almost completely. Today, however, *kyūdō* is viewed as a sport, and with the establishment of new archery grounds there is a steady demand for this traditional type of bow.

There are four types of plain-finished *shiraki yumi*, graded for use ranging from beginners to the most skilled archers. The main material is bamboo, but *hazenoki* (sumac) wood is also used. The bow itself is laminated, composed of a core and outer layers. In the first stage of production, the bamboo is prepared by being split into flat strips of the required width and having its oils removed. Once it is thoroughly seasoned, the skin is removed from the bamboo to be used for the outer layers. The surface of the bamboo for the core is then burned on both sides and planed to a standard thickness. The sumac for the core is seasoned, then similarly planed. Two strips of the charred bamboo are placed between three strips of sumac to form the core, and then laminated together with a glue made by boiling deerskin. Final adjustments in the thickness of the core are made by using a special hand plane before the outer layers of bamboo are laminated to the core with the glue. To finish, a plane is run over the surfaces to render them smooth, and if necessary a grip of rattan is wound around the bow in the appropriate place.

There are at present twenty-three master bow-makers at work in Japan, nineteen of them in Miyakonojō. While their customers are no longer the warriors and noblemen of old, popular interest in the art of archery will certainly keep them busy for many years to come.

W A S H I

Japanese Paper

Introduction

The traditional handmade papers of Japan are collectively called *washi*, literally "Japanese paper," to distinguish them from papers produced elsewhere. Paper was invented in China, and the technology was introduced to Japan via Korea by the sixth, perhaps as early as the fifth, century. At first, production was centered in the capitals to meet the growing demands of Japanese government officials and Buddhist temples, but from the eleventh century on it shifted to those rural provinces where abundant supplies of the necessary raw materials were available. Much of Japan's *washi* continues to be made in historic papermaking regions.

Papermaking has long been a family occupation in Japan, and in some areas entire villages are involved. It was traditionally carried out during the winter months, in between the harvesting and planting seasons. Paper is made in many different areas of Japan, with only slight variations in the basic methods. The principal raw materials are bast fibers, taken from the inner bark of the *kōzo* (paper mulberry), *mitsumata*, and *ganpi* shrubs, although hemp, bamboo, and rice straw are also sometimes used. Western-style papermaking using wood pulp was not introduced to Japan until the second half of the nineteenth century.

Ganpi bark can be stripped directly from the branches, but *kōzo* and *mitsumata* must be steamed before the bark can be removed. The bark is then dried for storage. To prepare the bark for papermaking, it must be soaked in water and the dark outer bark scraped away. The bark is then cooked in a strong alkaline solution to release the starches and fats. The fibers are next washed until clean and impurities are meticulously removed. The resulting white fiber is beaten to further loosen and separate the fibers, which are then put into large papermaking vats filled with water. A formation agent (sometimes referred to as mucilage) is normally added. This serves to keep the fibers suspended in the water, to slow down the drainage of water from the mold, and to allow newly formed sheets to be stacked, pressed, and separated without the use of interleaving felts as in Western papermaking. The most commonly used formation agents are extracted from the *tororo-aoi* (*Abelmoschus manihot*) and *nori-utsugi* (*Hydrangea paniculta*) plants.

The Japanese papermaking mold generally consists of a hinged wooden frame enclosing a removable screen of bamboo splints, although molds can also be made of reed, metal, or gauze. Many papermakers employ the *nagashisuki* method in which the contents of the vat are stirred until the fibers are evenly dispersed. The mold is dipped into the vat and then lifted out and rocked rhythmically back and forth so that the fibers intermesh as the liquid drains away. Thin, even sheets are particularly difficult to form. Further dips of stock may be added to increase the thickness of the sheet, and excess stock can be thrown out of the mold when the required thickness has been achieved.

Another method called *tamesuki*, similar to that used in Western papermaking, is employed less frequently. The mold is first dipped into a vat and lifted out full of stock. The papermaker then skillfully shakes the mold, causing the water to drain off and the fibers to settle evenly on the screen. In both methods, the screen is removed from the mold and the tender, wet sheets are placed one on top of another in a stack. The stacks are then slowly pressed to drain away excess water, after which the sheets are separated and dried by one of two methods. In the method called *itaboshi*, the wet sheets are brushed out on wooden boards, which are then stood outside to let the paper dry in the sun. The other method is called *teppan kansō*, whereby the sheets of paper are brushed onto hot metal plates that are then heated with steam or electricity.

In addition to paper for writing and painting, *washi* is used for a diversity of things including fans, lanterns, toys, sliding *shōji* panels, and even clothing. Unlike most mass-produced Western-style papers, it has a subtle beauty and tactile richness that have earned it international acclaim. Over the centuries, papermakers in different prefectures have developed local variations of *washi*, some of which are discussed below.

Dipping the mold into a vat of stock.

ECHIZEN PAPER
Echizen Washi

Imadate-chō, Fukui Prefecture

Papermaking in Fukui Prefecture historically has been closely connected with the ruling authorities, perhaps because of its proximity to Kyoto. The *Engishiki*, a collection of fifty volumes on official procedures and ceremonies compiled in 927, records that paper from Echizen was at that time being supplied to the court. By the Muromachi period, much of the paper on which official orders (*hōsho*) were written was made in Echizen. Papermaking came under the protection of Ōtakiji temple, and a *kamiza*, or paper guild, was established. Some of the region's most distinctive papers were developed during the feudal period. As official ties continued, the Tokugawa government gave its approval for paper from Echizen to be stamped with a seal signifying that it was used by the shogunate. At the end of the nineteenth century, paper for official bank notes was made in Echizen, and during the late 1920s and early 1930s, Imadate-chō also supplied paper for the bills of the Joint Chinese Reserve Bank. In 1951, the Ministry of Finance began to commission paper for bank notes, and to this day, paper from Echizen is still part of the country's official operations.

The wide range of *washi* currently produced in Echizen includes paper for sliding door panels (*fusuma*) and wallpaper, special papers for calligraphy and painting, fine-quality papers for official documents such as bonds and checks (*kyokushi*), and various other forms of stationery and envelopes. Decorated papers dyed a range of colors and patterns—some displaying designs harking back to the Heian period—are also made in Echizen.

Echizen papermakers employ all of the traditional raw materials: *kōzo*, *mitsumata*, and *ganpi*. No formation agent is used in the production of the *kyokushi* official papers, which are produced by the *tamesuki* method in molds with metal net screens. The *nagashisuki* method is employed for all the other papers. The formation agent is *tororo-aoi* and the screens are made of either bamboo or gauze.

Because of its long history of association with Japanese authorities, Echizen *washi* could almost be called part of the establishment. It nevertheless produces a fine range of Japan's most traditional papers.

AWA PAPER • *Awa Washi*

Yamakawa-chō, Tokushima Prefecture

It is said that the origins of Awa *washi* can be traced back to the Nara and Heian periods, to the paper made by people of the Inbe clan who inhabited Tokushima at the time. This clan was in charge of imperial court ceremonies. There is a shrine in Tokushima today named Inbe Shrine in which Amenohiwashi no Mikoto is enshrined as an ancestral god of paper. A document dated 774 records that Awa had failed to make a payment of some 40 *kin* (approximately 24 kilograms) of hemp, which was used for making paper. It is therefore assumed that papermaking was being carried out in this region by this time.

Papermaking seems to have spread rapidly along the banks of the Yoshino River after the construction of two temples near the river's mouth. As one of the earliest centers of production, the paper industry here seems to have contributed to the advancement of papermaking further afield in Shikoku. In an effort to develop agriculture in his domain, the feudal lord of Awa encouraged the cultivation of *kōzo* for papermaking along with tea, indigo, and mulberry for silkworms. Under strict regulation, the manufacture and retailing of paper in this area continued to develop, bringing an increase in the variety of papers available.

The raw materials used for Awa *washi* are *kōzo, mitsumata,* and *ganpi*. The formation agent is *tororo-aoi* or *nori-utsugi*. The mold screens are made of either bamboo splints or reeds, and the paper-forming method is *nagashisuki*. If a paper is to be dyed, the paper stock is put into a dye vat before being formed.

Some calligraphy, craft, and painting papers are being made today, yet much of the production in the Saji mills consists of printing, publishing, and wrapping papers. The indigo-dyed paper of Awa, however, is definitely the most distinctive of the Yamakawa-chō output.

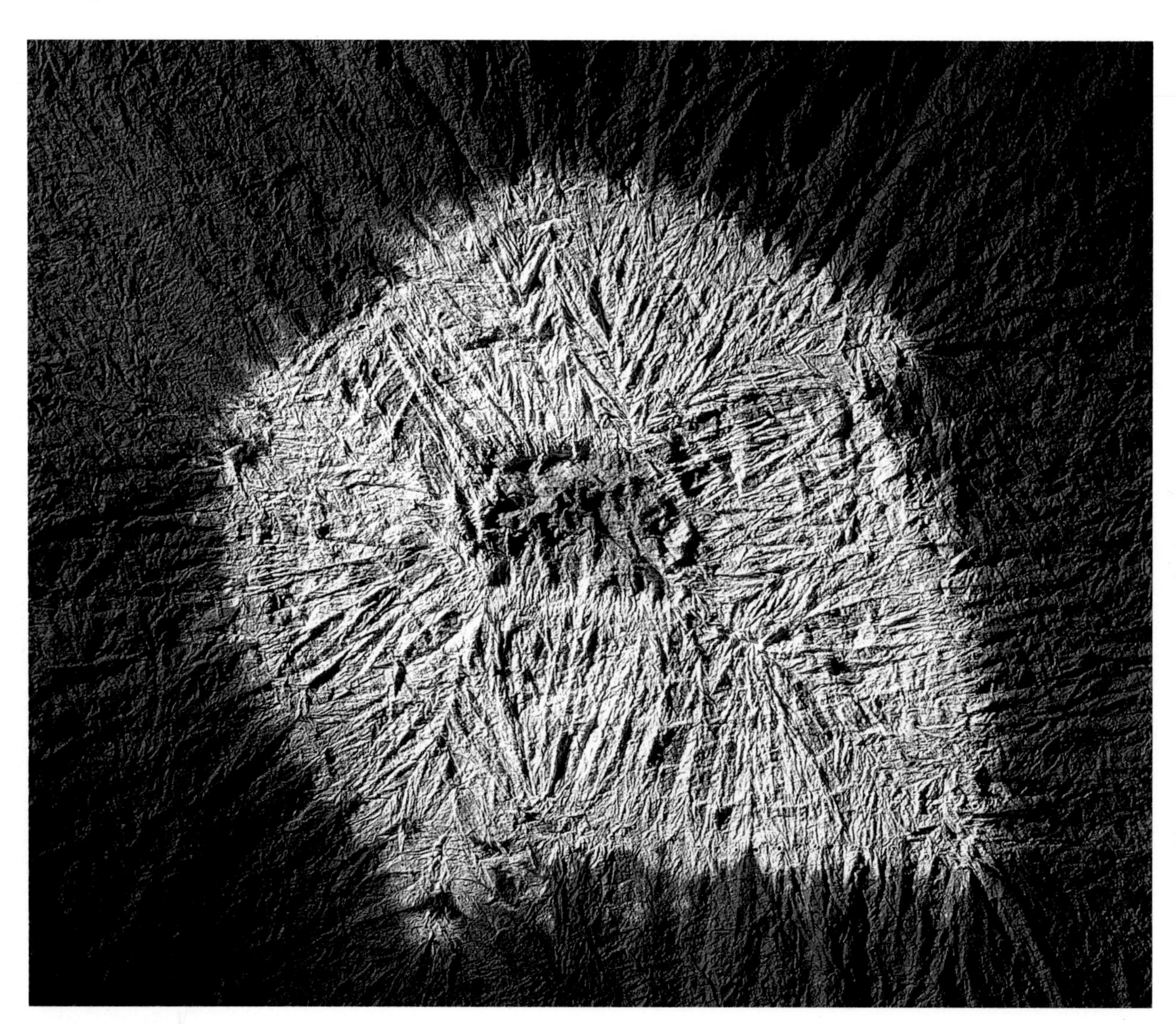

TOSA PAPER • *Tosa Washi*

Ino-chō, Kōchi Prefecture

Of all the traditional craft papers in Japan, Tosa *washi* is perhaps the most famous. As in the case of other Japanese papers, its origins are unclear, but the *Engishiki* relates that paper for religious purposes (*hōsho gami*) and a similar *kōzo* paper named *sugihara gami* were supplied to the court by this region. It therefore seems safe to assume that paper was being made here by the tenth century at the latest. The origins of colored Tosa paper are better documented, for it is recorded that Aki Saburōzaemon successfully produced some dyed papers around 1592. Recognizing the importance of the local papermaking industry, Yamanouchi Kazutoyo, who became daimyo of Tosa in 1601, lent his protection and actively encouraged its production as the "clan paper." The export of colored paper outside the domain was prohibited.

Papermaking continued in the traditional vein until a method of producing large sheets was invented in 1860. This led to the making of paper in large quantities and the creation of various other papers, including a resinous paper for writing with fountain pens that does not smudge, paper for postage stamps, and a paper suitable for map making. The techniques developed in Tosa were passed on to papermakers in other regions. Tosa is also the hub of papermaking machinery and tools, as well as a supplier of the raw materials needed for making *washi*. In addition, more than half of the people presently producing screens and molds work in Kōchi. The thread and the bamboo splints necessary for making the screens are made nowhere else. In sum, Tosa has played a major role in the development of Japan's papermaking industry, and seems likely to lead the way for many years to come.

All the paper made in Tosa is characterized by its fine-quality *kōzo*. *Kōzo* is not the only raw material used, however. In addition there are *mitsumata* and *ganpi*, as well as hemp, bamboo, and rice straw. The formation agent is either *tororo-aoi* or *nori-utsugi*, and the screens are made of bamboo, reeds, or gauze. Tosa paper is formed by both the *nagashisuki* and *tamesuki* methods.

A vast range of papers is still being produced in the area. Some, like *shōji* window-screen paper and calligraphy paper, are of a traditional nature. Others are designed for more contemporary uses, such as fine-quality printing and gift wrapping.

UCHIYAMA PAPER • *Uchiyama Gami*

Iiyama, Nagano Prefecture

The history of making Uchiyama *gami* extends back to the late seventeenth century. By 1698 there were already more than twenty papermakers in the area, and the numbers increased steadily to the beginning of the eighteenth century. The method of production remained more or less unchanged up until the end of the Edo period, but in 1872 bamboo screens came into use, and from 1902, paper-forming techniques introduced from Tosa (Kōchi Prefecture) were applied for making sheets two and four times the size of a standard sheet. Further improvements during the 1920s made it possible to form sheets six and eight times larger. The introduction of such machinery as stampers, hollander beaters, and compressed air machines started before World War II, and after the war papermakers began to use chlorous acid ($HClO_2$) in bleaching.

Uchiyama *gami* is particularly noted for its strength. Only *kōzo*, which grows profusely in the surrounding mountains, is used. Iiyama's abundant streams, clean air, cold winter temperatures, and heavy snowfall also play important roles in the formation of this paper, which is made by the *nagashisuki* method. Techniques unique to the papermaking of this region include freezing the bark fibers by laying them out on the snow, and snow bleaching (*yukizarashi*), which imparts a soft whiteness to the sheets. Iiyama's white, translucent *shōji* paper is perhaps the best known, but paper for important records and documents is also produced in this region, along with special papers for calligraphy and accounting books.

INSHŪ PAPER • *Inshū Washi*

Village of Saji, Yazu-gun, Tottori Prefecture

Although the *Engishiki*, the tenth-century publication on official protocol, recounts that the court was being provided with Inshū paper, the early history of papermaking in this region is not completely clear. According to Edo-period records, paper was made here for the local feudal lord—first at Aoya-chō from 1624, then in the village of Saji from 1726. The range of Inshū papermaking expanded during this period by adapting technologies introduced from Echizen (Fukui Prefecture), Mino (Gifu Prefecture), and Harima (Hyōgo Prefecture). Production continued to increase, and at the beginning of the twentieth century the prefectural government moved to develop the local industry further by inviting a papermaking specialist from Tosa (Kōchi Prefecture) to provide instruction in new techniques. As papermaking became modernized, Inshū *washi* won recognition as a fine-quality paper. Some mechanized equipment was introduced after World War II, but papermakers switched to making fine-grained artists' papers in 1954.

Inshū papermaking employs the basic raw materials of *kōzo*, *mitsumata*, and *ganpi* and the *nagashisuki* method. The formation agent is *tororo-aoi* and either bamboo splints or reeds are used for the mold screens.

In addition to art and craft papers, special papers for important records and documents as well as wallpaper, paper for *fusuma* sliding screens, translucent paper for *shōji*, and some colored papers are produced. Calligraphy paper as well as regular stationery and envelopes are made in the mills in Saji. Among these it is the artists' papers that are the most distinctive, especially one known as Inshū *mitsumata* paper.

MOKKŌHIN

Woodcraft

Introduction

There is a uniquely unpretentious textural quality to Japanese wood, especially when compared to some of the figured hardwoods from Southeast Asia. As a material, Japanese wood has an unmatched warmth and responsiveness, and the large number of traditional woodcrafts it has inspired have found a correspondingly special place in the Japanese heart.

A number of distinct ways of handling wood can be singled out for attention. What are known as *sashimono* (literally, "assembled items") are, in the broadest sense of the term, small pieces of furniture and other household articles assembled by jointing together a number of components. By a stricter definition, *sashimono* are objects assembled from boards of wood, rods of wood, or combinations of the two. The craftsmen who make *sashimono*, known as *sashimonoshi*, were originally makers of standard pieces of furniture. During the second half of the seventeenth century they came to be recognized as a distinct group, separate from the craftsmen who made architectural fittings such as *fusuma*, windows, and other removable items, or from the architectural carpenters such as the *miya daiku*, who were mostly concerned with building wooden temples and shrines. More often than not, the term *sashimono* is used today in its more narrow sense, but in the case of Kyō *sashimono* the more general meaning is understood.

Other techniques include *hikimono*, articles that are mostly turned on a lathe, and *magemono*, which are produced by bending wood. At a more conventional level, many woodcarvers produce what are known simply as *horimono*, or "carved goods."

The techniques used to carve sculptures and patterns in wood probably developed in Japan following Korean models. By the beginning of the Heian period, the carving of wooden statues in Japan was being carried out as a matter of course. Over the centuries, the art and techniques of woodcarving developed and expanded to include such items as *netsuke* and Noh masks, as well as architectural decorations. Like other crafts, carving too had its master craftsmen. In recent years, however, demand for quality craftwork has decreased as new and economical materials have become more readily available. Most of the work done in wood today is thus of an artistic nature: other than the repair of old pieces and the making of religious statues and other paraphernalia, contemporary craft-oriented woodcarving is applied to things that will become forms for lacquer ware, ornaments, or decorative architectural fittings. The last-named most frequently involves the carving of *ranma* (transoms).

While most techniques of tree conversion and many ways of finishing wood used in Japan are similar or identical to those used in other parts of the world, one method of preparing wood is worthy of special mention. Called *masame dori,* it is roughly equivalent to quartering or quartersawing. Put simply, this yields wood with straight grain. Boards and planks of wood are usually obtained by plainsawing across the grain, exposing the annual rings as wavy lines on the surface of the cut pieces; in Japan this method is known as *itame kidori.* By contrast, *masame dori,* or *masame kidori,* produces wood in thin and relatively narrow strips by cutting or splitting them in a radiating pattern, quite often from a short log. The word *masame* itself really refers to the close, straight grain obtained. Such quartersawn or split wood expands and contracts equally in all directions and is therefore less likely to warp.

Joinery, as practiced by the cabinetmakers, woodworkers, and carpenters who construct buildings in wood, is almost an art in its own right. There are literally dozens of ways of joining wood, some of which produce what look more like puzzles than joints. In reality these are ingenious adaptations of the same or similar kinds of joints used by craftsmen the world over.

Hammering out the design on metal fittings for an Iwayadō chest.

IWAYADŌ CHESTS
Iwayadō Tansu
Morioka and Esashi, Iwate Prefecture

Japanese furniture as a whole is not especially well known abroad, however the traditional *tansu* is not only known but eagerly sought after. Iwayadō *tansu* are particularly prized for both the beautiful grain of the zelkova wood and the lacquer-coated metal fittings on their surface.

Such *tansu* were originally used as safes, and even today they have a lock. In the past, the chests would have occupied the lower compartment of the special cupboard where foldaway bedding was stored during the day. Another type sat firmly on a palette fitted with wheels so that the chest and its contents could easily be wheeled in and out of a storehouse, or moved to a safe place in the event of fire.

The Iwayadō-style *tansu* is said to date back to the end of the eighteenth century, around 1783. Until the end of the nineteenth century, the same style continued to be made in this northeastern corner of Japan in the traditional manner, with one person doing all the woodwork and metalwork necessary; after this, there was some division of labor between metalwork and lacquerwork specialists. Production, mainly for the agricultural community in the southern part of the prefecture, reached its highest levels in the first two

decades of the twentieth century. Since then, Iwayadō chests have been in demand throughout Japan and, more recently, have been exported.

Although similar in feeling to those made in Sendai, in neighboring Miyagi Prefecture, the chests of Iwayadō are simpler in appearance, as are the other pieces of furniture produced today such as sideboards, tables, and ornamental shelves. Nevertheless, the chests in particular retain their strong-looking metal fittings and all-round solid appearance.

Zelkova and paulownia are used for drawer facings, doors, tops, and visible panels; cypress and paulownia for other, less visible parts. Instead of metal nails, wooden pins are employed. The wood is finished with lacquer. The metal fittings are cut and shaped by hand from steel, sometimes from sheets no more than a millimeter thick. They too are given a coat of natural lacquer, which is warmed until it smolders and turns black to create the vivid contrast so characteristic of these appealing chests.

KAMO PAULOWNIA CHESTS
Kamo Kiri Tansu

Kamo, Niigata Prefecture

For many Japanese, paulownia wood (*kiri*) is synonymous with traditional chests. In the past, a paulownia *tansu* was an essential part of a bride's dowry; indeed, it was customary to plant a paulownia sapling when a baby girl was born to ensure that she would have a *kiri tansu* when she became a bride.

Paulownia wood is blessed with a beautiful, straight grain, together with a warm color and appealing texture. Particularly light in weight, it is not subject to warping like some other woods, and is also easy to care for. What is more, old wood can be restored so that it looks as good as new—a definite advantage in the case of an expensive and perhaps sentimentally valuable *kiri tansu*.

Kamo, a city in the center of Niigata Prefecture with a traditional atmosphere sometimes likened to Kyoto's, is the country's number-one production center for *kiri tansu*. It has always had an abundance of natural paulownia to support the industry, which started in the late eighteenth century but only became known nationwide toward the end of the Edo period. By the end of the nineteenth century, the industry was thriving, and during the 1910s and 1920s production reached unprecedented levels.

The boards chosen vary in thickness, but all are of solid paulownia. Since *kiri* boards are narrow, they must be jointed together to achieve sufficient thickness for building a chest. To do this, Kamo craftsmen use pins fashioned from the stem of the Japanese sunflower, *utsugi* (*Deutzia scabra*), which are driven in at an equal depth to both boards being joined. The tops and shelves are usually more than 19 millimeters thick, but back panels and the bottoms of drawers are normally no thicker than 7 millimeters. Depending on the pieces of the chest to be joined, a variety of joints are employed: ox joints, tenon joints, and wooden pegs, but never nails. Fittings of copper, copper alloy, or steel are attached. To finish, the surfaces are polished with an *uzukuri*, the end of a bundle of straw, then the wood is given a coating of *yashabushi*, a liquid made by boiling the seeds of the *yamahan* (Alder family; *Alnus hirsuta* var. *sibirica*). Finally, the wood is polished with wax—ending a processing of surprising simplicity for what is, when completed, one of Japan's most refined pieces of furniture.

KYOTO WOODWORK • *Kyō Sashimono*

Kyoto

This beautiful work, closer in feeling to an artwork than to a simple piece of furniture or joinery, can be broadly divided into two types. *Chōdo sashimono* includes various types of furniture and household goods, such as chests (*tansu*), shelving, desks, storage for stationery, various kinds of boxes, trays, freestanding screens, and light fittings. The other group, *sadō sashimono*, comprises a number of small items of furniture and utensils used during the tea ceremony. This group in fact comprises objects that are true *sashimono*, such as shelving, simple boards, and trays, water containers, and other items made of wood bent into shape (*magemono*), and some turned items (*hikimono*). There are also carved articles (*horimono*) and similar items called *kurimono*.

Typical woods used are Japanese cypress, zelkova, paulownia, mulberry, cherry, and Japanese cedar. It can take anything from two to ten years to naturally season and process the wood. Particular care is taken to ensure that paulownia is rid of all its impurities before it is used.

Methods of jointing and decoration employed for *chōdo sashimono* are largely laid down by convention. In the case of *magemono* posts and bowls, wood with a fine, straight grain is needed. It is bent either by making a series of saw cuts in the back of the part to be bent or by steaming the boards first to render them flexible. Carved pieces (*horimono, kurimono*) are fashioned from one piece of wood. In the case of *sadō sashimono* bowls, bamboo is used for pinning joints; otherwise Japanese sunflower (*utsugi*) is used. Metal fittings are of copper, copper alloy, gold, silver, or steel.

The finish applied to *sashimono* depends partly on the type of wood used. Chinese wax (*ibotarō*) is used to polish pieces made of paulownia. With mulberry, however, the color is first brought out with lime, then the wood is wiped with oil or lacquer. Lacquer is also used for zelkova, but for other woods either scouring rush (*tokusa*) or the leaf of the *muku* is used to polish the wood. Decorative techniques employed include painting with lacquer or pigments, *makie*, marquetry, gilding, and the use of gold and silver powders or gold and silver paints.

Undoubtedly, the influence of the tea ceremony and its aesthetics on *sashimono* has been considerable. These finely crafted pieces of furniture are in a sense an expression of the tea spirit and sensibility, reminiscent of the controlled stylization of the ceremony itself.

ŌDATE BENTWOOD WORK • *Ōdate Magewappa*

Ōdate, Akita Prefecture

Generally speaking, the natural plasticity of wood can be increased either by steaming or by making a series of cuts with a saw on the inside of the section that is to be bent. In Japan, the former technique is usually applied. Some *magemono* (bent pieces) are used for furniture, but more often they are small pieces that make use of the attractive grains and textures of Japanese cedar, Japanese cypress, white cypress, and fir. When preparing *magemono* the wood is first quartersawn or split into narrow boards to expose the straight grain. Each board is immersed in boiling water for up to ten minutes, then removed and bent around a template of firm canvas. The process is repeated many times before the board is pegged into shape and left to dry. After several days, the bentwood is able to be worked.

The *magemono* from Kiso (in Nagano Prefecture) and Akita are very well known. Ōdate, in particular, is famous throughout Japan for its *magewappa*. *Wappa* is a local variant of *menpa* or *meppa*, the word for a bentwood lunch box. The elegant simplicity of Ōdate lunch boxes is enhanced by the scent of the wood, which mingles with the aroma of rice to create a very special meal.

In Akita, this particular style of bentwood box is believed to have been in existence for more than a thousand years. The abundant local supply of cedar has always sustained a lot of woodcraft, work that can be done during the long, cold winters when labor outside becomes impossible. From the early seventeenth century, however, the lower-ranking samurai of the clan were encouraged to take up the making of these bentwood boxes as a way of supplementing the rather meager budget of this rural province. Eventually, instruction in the making of *magewappa* came to be given with the full support of the clan, and in the early nineteenth century, the making other products began.

Demand for *magewappa* dropped after World War II, but it gradually recovered as a result of efforts to modernize production methods without destroying the handmade quality of the goods. Other goods being made today that utilize the bentwood principle are a kind of slop basin (*kensui*) used in the tea ceremony, a basket (*seiro*) for steaming rice and other foods, a container for saké known as a *tokkuri hakama* used on official or auspicious occasions when saké is presented as a gift or offering, and a variety of trays, coasters, and bowls with lids for small Japanese-style cookies and cakes. The *magewappa*, however, remains the most representative and sought-after example of Ōdate bentwood.

AKITA CEDAR BOWLS AND BARRELS
Akita Sugi Oke Taru

Akita, Akita Prefecture

Taru—barrels or kegs—are containers for storage, usually of liquids such as saké and soy sauce. *Oke*—tubs, bowls, or pails (often with a handle)—are frequently used for making Japanese pickles or as general kitchen utensils; they are washed out after use and put aside until needed again.

Both the bowls and the barrels are round and made from a number of narrow slats of wood fixed on a circular baseboard; both also use the same essential technique for assembly. Without use of any adhesive, a number of staves (*kure*) are bound tightly together with a *taga* or band, which is made either of interwoven strips of bamboo or of copper. Two or more bands are necessary for an *oke*, four or more for a *taru*. The bottoms of the *kure* are simply slotted into the baseboard: a perfect fit is essential if the finished article is not to leak.

Despite the similarities in assembly, there are significant differences in the choice and preparation of woods used for *oke* and *taru*. The Japanese cedar chosen for the former is usually of the fine-grained *masame* type, while that for *taru* is of the *itame* type. When they are cut, the staves for *taru* are made to taper at both ends, providing the requisite strength in the middle of the barrel; staves for *oke* are of uniform width. Finishing varies according to the craftsman's preference. The most usual finish is a light coating of wax, but sometimes persimmon juice is used, or *urushi* is applied in the *tame nuri* method.

Even though many of the bowls in particular have a utilitarian purpose, the cedar used in Akita woodcraft has a lovely grain and a scent that flavors the stored objects, endowing these containers with a lasting aesthetic appeal.

HAKONE MARQUETRY • *Hakone Yosegi Zaiku*

Odawara and Hakone-machi, Kanagawa Prefecture

The art of marquetry has a long tradition in Japan. Wood craftsmen living in the area of Yumoto, in the foothills of Hakone, have been carrying out the art for hundreds of years. The mountains of Hakone rival Arashiyama in Kyoto or Hōki-daisen in Tottori Prefecture in the abundance and variety of trees that they offer craftsmen. Marquetry pictures and other small items have over the years become a valued purchase for passing travelers or visitors to the local hot springs. The craft, originally known as Yumoto *zaiku,* has been loosely referred to as Hakone *zaiku* since the end of the nineteenth century.

Hakone marquetry is based on geometric designs that can combine upward of fifty patterns of different-colored woods. After selection, these woods are sawn or split, then thoroughly dried and seasoned over a period of months. Various of these sheets are next bonded together with a synthetic or animal glue to make unit patterns, then cut and reassembled in a block to form *taneita* mosaic patterns. By planing extremely fine cross sections from the block—a delicate and demanding process—the craftsman obtains the end-grain veneers (*zuku*) for the marquetry. Once they have been backed with paper, these veneers are applied to the base wood of the utensil or accessory.

The geometric line patterns on the *taneita* can be either bilaterally symmetrical or spread out in a patchwork of totally symmetrical forms. Apart from *taneita* that have been made by joining pieces of wood together at a chosen angle, less systematic effects are also possible. *Midare yosegi* has pieces joined at various angles at random, while *suji yosegi* is the reversal of pieces in a line. Typical patterns used are a check (*ichimatsu*), a twill-like weave (*ajiro*), grinds (*kōshi*), boxes (*masu*), scales (*uroko*), and a hemp-leaf pattern (*asanoha*), but these are only a few of the many hundreds of variegated patterns that can be created in this intricate and appealing craft.

NAGISO TURNERY • *Nagiso Rokuro Zaiku*

Nagiso-chō, Nagano Prefecture

Flanked by well-forested mountains, the old Kiso Highway allowed access to timber of the Kiso region in Nagano Prefecture. Nagiso lies toward the southern end of this region, where forestry was promoted by the Owari clan throughout the Edo period. Records tell of plain wooden trays and bowls made by woodworkers in the village being sent to Nagoya and Osaka around the beginning of the eighteenth century. At the end of the nineteenth century, the woodworkers were joined by craftsmen skilled in the use of *urushi*, and the production of turned lacquer ware began. At the time, the lathes were, of course, driven by hand, but later the water wheel, then electricity, were to provide new sources of power, and the work of making these turned goods became more efficient. The introduction of inexpensive plastic products after World War II had a damaging effect on woodcrafts generally and turned goods in particular, but more recently renewed popular interest in the "real thing" has restored Nagiso turnery to some of its former glory.

The turned goods now being made in Nagiso include salad bowls, mixing bowls, tea caddies, and containers for Japanese-style confectionery, as well as smaller items such as coasters. All possess a natural beauty enhanced by the way the grain has been drawn out by the craftsman, the inherent warmth of handmade objects, and the appeal of simple, unpretentious forms.

The first stage in creating Nagiso turnery is selecting just the right wood; this is effectively a craft in itself, acquired through years of experience. Japanese horse chestnut, zelkova, castor arabia, katsura tree, and birch are most commonly used. Logs are cross-sawn, split, shaped, and then dried for up to a year prior to turning. Once the wood is ready, blocks of the required size are roughly turned on a lathe to give them their basic form. After further drying, the craftsman returns the block to the lathe for the turning proper, cutting with as many as twenty turning chisels, which he constantly sharpens while working. The chisel is held at right angles to the body and worked from the center of the wood to the rim and then back to the center. Both choice and application of the chisels are mastered only after a long apprenticeship.

At the completion of turning, the piece is finished off with sandpaper. Articles where the wood is to be left in its natural state are polished with *tokusa* (scouring rush; *Equisetum hyemale*) or the straw of the *suguki* (a type of turnip) and water: these provide a natural sheen. On other articles, natural *urushi* is used, the lacquer being repeatedly rubbed into the wood and wiped off until a characteristic gloss is achieved.

Similar turned goods are made in Imaichi in Tochigi Prefecture, Hitoyoshi in Kumamoto Prefecture, and Inabu in Aichi Prefecture, but the work of Nagiso, because of the area's long association with wood and woodcrafts, still retains a special place in the hearts of the Japanese.

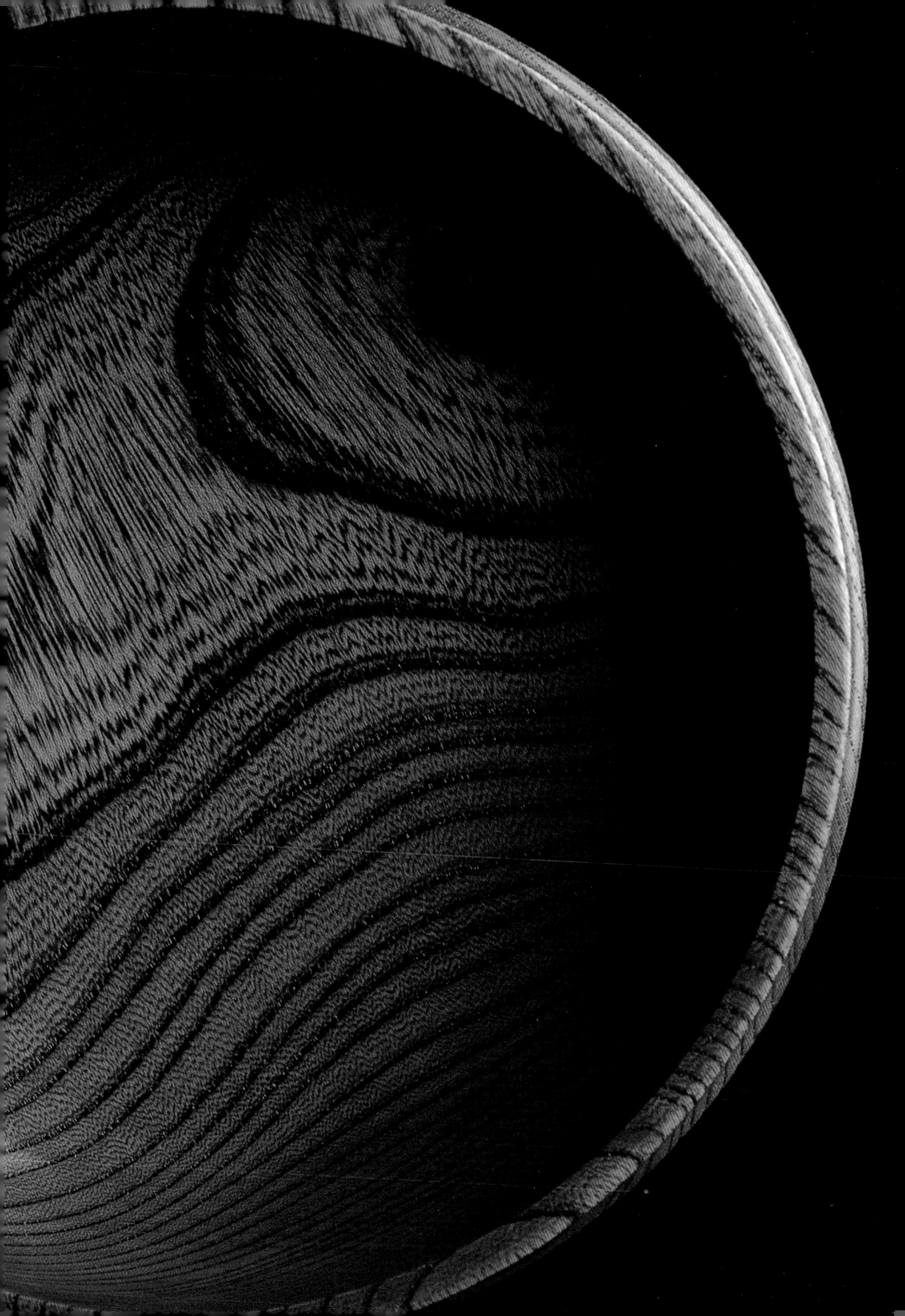

TAKAYAMA WOODCARVING
Takayama Ichii-Ittōbori

Takayama, Gifu Prefecture

The type of woodcarving known as *ittōbori* has been practiced for many centuries. The better-known kinds include Sasano *ittōbori* from Yonezawa (Yamagata Prefecture), Nara *ittōbori* from the city of the same name, Sanuki *ittōbori* from Takamatsu (Kagawa Prefecture) on the island of Shikoku, and Ise *ittōbori* from the city of Ise in Mie Prefecture.

The history of woodcarving in the Hida area where Takayama is located, almost due west of Tokyo, also dates back many centuries. In the seventh century, the Hida officials, instead of paying taxes, would dispatch ten skilled craftsmen to the capital of Nara, where they made a significant contribution to the building of large temples undertaken during the Nara period and the succeeding Heian period. Some of these men—unusual for an artisan at the time—even became famous enough for their names to be remembered by posterity. During the Edo period, the area produced a number of famous carpenters who dedicated themselves to the building and decoration of shrines and temples. Among the most celebrated of these carpenters was Hidari Jingorō. In addition to such craftsmen, who were both carpenters and sculptors, others from the area specialized in wood sculpture alone. One such was Matsuda Ryōchō (1800–71), who went to Edo (now Tokyo) and worked under one of the most famous *netsuke* craftsmen, Hirata Ryōchō. On his return to Takayama, he too began to produce fine *netsuke*, but at the same time he revived the *ittōbori* technique of carving using *ichii*, a type of Japanese yew, to carve masks, elephants, and other ornaments in a characteristically direct and simple style. It is this work that is thought to have marked the beginnings of the craft and style of carving.

The wood is obtained either locally or from Hokkaido. Japanese yew, with its beautiful grain, is particularly good for carving: being neither too hard nor too soft and containing a lot of oil, it is much easier to carve than some other kinds of wood. The wood is thoroughly dried and seasoned before the initial rough carving (known as "trimming"). In a process demanding unbroken concentration, the craftsman studies the natural form of the wood until he settles upon a design, which he draws in pencil. A rough carving is made following these outlines. For the final carving, which is also done by hand, the craftsman selects from as many as one hundred different chisels. The choice is significant, it being a characteristic of *ichii-ittobori* that the carver aims to leave deliberate traces of the chisel behind on the finished work. Throughout, care is taken to bring out both the natural form of the grain and the contrast between the reddish heartwood and the lighter sapwood that surrounds it.

The masks, tea-ceremony utensils, animal carvings, and other ornamental pieces of *ittōbori* are all refreshingly simple and, with time, acquire a natural gloss, a credit to the craftsmen and to the fine wood from which they are made.

INAMI WOODCARVING
Inami Chōkoku

Inami-machi, Toyama Prefecture

The traditional style of Japanese house is still being built today. Some modern techniques and mannerisms have of course been adopted, but the general style can be traced back over four hundred years. One of the major decorative features of this style—which on the whole avoids decoration—is the *ranma*. More or less equivalent to the transom, these decorative screens have a special place among the various types of woodcarving done by the craftsmen of Inami.

The sponsorship of powerful Buddhist authorities gave a strong impetus to the advance of Inami woodcarving. Particularly during the peaceful years of the Edo period, there was much work to be done not only for Buddhist temples but also for their associated shrines and other independent shrines. Inami woodcarvers were therefore guaranteed a livelihood and were able to develop their craft with relative freedom. Over the years, their repertoire expanded to include pieces of domestic carving such as *ranma*, the freestanding screens know as *tsuitate*, and carvings of animals, *daruma*, and mythical figures. Today, when demand for *ranma* is steady though limited, the craftsmen are thus able to supplement their incomes by making more domestic-oriented pieces such as ornaments and *shishi* (carved lion heads), as well as commercial items like picture frames.

The designs for the *ranma*, traditionally based on the wall paintings of the Kanō school and the landscapes of the Shijō and Maruyama schools, become grand, elegant pieces of interior decoration, sometimes enhanced by a technique known as *fukurin* whereby the relief is made to appear to be deeper than it actually is by extending the carving outside the frame.

For the *ranma*, woods such as camphor and zelkova are used, and for the ornaments paulownia is used as well. The boards from which *ranma* are carved are usually at least 45 millimeters thick. The wood is cut to size and seasoned naturally for between six months and a year, then the design is drawn on the wood and those parts not needed are cut out. Carving is first rough and then more detailed, the board being carved from both sides until the wood is cut through. The final detailed work is done with a sculpturing knife.

The carving of *ranma* is also carried out in the Osaka area. Together, Inami and Osaka contribute significantly to the continuance of a craft that over the years has adorned the places where the Japanese worship and that continues to grace many homes.

KISO OROKU COMBS
Kiso Oroku Gushi

Village of Kiso,* Kiso-gun, Nagano Prefecture

In Japan *kushi* (*gushi* in some cases) or combs, have traditionally been used not just to comb the hair but also as a hair ornament. Originally they had long teeth and were made of bamboo, but during the Nara period, more conventionally shaped combs were introduced from China. During the Heian period, the characteristic Japanese half-moon comb became the norm, and from about the same time, boxwood was a common material. By the beginning of the Edo period, however, combs made of other woods—rare woods from abroad and homegrown paulownia—came into favor, especially among the prostitutes, who were in a sense leaders of fashion at the time. The combs they used were rounded at the top, straight at the sides, and thinner than conventional combs of the period. Subsequently, combs of various shapes appeared, including some with a large flat area above the teeth that provided space for decoration in materials such as mother-of-pearl or in *makie*. Combs of tortoiseshell and ivory also appeared, but were mostly used by the women of samurai families.

Until the years preceding World War II, Oroku combs, the most representative of all the combs made in the Kiso area, could be found in almost every home in the country. Although less widespread today, they are still being made, often with material imported from Thailand. Large pieces of timber are first cut to a manageable size, then dried naturally for one to three months. The comb-shaped pieces of wood are then trimmed to the correct thickness and size by machine and, in the case of Oroku combs, the teeth are cut by hand. This is no easy task—there are ten to fourteen teeth per centimeter, which means that unless the comb is held up to the light, they cannot be clearly distinguished. The back of the comb is then shaped before the combs are buffed and finished to a natural sheen.

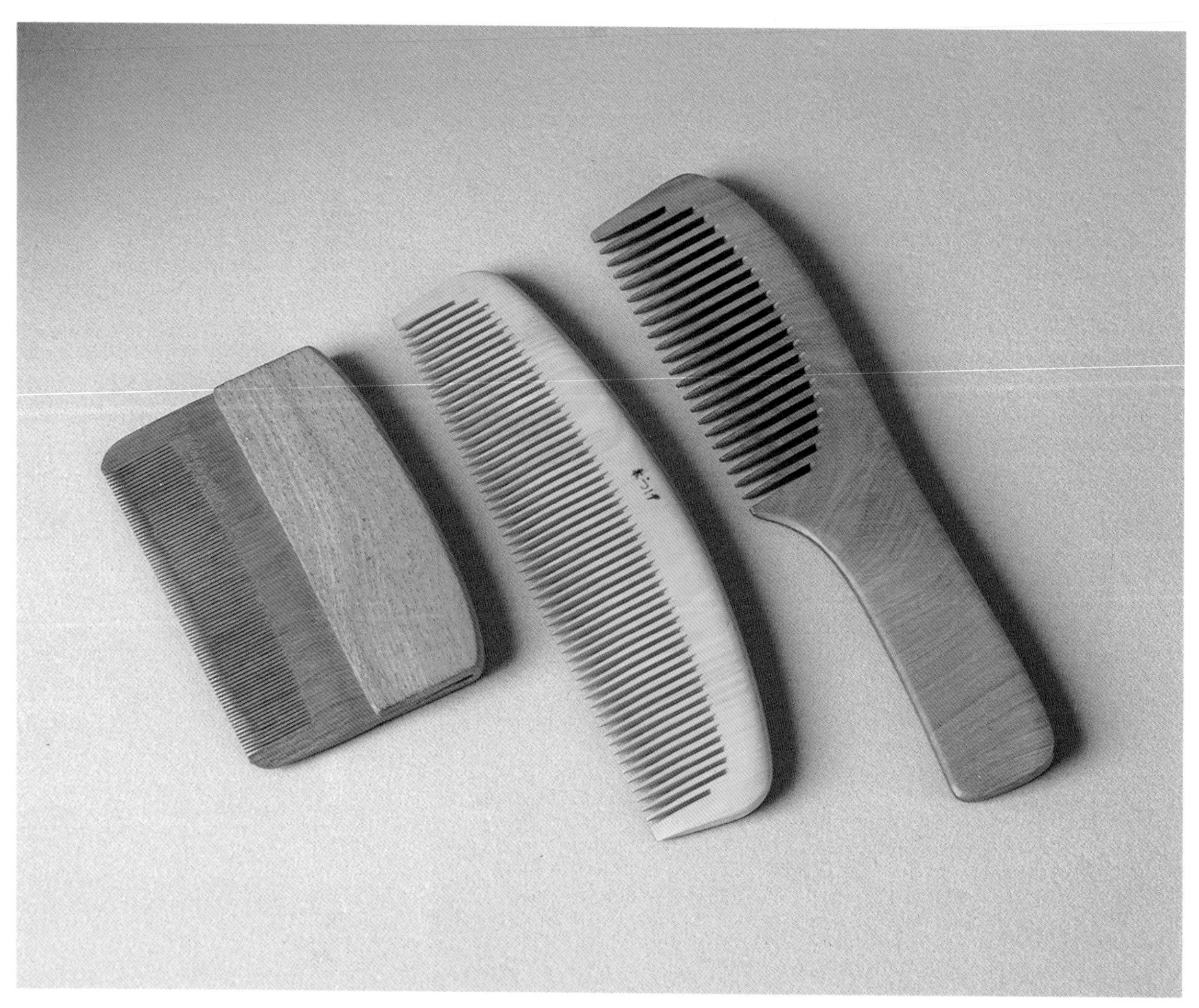

AKITA CHERRY-BARK WORK
Akita Kabazaiku

Kakunodate-machi, Akita Prefecture

The wild mountain cherries that eke out an existence in the poor soil among the cold rocky crags of Japan develop a tough fibrous bark. Although the bark can easily be peeled off in strips around the trunk of the tree in a horizontal direction, it is extremely strong in a vertical direction, and is also impermeable to air. These are the properties exploited in *kabazaiku*, cherry-bark work.

In the past, this type of work was produced around Hida Takayama in Gifu Prefecture, in the vicinity of Kōfu in Yamanashi Prefecture, and in various places in both Iwate Prefecture and Akita Prefecture. Even now, cherry bark is still used to "sew-joint" bentwood goods, and to make sheaths and handle covers for large mountain knives used in the lumber industry. But Kakunodate is the only place where cherry-bark work is still carried on in earnest.

The techniques of *kabazaiku* did not originate in Kakunodate, but were introduced in the 1780s from the Ani area north of the town. Developing into a side business for lower-ranking samurai, *kabazaiku* became widely known during the early part of the nineteenth century, and cherry-bark craftwork was often bought as a gift item. With major technical developments toward the end of the nineteenth century, it became an industry in the real sense of the word. In the years before World War II, the idea of using slivers of willow as a base on which to veneer cherry wood bark emerged, and in time even the use of tin became acceptable in the attempt to produce cheaper and more practical goods.

Today, a number of goods are being made alongside the more traditionally produced pieces. The work falls into the following basic categories: objects that use a tin or willow-sliver form, notably tea caddies, cigarette cases, and *natsume* (special small tea caddies for the lighter powdered tea of the tea ceremony); those relying on a wooden base, and including such objects as stationery boxes, jewelry cases, tables, small cabinets, and trays; and objects without a form such as *netsuke*, pendants, and brooches. Whatever the technique, these products have a durability and charm stemming from the strength, impermeability, and natural patina of the bark, which deepens with use.

A number of different woods are used for the bases, including cucumber tree, magnolia, Japanese cedar, Japanese cypress, white cypress, and paulownia. Before application of the cherry bark, the wood is covered with an animal glue. Cherry bark is applied to the inside surfaces of boxes and other articles; the outside is then veneered with one or more layers of the bark.

In the case of *katamono*, which use a tin- or willow-sliver form to make such items as tea caddies, the material is first bent into the required shape and fitted to a circular wooden base and top. The form is then veneered with the cherry bark, again using an animal glue as an adhesive. *Tataimono* are made from between ten and fifteen layers of cherry bark, bonded together with an animal glue and then carved into the required shape.

There is little doubt that on the trees, the bark will always be overshadowed by the cherry blossoms. But this relatively new craft is still thriving and producing forms that could one day rival the beauty of the blossoms.

KINKŌHIN

Metalwork

Introduction

Having had its foundations laid in antiquity, metal craft using both precious and nonprecious materials has achieved a high degree of refinement in Japan. This superb quality notwithstanding, metalwork does not enjoy the renown of other Japanese crafts. Yet it is well worth closer consideration.

From the representative metalworking techniques in Japan, two stand out as warranting special mention. In the technique of *hera shibori*, the base metal is placed against a wooden form and shaped to it by hammering with a short staff, normally of wood, as the form is turned. In *kirihame*, a design is drawn on the base metal, cut out, and then inlaid with a gold-copper alloy.

The techniques and styles discussed here represent some of the best examples of contemporary metalwork in Japan and cover a range of different metals and regions.

Kyoto metalwork.

NANBU CAST IRONWORK • *Nanbu Tekki*

Mizusawa and Morioka, Iwate Prefecture

Iwate Prefecture possesses all the main ingredients necessary for casting—iron sand, sand, clay, and charcoal. The area has thus been known for its casting skills for many hundreds of years. Nanbu *tekki* is just one example of the traditional casting techniques extant today.

The beginnings of Nanbu *tekki* are said to date back to the Momoyama period, when the feudal head of the Nanbu clan invited some craftsmen skilled in the techniques of casting to come to Morioka from Kyoto. There they produced *chagama*, or pots in which to heat water for the tea ceremony. Some time later, in 1626 (Kan'ei 3), the famous *Kan'ei tsūhō* was made in the region. Later, four more skilled craftsmen were invited to the area from Tokyo (then called Edo) and elsewhere. Descendants of these four jealously guarded and preserved the traditional skills and techniques of casting. However, in 1645 (Shōho 2), the local administration invited other metalworkers and casters from Tamoyama (now Ōfunato) to come to Mizusawa, which was then part of the Daté clan. A large amount of metalwork was produced here during the Tenna era, between 1681 and 1684. The design of the *chagama* was revised and more conventional-looking teakettles were made, along with cooking pots. The production of such items continued almost unabated until after World War II, when a drastic change in the lifestyle of the Japanese meant that iron teakettles and other traditional metal pots were no longer in such great demand. Diversification of production was therefore pursued, including decorative items, ashtrays, and wind bells.

Present-day Mizusawa is best known for the production of machine parts and for its wind bells. This does not mean, however, that metal cooking pots have disappeared. Apart from the wind bells, traditional metal teakettles, *chagama*, and sukiyaki pans are still being produced, along with some smaller decorative items. Many of these products are characterized by the raised spot pattern known as *arare moyō*, the dull luster of the metal, and a feeling of strength and substance.

The production techniques have changed little over the years. Iron sand or casting pig iron is poured into the cast. When it has cooled the cast is split open and the casting removed. The making of casts themselves is worthy of note. Clay that has been fired is pulverized and kneaded together with clay slurry. The design is put on the inside of the cast, which is then fired and hardened. The core is made in the same way as the mold. Placed within the outer mold, it leaves a ridge on the casting. Because the parts of the cast must be made each time, this method is not suited to mass production. Dry molds are therefore employed for mass-produced items. In order to make one of these, the original form is first made from stone chips. An outer cast is made by using a mixture of sand and clay slurry and is then put into a drying furnace. The same method and casting process are employed to make the core.

Nanbu *tekki* is found in use all over Japan. It brings an air of quality and tradition to many meals, not the least of which is sukiyaki, one of Japan's foremost contributions to the dining tables of the world.

TSUBAME BEATEN COPPERWARE
Tsubame Tsuiki Dōki

Bunsui-machi and Tsubame, Niigata Prefecture

Beating a metal sheet into shape with a hammer or staff is a technique that can be found in most parts of the world. It is known alternatively as raising or sinking; in Japan it is called *tsuiki*. Copper is the preferred material as it is extremely malleable.

The art of metal forming was brought to Tsubame in the 1760s from Sendai, the provincial capital of what is now Miyagi Prefecture. The techniques were mastered by the craftsman Tamagawa Kakubei, whose work remains the basis for today's craft. Although production first centered on utilitarian items such as cooking pots and water pitchers, as the craftsmen acquired more expertise, they gradually began to create pieces of a refined and elegant nature. Plating and the technique of inlaying gold and silver were introduced at the end of the nineteenth century, and these served to popularize fine copperware.

In the 1920s, however, the introduction of aluminum goods severely eroded the market for copperware. The situation declined further with the outbreak of war. The price of copper soared, and most of the skilled staff left or were conscripted. Fortunately, since the end of the war, demand has returned more strongly than ever. Tsubame copperware is much loved for its beautiful coloring and marking, as well as for its durability. There is a greater emphasis now on decorative items, yet finely crafted household goods and tea-ceremony utensils remain the mainstay of production.

After the copper has been heated so as to anneal, it is formed with a hammer or wooden mallet either by sinking (*uchiage*), in which the copper is beaten to conform to a sunken form made of zelkova wood, or by raising, in which the metal is gradually reduced by beating it around a steel form (*toriguchi*). A third technique is *itamaki*. The metal is bent into a cylindrical shape, and the bottom and side seams are fastened by a claw joint or by soldering. Decorations, if any, are either beaten into the metal, chased, or inlaid (*kirihame*). The surface of the copper is treated in order to prevent oxidation. It is possible to color the metal a warm yellow, scarlet, or amethyst by soaking it in solutions mixing copper, copper sulfide, and calcium sulfide, with burnishing or polishing to determine the final hue.

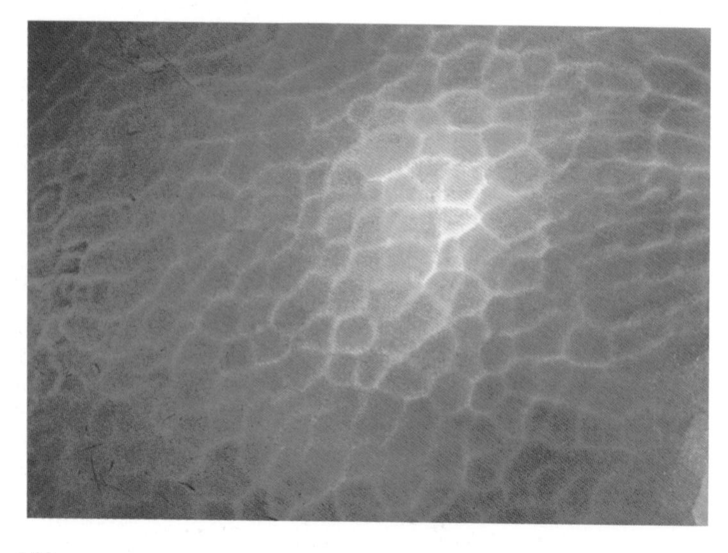

TAKAOKA CASTING • *Takaoka Dōki*

Takaoka, Toyama Prefecture

The primary material for casting is bronze, but because it is an alloy of copper, articles made of bronze and other copper alloys are categorized as *dōki* (copper articles). Copper itself is often referred to as *akagane*, meaning "red metal," or simply as *aka*. On the other hand, bronze, which has a somewhat greenish cast, is called *seidō* (green copper).

Located in Toyama Prefecture not far from the Japan Sea, Takaoka is the nation's top producer of cast bronze. Casting began around 1611 when the Takaoka clan summoned seven casting experts from the Kawachi area of Osaka Prefecture, where casting had first developed, in order to set up an income-producing industry for this rather poorly endowed castle town. In its early days the foundry produced cast-iron kitchen ware and other household articles as well as agricultural equipment such as plows (*suki*) and hoes (*kuwa*). Later, religious articles, incense burners, vases, and other finely crafted items were cast in bronze. Although demand for these bronze pieces rapidly increased during the thirty-year period from 1751, production remained a cottage industry until 1819, when a distribution system was organized. Specialization resulted in a greater division of labor, which in turn led to larger and more efficient production. After World War II, factory-style management was introduced, and production was mechanized.

The *karakane* (Chinese metal) alloy that is used for these decorative objects contains copper, zinc, lead, and 10 percent or less tin. By contrast, temple bells are made of bronze having 15 to 20 percent tin. Casting is done with either a dry-sand mold, another type of dry molding material, or lost-wax. Molds for round objects are called *hikigata*, those for irregular shapes *komegata*. The metal chasing is all done by hand. The coloring of the metal is achieved by immersion in a chemical bath or by painting a mixture of saltpeter and rice bran onto the surface and then heating it in a furnace. Although cast bronze industries also exist in Niigata, Yamagata, and Hiroshima Prefectures, Takaoka is noted for being able to cast exceptionally large pieces as well as delicate work.

OSAKA NANIWA PEWTER WARE
Osaka Naniwa Suzuki
Osaka

Tin is easy to work and neither corrodes nor tarnishes. In Japan it is reputed to improve the taste of water and saké, and is for this reason considered an ideal material for teapots and vessels for serving Japan's national liquor.

The earliest tin artifacts to have been found in Japan date back thirteen hundred years and are kept at *Shōsōin*, the imperial repository in Nara. The pieces are of Chinese origin, however, and this would have made them too precious for use by anyone not of a privileged status. Indeed, it was not until the eighteenth century that tinware came into general use.

Tin mining in Japan began in 1701 at Taniyama in Kagoshima. The Taniyama tin was of good quality, and rather than being processed locally, a large proportion of it was sent for sale in Osaka. Tinware had already become a viable industry in Osaka by 1679 at the latest and in Kyoto by 1690. In the second half of the nineteenth century, a way of purifying tin was mastered and lathes were improved, enabling great advances in the craft. In 1872, the *ibushi* method of coloring the metal was discovered. This popular effect is achieved by exposing the formed piece to the smoke of smoldering wood.

The tin that is being used today is 97 parts pure; lead is added to bind and strengthen the material. Molten tin is poured into molds and cast. Some of the articles are turned on a lathe while others are beaten and polished. The body, lip, and base are either welded together or joined using a solder known as *rogane*. Designs or motifs are painted on in natural lacquer, which acts as a resist, and then the surface is etched with nitric acid. Lacquer is rubbed into the surface, and final finishing is carried out.

Tinware is now being produced only in Osaka and Taniyama, where it was first mined. Competition from other materials is such that most contemporary pewter ware is turned, since mass production is easier by this method. Product ranges have also expanded to include jugs, tumblers, coasters, and novelty items.

TOKYO SILVERSMITHERY • *Tokyo Ginki*

Tokyo

Tokyo is one of the few places in Japan where silverware is made. The craft took root and developed in Tokyo just as the city was beginning to grow as the military capital of the country in the early seventeenth century.

Historical records of the period between 1624 and 1644 tell us that silver items made in Edo (present-day Tokyo) were presented to the emperor and the shogun. They also describe the types of tools being used in the silver craft. Although at first only the higher-ranking members of the warrior class could afford articles made of this precious metal, by the end of the seventeenth century the clientele had grown considerably.

Three of the techniques still in use today—*kirihame* inlaying, *chōkin* chasing, and metal forming—can be traced back to individuals who worked during the Edo Period. They are employed for a large range of accessories and household goods. The silver is a very pure 925 parts to 1,000. It is formed by either raising and sinking, in which case the surface is always decorated, or by spinning (*hera shibori*). Parts are joined by soldering with silver, fusing, or riveting.

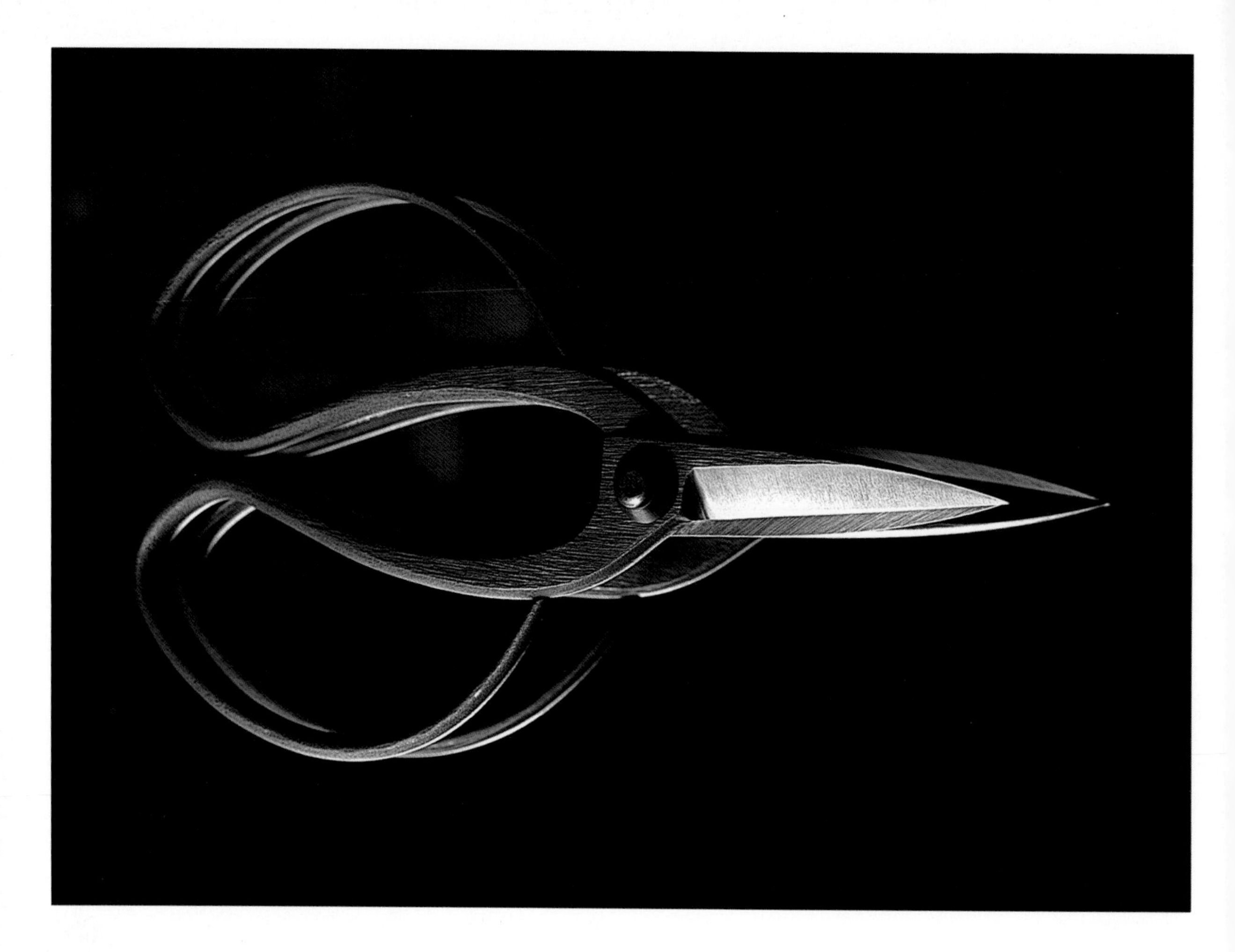

SAKAI FORGED BLADES • *Sakai Uchi Hamono*

Sakai and Osaka, Osaka Prefecture

In Japan the techniques of forging became increasingly sophisticated from the eighth century onward. Those of the Osaka area were known as Kawachi forging, after the name of that district. The import and export of iron and steel itself was being handled through Sakai, which was a prosperous port doing trade with other Asian countries. Then, in 1543, an event occurred that changed not only the forging industry in Sakai, but the whole history of the country. A Portuguese ship arrived at Tanegashima Island at the tip of Kyūshū, bringing with it the first firearms known to the Japanese. Among the metalworkers who made copies of these weapons was a man from Sakai. He returned to Sakai and with the financial backing of one of the wealthy Sakai merchants began producing firearms, which came to be known as Sakai *teppō*.

The skills and techniques used in the production of Sakai *teppō* were later put to use in the making of knives and other blade tools. The first of these was a tobacco knife introduced between 1570 and 1590. It was given the name *okata* (after the wife of the metalsmith) *hōchō* (knife). The general name for the blade tools of Sakai was Sakai *uchi* (forged) *hamono* (blades). The techniques were based on those employed by swordsmiths, but in the first half of the eighteenth century a blacksmith created a very sharp knife called *yamanoue hōchō,* literally "mountaintop knife." This led to the production of various other knives for cooking and carving, scissors, razors, sickles, and a wide range of carpentry tools. To this day, the sharpness of the Sakai blade tools sets them apart.

The base metal is a very soft steel, and the cutting edge is a high-quality blade steel. The two are forged together with the hard blade steel sandwiched between the soft steel that forms the back of the blade. The method of tempering is unique to Sakai. The blades are first completely covered with mud, then placed in a furnace. When the red-hot blades are removed, they are immediately plunged into water and cooled. The blades afterwards are sharpened and fitted with wooden handles.

Miki and Ono, both also in Hyōgo Prefecture, produce more knives and blade tools than Sakai, yet the city maintains a thriving industry renowned for high-quality goods.

KYOTO METAL INLAY • *Kyō Zōgan*

Kyoto

The technique of inlaying metals was probably brought to Japan from Korea with the great influx of continental culture during the Asuka period, which lasted from the late sixth century to 710. In the eighth century metal inlaying was already taking place in Kyoto.

Two families of metalsmiths, the Shoami and the Umetada, had become the leading experts in the craft by the time the provinces of Japan fell under the control of powerful feudal lords in the sixteenth century. Craftsmen who had been apprenticed to these two families attached themselves to these feudal lords, and in this way the techniques of *Kyō zōgan* spread throughout the country. Although inlaying was first employed for decorating religious objects, sword hilts, and other articles for the nobility and warrior class, the prosperity and peaceful environment brought by the consolidation of Japan at the beginning of the seventeenth century saw *zōgan* also being used for objects of everyday life such as hibachi braziers, the thin tobacco pipes called *kiseru,* and purse closures. Toward the end of the nineteenth century, almost all the metal inlaying and cloisonne was being done for export. Today, 80 percent of Japanese *zōgan* is produced in Kyoto, with the remainder being made in Kanazawa in Ishikawa Prefecture and Kumamoto on the island of Kyūshū.

Some *zōgan* still is employed on work for religious purposes; however, there is a greater demand for its use in accessories such as brooches, tiepins, buckles, and badges.

The techniques are the same as those once used to decorate the samurai swords: *ito zōgan* which resembles a fine thread; *nunome zōgan,* which resembles a rough weave; *hira* (flat) *zōgan*; and *taka* (relief) *zōgan*. The base metal is usually polished steel; the engraving tool is called a *tagane* or a *burin*. The engraved lines are filled with gold or silver that has been hammered into thin sheets or cut with a die. This inlay process involves some sixteen, painstaking stages. Corrosion is prevented by immersing the metal in a bath of hot tannin obtained from Japanese tea. Natural lacquer is then applied to the surface and warmed to fix it before the piece is polished to its final sheen with wood charcoal or a metal spatula.

SHOKŌGEIHIN

Other Crafts

Introduction

Of the vast array of crafts still being made in Japan, a number of them fall outside of the major material categories presented in the previous pages. Ogatsu inkstones and Izumo stone lanterns represent the stone-carving crafts, and the luxurious lacquered work of Kōshū deerhide pouches constitutes a rare example of an object fashioned from animal products.

Many of the crafts in this section rely on more than one material. Such traditional objects as brushes, fans, umbrellas, and lanterns require a second material to be used in tandem with a skeletal base of bamboo, a material that found its way into all walks of life in olden times. Other objects, most notably the portable shrine and the Buddhist altar, require a team of skilled craftsmen working in various genres to complete.

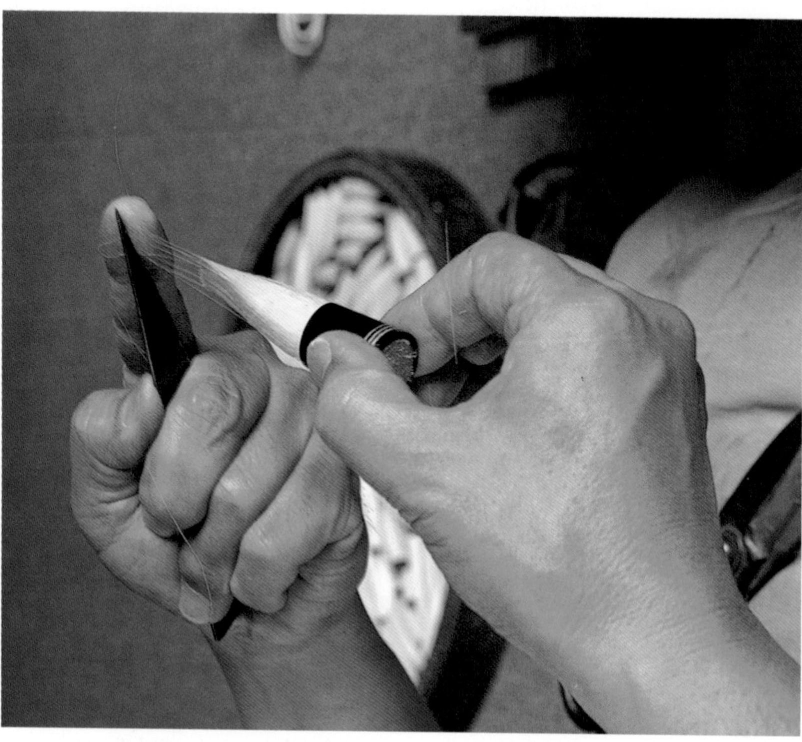

Trimming the hair of a Nara brush.

A Miyagi *kokeshi* doll on the lathe.

BANSHŪ ABACUS
Banshū Soroban

Ono, Hyōgo Prefecture

As a way of making calculations, the abacus is one of mankind's most ancient instruments, and only in recent times has it been superseded by the electronic calculator. A good abacus operator in Japan can beat someone using a calculator, a feat that has been demonstrated on several occasions in competition. Demand for the abacus has certainly dropped as electronic calculators have become cheaper and more widely available, yet it is still used to teach children the principles of mathematics, something that has caused great interest in America, Europe, and Russia. Even today, many shopkeepers and accountants still use an abacus to double-check their electronic calculations.

It is thought that the abacus reached Japan during in the 1590s via China. Having been introduced, it was adapted to meet local requirements, the form it took being the one that is used to this day. It was the settled Edo period that saw the generalization of their use for everyday calculation and, particularly in Western Japan, the *soroban* became known as both the "life" and "spirit" of the tradespeople who prospered during this peaceful time.

Making a good *soroban* is, of course, an intricate job requiring much patience. The frames are made from a variety of woods, including ebony, oak, and birch. The shafts for the beads are generally made from bamboo, and the beads are fashioned from such woods as boxwood, birch, or holly. The beads are usually polished, but if they are made of birch they are stained. The bamboo shafts are also stained, unless they are made from *susudake*—a type of smoke-stained bamboo traditionally obtained from inside the roofs of old dwellings. A small white dot, called a *hatome* (literally, a "pigeon's eye"), is added to the upper edge of the top and bottom frames as a guide in calculations. The frame itself is then jointed together using either round or triangular mortise and tenon joints. These are employed because they tend not to loosen as the abacus is used.

Banshū *soroban* now account for approximately four-fifths of national production, the remainder coming from Unshū in Shimane Prefecture and Ōsumi in Kagoshima Prefecture. It is difficult to say how long the *soroban* can hold out against the electronic revolution, although it is interesting to note that at one time it was possible to buy a combination abacus-calculator, an attempt to combine the best of both worlds.

FUKUYAMA PLANE HARPS
Fukuyama Koto

Fukuyama,* Hiroshima Prefecture

The *koto* appears simple and unassuming when compared to its upright Western cousins; it is somehow so characteristic of Japan, yet it is not originally a Japanese instrument. Its prototype was Chinese, and only after a long and gradual period of adaptation and development has it become, along with the *shakuhachi* and the *shamisen,* one of Japan's representative musical instruments. During its history, the *koto* has taken various different forms: a 25-string instrument called a *hitsu,* a 13-string *sō,* a 7-string *kin,* and the 6-string *wagon* (sometimes also called the *yamato goto*). But since the Kamakura period, the 13-string *sō* has become the instrument known under the generic term of *koto.*

Production of *koto* started in Fukuyama in the Edo period, but large quantities were not made until the beginning of this century. The city has subsequently become Japan's center for production of *koto* of the highest quality and now produces four-fifths of all these instruments.

Basically speaking, the *koto* is composed of a bowed and barreled wooden soundbox over which silk strings are stretched and raised on individual bridges. The upper part looks something like a flat, upturned U-shape in section, being cut from a log of paulownia (*kiri*). The original cutting from the log is done with sophisticated circular band saws, and it is then primitively hacked and mostly planed into shape. After this sounding board has been thoroughly dried, a herringbone pattern known as *ayasugibori* is carefully inscribed across part of the back with a chisel, the effect being to improve the quality of its acoustics. This painstaking work becomes more or less hidden, as a similarly well-seasoned board of the same wood is attached behind it, with spacers known as *seki-ita* and *hari-ita* fitted between. Next, all the surfaces are charred. A red-hot iron heated to 1800° F (1000° C) is dragged across the surface of the wood, which immediately flares and is naturally extinguished. The required degree of charring can only be obtained by carefully controlling the pressure applied to the iron as it cools. The final finishing of the surface to give it a "salt and pepper" hue is done with a wire brush called an *uzukuri,* made of a bundle of roots from the Japanese yew, and a light polishing is done with a beetle wax. The wax is put into a bag of loose material, dusted onto the surface of the wood, and then polished up with a cloth. To complete the instrument, decorations of ivory and rare wood are added.

While there are two basic types, the only real difference is that the Yamada *koto* has a slightly more swelling body than the dead-straight Ikuta *koto.* Both have the same gently bowed form when seen from the side. What they also share, of course, is an elegance and enigmatic air fostered by years of development. The end result of this truly Japanese mix of techniques—from the highly crafted yet usually concealed pattern on the back of the sounding board to the simple, almost primitive finishing with fire and beetle wax—is an instrument as sophisticated as those in the West, but unmistakably Japanese in character and musical tone.

KŌSHŪ LACQUERED DEERHIDE
Kōshū Inden

Kōfu, Yamanashi Prefecture

Situated off the eastern edge of the Asian continent, Japan has inevitably been a cul-de-sac for various customs and technologies originating elsewhere. Even today, Japan is widely regarded as the best developer of ideas in the world; Kōshū *inden* is but an ancient example of this phenomenon.

Thought to have originated in India, Kōshū *inden* is basically a technique of coloring and applying a pattern to deerskin. According to the *Engishiki*, an ancient book on official protocol, the process by which the skin is dressed was brought to Japan by Korean tanners in the Asuka period. Much later, a monk from Kyoto's Myōshinji temple brought back a tiger-skin accessory from India for the feudal lord, Takeda Shingen, and this triggered the production of a bag made from deerskin. During the civil war period from 1490 to 1600, the Kōshū *inden* technique was utilized on the armor, helmets, and other military goods of the warriors. This gave way in the Edo period to its use on some formal pieces of clothing and especially on pouches, money belts, and tobacco pouches, since many people were moving about the country and needed small, light, but durable receptacles for their possessions. Nowadays, its use has been extended to such things as belts, ladies' handbags, and business-card holders, all of which take ample advantage of its ability to breathe, water resistance, durability, and almost everlasting softness and smooth texture.

As with many other traditional techniques, the process of preparing the hide relies on both simple methods and the effects obtained by using natural materials. After the hide has been tanned, any remaining hair or other extraneous matter is removed by singeing it with a hot iron, and the smooth, suedelike hide is smoked or stained with a ground color by using straw and pine resin. To soften it further, it is rubbed down with pumice (a sander is normally used nowadays). Next, the hide is cut into appropriate sizes. The application of the pattern is mainly done using natural lacquer into which a fine-powdered polishing stone (*tonoko*), pigment, and egg white have been mixed. This mixture is applied to the prepared hide with a stencil and squeegee. Various patterns are used, most of them small, overall repeats, featuring wave motifs, chrysanthemums, small cherry flowers, and geometric designs.

The products utilizing Kōshū *inden* have the kind of appeal that transcends cultural and generational barriers with the agility of the deer from whose hide they are made.

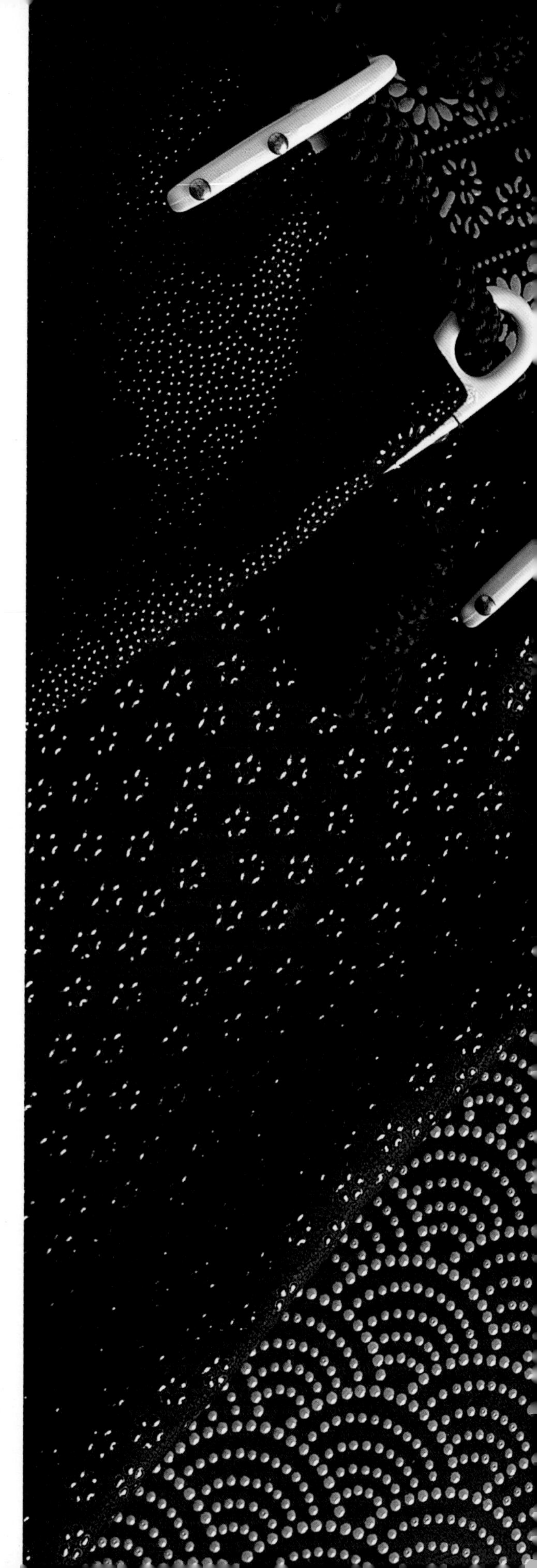

NARA SUMI INK • *Nara Zumi*

Nara, Nara Prefecture

A good ink stick may be as highly treasured by calligraphers and artists as the ink stone on which it is ground. The ink is made by pressing the stick onto the surface of the stone and moving it back and forth in a small pool of water. Nowadays, traditional ink comes in stick, liquid, or paste form, and in black or red.

The actual origins of Nara *zumi* go back to the ink made at Kōfukuji temple in the city of Nara between 1389 and 1427. The year 1739 was a turning point: with a view to raising the quality of the ink, further instruction was given on production methods by some Chinese living in Nagasaki at the time. Ever since, Nara *zumi* has been well known.

Even to the uninitiated, production is simple. First, soot is collected from smoke deposits left by burning oil or pine logs. Some scent, such as musk, is added along with an animal glue, and the mixture is kneaded into a paste that is then pressed into molds. Traditionally, the sticks are hung out to dry for two to four months indoors, in low temperatures. The dried sticks, which become very hard, are then polished with a shell to bring up a gloss. Lastly, gold, silver, and colored powders are applied over the embossed characters and designs on the sticks. The result is a highly crafted object that is nonetheless fated to disappear, worn away to nothing as it is ground down for another work of art, or just a letter to a friend.

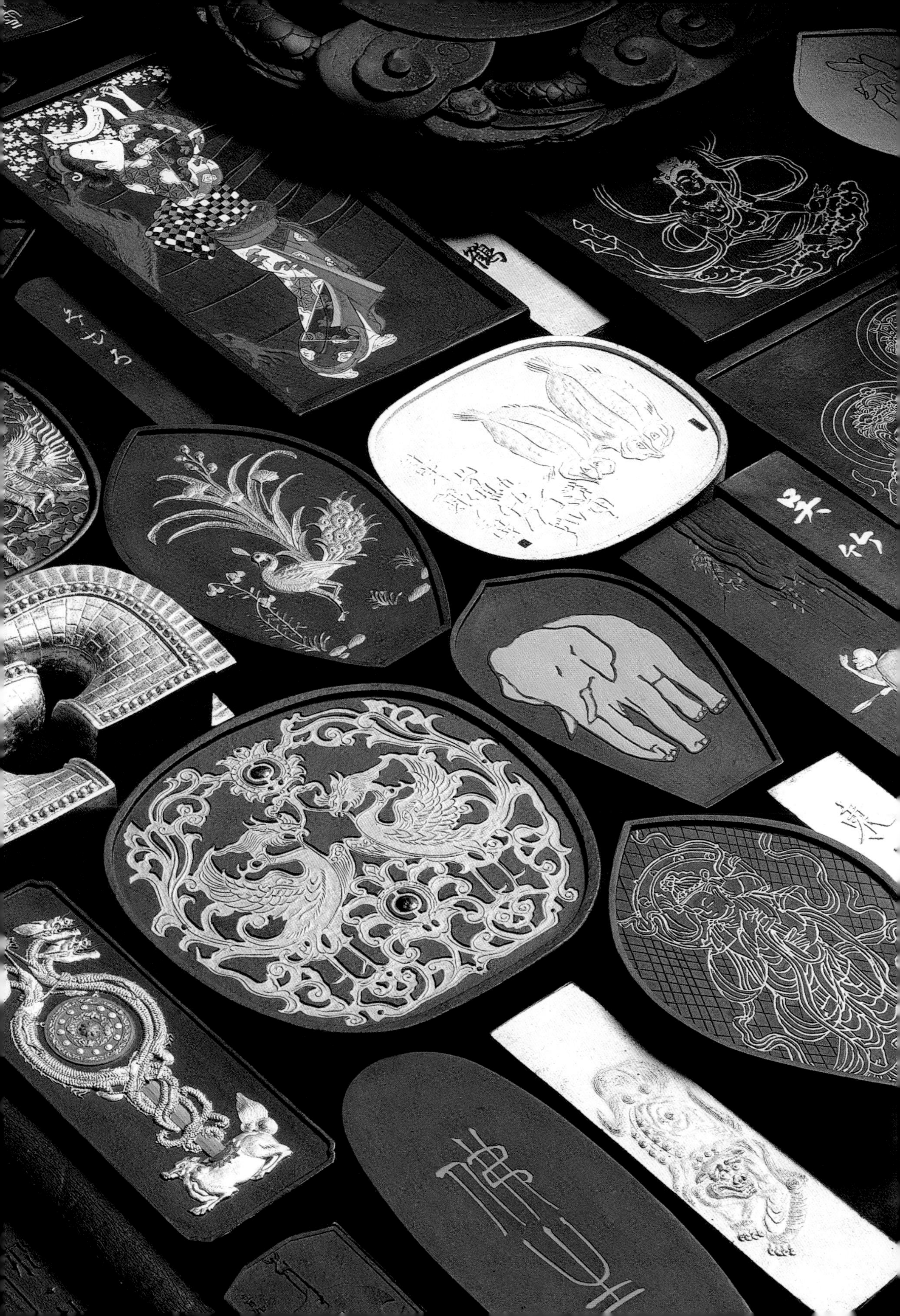

NARA BRUSHES
Nara Fude

Nara, Nara Prefecture

The writing brush, or *fude*, became an essential writing tool on the Japanese archipelago during the Kofun period, when there were active dealings with China. The origins of Nara *fude* are said to go back to the eighth century, when the introduction of Buddhism led to the establishment of places to copy the sutras. Brushes were essential for the many students of Buddhism to reproduce the holy scriptures, which were the essence of their teachings and devotions. The Matsunaga Revolt of 1567 caused an abrupt drop in demand for brushes from temples and shrines alike, obliging the brush makers themselves to take to the roads and travel the country selling their wares. Stability returned in the Edo period, and *fude* came into general use with mandatory education. They have, however, fallen into relative disuse since the end of World War II. At present, production figures for Kumano *fude* from Hiroshima Prefecture are the high-est in the country, but Nara *fude* are the oldest and, together with those produced in Toyohashi, in Aichi Prefecture, they are acclaimed as brushes of the highest quality by specialists both in Japan and overseas.

The hair for these brushes is taken from lambs, horses, the Japanese raccoon (*tanuki*), or even cats. The shafts are usually made of bamboo, but are sometimes fashioned from wood. The selected hair is first boiled to remove any kinks and to soften it. Enough hair—according to the type of brush to be made—is cut to the same length and bound at one end with a linen thread. This bundle of hair is then introduced into its shaft and fixed with an adhesive. The tip of the brush is either left as it is or is given a coat of gum or funorin to bring the hairs together into a point.

Each kind of brush, be it wide or narrow, has a purpose in calligraphy or painting. But whatever its purpose, it becomes an extension of the hand of the person using it, giving expression to their words and thoughts, and lending a distinctive style and expressive quality to the characters and images that it is used to write or draw.

OGATSU INKSTONES • *Ogatsu Suzuri*

Ogatsu-chō, Miyagi Prefecture

For better or worse, word processors are the writing tool of the age. But in days gone by it was the writing brush, inkstone, ink, and paper that were the essential tools of the written word in Japan and other Asian countries. The inkstone, in particular, was considered of great importance in Japan, being likened to the sword of the samurai and the mirror of a lady, according to one old saying.

Apart from the Ogatsu inkstone—the most well known in Japan—the Amahada *suzuri* from Yamanashi Prefecture and the Akama *suzuri* from Yamaguchi Prefecture are of particular note. Those from Ogatsu are first mentioned in records from 1396. After 1591, the techniques of making the stones were handed down from father to son as a trade secret by the Okuda family, who were the recognized craftsmen (*suzuri kirikata*) of the times. In the later Edo period, the Daté family took over as the keepers of this art. The number of places making them increased as the literacy rate rose, yet today Ogatsu alone produces 90 percent of the country's natural stone for *suzuri*. Of course, it must be said that ever fewer inkstones are in daily use.

The slate used for Ogatsu inkstones is prized for its luster and grain quality. It is resistant to extreme temperatures, wear and tear from long usage, and warping. In addition, the copper, iron, and quartz in the stone create the ideal surface on which to rub ink sticks.

After the slate is mined, the substandard stone is discarded. The remainder, about 10 percent of the total, is separated sheet by sheet and cut to the required size and thickness. The surface is then smoothed out with sand and water and the inkwell is carved. Once the carving is concluded, the polishing commences. The interior is done first, followed by the exterior, and a final polish brings out the stone's color and sheen.

Some Ogatsu *suzuri* are collectors' items; others are in general use. They are prized equally by those who actually use them and by those who just admire them for their deep, dark indigo or sooty coloring, their smooth surface, and their hardness, which makes them ideal for preparing ink. The better-carved examples are considered works of art.

GYŌTOKU ŌKARAHAFŪ PORTABLE SHRINE
Gyōtoku Ōkarahafū Mikoshi

Ichikawa,* Chiba Prefecture

While the Japanese are not usually considered to be a devout race of people in the Western sense of monotheistic belief, they are nevertheless religious believers. There is a type of religious "parallelism" in Japan: the actual population is outnumbered by the official number of believers, possible only because so many people recognize at least two religions. This mania for religious adherence helps to sustain a number of highly refined craft items, particularly those associated with the main beliefs of Shintōism and Buddhism. The portable shrine and the household altar are two significant examples. As anyone visiting Japan quickly realizes, the Japanese like festivals. Originally, festivals were religious rituals at which offerings were made in a ceremonial manner to welcome divine spirits. Although their religious significance may have dimmed, these festive occasions are still begin held up and down the country and are attended by hundreds of people celebrating and commemorating a particular event or happening.

The carrying of a *mikoshi* is just one of the things that constitute such a religious festival. A *mikoshi* is a portable shrine in which the embodiment of a god or a divine spirit is enshrined. They are carried enthusiastically by young men and women who were born in the area where the enshrined god or spirit resides. The *mikoshi* are carried around the immediate neighborhood while bearers chant *"wasshoi,"* a word that according to one explanation, was simply a greeting to the gods in the past. In the grounds of the shrine where the festival begins, there is chanting (*kagura bayashi*) and dancing to music played in praise of the god. On the streets the sight and sound of the *mikoshi* bouncing in time with the chanting draws onlookers away from the real world and into another bound by a common spirit of participation. While many modern Japanese consider festivals simply as an excuse for a lot of noise and merrymaking, the events still foster a sense of community.

Historical records tell us that the first use of a *mikoshi* was in 749, a purple-colored palanquin that was used to carry the divine spirit of the Usa Hachiman shrine at the time of the founding of the great temple of Tōdaiji, in Nara. Not until sometime later, during the middle of the Heian period, were *mikoshi* used at festivals dedicated to local gods in small settlements and villages. By the end of the twelfth century, references were being made to *abare mikoshi*. This is the raising and lowering, or bouncing, of a *mikoshi* as it is being carried. Prior to this they had simply been carried in a solemn and dignified manner. This more boisterous style of festival became increasingly popular during the Edo period.

The Gyōtoku *ōkarahafū mikoshi* are characterized by the brave, elegant line of the gable. This is just one of the elements that is borrowed from architecture in the construction of what is essentially a miniature building. The kind of gable employed on these particular *mikoshi* was derived from Chinese architecture. These often form the end of a "porch roof," or secondary gable set into the sweep of a main roof.

The woods generally used to make *mikoshi* are Japanese cypress, evergreen oak, and zelkova. The zelkova is used for the posts and beams; oak and zelkova are used for those parts of the construction that need to be strong; and the cypress is used for those parts that are coated with lacquer. All the wood is well seasoned; heartwood is avoided.

Most of the pieces are made individually and then assembled. Construction begins by building a frame onto which four posts are erected. These support the body and roof of the *mikoshi*. Fences and *torii*, or characteristic Shintō shrine arches, are then fixed. The roof is decorated with such things as the mythical *hōō*, or phoenix. All of the woodwork requires frequent use of various kinds of chisels, saws, and planes. Moreover, the entire work—carving, woodwork, lacquering, and application of decorations—is done by either one person or a small group of people, and therefore demands strong all-round skills.

One of Japan's major festivals is the *Sanja Matsuri*, which is held every May in the Asakusa district of Tokyo. Hundreds of people are involved in the bearing of the one hundred *mikoshi* that are paraded over three days and, as can be expected, this draws huge crowds of onlookers. Latterly the festival itself has become quite international, having been held in such places as Venice, Hawaii, Düsseldorf, Nice, and Cannes. Yet whether the festival is held in some small mountain village or in some far-flung corner of the world, it would be nothing without the *mikoshi*.

KYOTO HOUSEHOLD BUDDHIST ALTARS, FITTINGS
Kyō Butsudan, Kyō Butsugu

Kyoto

Buddhist family altars and altar implements are ordinarily considered separately. The essential difference is that family altars are used in the home, whereas the other items are used at official places of worship. In the simplest possible terms, the cupboard-like *butsudan* is used to house memorial tablets inscribed with the secular names of ancestors. There is also usually an image of the Buddha so that worship can take place in front of a *butsudan* on a daily basis.

The making of implements used on Buddhist altars in Kyoto dates back to the very introduction of Buddhism, sometime around 552. During the second half of the Heian period, the building of esoteric temples and the making of religious articles was being actively pursued, nowhere more so than in Kyoto, which was already the center of religious crafts. The Kamakura period was to see a further increase in the making of Buddhist statues and metalwork altar fittings. It was not until the beginning of the seventeenth century, however, that home altars came into use. The catalyst for this change was a declaration of faith made obligatory at the time of a census in order to suppress the Christian religion, and sometime after 1641, the altars found a place in every home as proof of religious belief.

Today, home altars are made in various places, each with their own particular characteristics. Gilded altars from such places as Nagoya and Hikone (Shiga Prefecture) are known as *nuri butsudan*. Rare wood altars (*karaki butsudan*) are made in Tokyo. Those fashioned in Tokushima are known for their large size. Conversely, the number of locations in which articles for altars are being made is rather small. In fact, Kyoto produces 85 percent of the country's total requirements, mostly for use at temples and monasteries.

An extensive range of religious paraphernalia is produced in Kyoto. Used when the sutras are chanted, the *mokugyo* is either a wooden board shaped like a fish or a hollow, shell-like piece of wood decorated with fish scales. It is struck rhythmically with a stick, the end of which is covered by a piece of cloth or leather. Placed in front of the Buddha, the *takatsuki* is a long-legged disk in which fruit or confectionery are offered. Looking something like a string of precious stones, the *yōrai* are placed around the neck, on the arm or at the elbow, and on the leg of a Buddha statue as a form of decoration. Made to hold the incense that is burned at Buddhist services, metal and pottery *kōro* are produced in a number of styles. Called a *kin* but usually known as a *rin*, this metal chime is struck as a sign that a *gongyō* service is going to begin. Again struck like an instrument, the *dora* is used at a *hōyō* memorial service. Decorative in nature, the *keman* are used to create an air of solemnity in the grounds of the temple. These flowerlike garlands made of various materials are hung over railings and other appropriate places. In addition, there are the statues of the Buddha and paintings depicting Buddhist subjects. But whatever the article, all are instilled with an essence of Buddhism and a sense of Kyoto craftsmanship.

The making of a *butsudan* involves a number of specialized craftsmen. There are woodworkers who cut and prepare the necessary material, woodcarvers, craftsmen who deal with the decorations, lacquerers, gilders, and specialists in the art of *makie*. Even the final assembly of a home altar is left up to a skilled person.

The fabrication of a *butsudan* is a process of assembling parts that in many cases have been finished beforehand. The woods used are Japanese cypress, zelkova, pine, and Japanese cedar. Such woods are used for the making of the roofs, doors, plinths, and inner sanctuary. After the wood has been carved it is coated with natural lacquer. Decorative applications such as *makie* and gilding are then carried out as required. Next the copper fittings are attached. The whole thing is then assembled and the final finishing is done.

Lacquer is used for most of the coloring and decorating of the metal and wooden articles produced, and the latter in particular are often gilded. Some of the articles carved out of wood are made by laminating several pieces of wood together (*yosegi*), whereas others are carved directly from a piece of wood (*ittōbori*). Copper or a copper alloy are often used for the metalwork. Casting, forging, and metal carving are the principal modeling techniques employed.

The artistry and craftsmanship that are exercised on these religious items are of the highest quality and yet are executed with apparent ease. The end result is a work of stature that still manages to glow from within.

KYOTO CANDLES • *Kyō Rōsoku*

Kyoto

Since time immemorial, people everywhere have sought a reliable source of artificial light. In Japan, the candle represents just one stage in this quest. Probably the first light sources used in the country were pitch-pine torches. Then came pine branches (*shisoku*) that had their ends scorched and soaked in oil. Later, tapers of paper were twisted up into a type of string and soaked in oil to form *shisoku*, which functioned much like a modern wick when lit. These in turn evolved into lanterns, themselves the precursors of the electric light.

Compared with other early light sources, candles have a long and unbroken lineage in Japan. They were first introduced into Japan mainly for religious purposes, although some were also used in upper-class households. These candles were made with imported beeswax until imports ceased sometime during the Heian period, when pine resin came to be used instead. Imports of beeswax resumed during the Muromachi period and, since the method of producing the wax was also introduced, greatly increasing the quantity of material available, the use of candles expanded to include the general populace. By the middle of the sixteenth century, both candles and candle wax were in production around Japan. The increasing use of candles spawned a variety of accessories for interior use, among them candlesticks (*shokudai*), small candlesticks (*te shoku*), and lamp stands (*bonbori*). Special candles decorated with festive scenes or flower designs were used at the Hina Matsuri (the girls' Doll Festival) or for ceremonial occasions with a religious significance.

Today, the ready supply of electricity has almost totally done away with the need for candlelight. The use of candles is mostly limited to Buddhist and other religious services. Yet so long as temples are functioning and these services are being held, there seems certain to be a demand for Japanese-style candles.

In fact, the candles that are being made in Kyoto today are almost exclusively for temple use, although some are also used for the tea ceremony. They do not produce soot, they last a long time, and they actually produce a lot of heat. Most come in one of two colors: white or deep vermilion. They are either conventionally shaped or of the distinctive *ikari* form. Described as "anchor" shaped, this means that they gradually become narrower toward their base and therefore present a tapering silhouette. The basic materials are *mokurō* or a vegetable wax sometimes known as Japanese wax.

Kyoto candles employ two methods of production. For the method known as *namagake*, the wicks are either *washi* paper or a reed known as *igusa*. To make a reed wick, the outer layer is peeled off and thoroughly dried; the paper wicks are made by coiling narrow strips of paper into a tube. One end of the finished wick is then pushed into a narrow bamboo tube so that most of its length remains exposed. Holding a bundle of these stuffed tubes in the right hand, the candlemaker steadies them against a stand and rolls them while, with the left hand, applying wax to the wicks. The bundle of tubes is then hung up to dry and the next bundle is treated in the same way. Once several bundles have been treated in this way, the candlemaker reapplies wax to the wicks of the first bundle, continuing this process until the required thickness for a candle is achieved. The number of applications needed to achieve the same thickness depends largely on the air temperature. Five or six coatings are generally sufficient, but fewer may be necessary in winter when the wax adheres better, and conversely, more may be required in summer.

The other way of making a candle is with a mold. Wax is simply poured into a wooden mold. Deep vermilion-colored candles can be made by mixing a pigment with some wax and applying it as a finish. Candles are also sometimes decorated by applying gold or silver leaf with natural lacquer. Molds are used for straight candles and for the "anchor" type, which can only be made in this way. For large candles too, a mold is required.

GIFU LANTERNS
Gifu Chōchin

Gifu,* Gifu Prefecture

To many people, paper lanterns are a symbol of the East, particularly of Japan. Their origins and first use, however, are not clear. Originally it seems that only fixed lanterns provided light. Lanterns that could be carried outside probably developed from a type of lamp called an *andon*. These were interior lamps that could be made portable by attaching a handle. (This development probably took place sometime after the Muromachi period.) In fact, it is difficult to distinguish between an *andon* and a *chōchin* of the times. But once lanterns made by covering baskets with paper entered into use, the *andon* became exclusively an interior lamp. The "basket lanterns"—as these early *chōchin* were known—were really improvised affairs, but toward the end of the sixteenth century, more substantial lanterns, foldable and made of narrow strips of bamboo covered with paper, were introduced. This type of *hako chōchin* was the first real portable lantern. During the peaceful Edo period, the making of lanterns became one of the jobs to which the lower ranks of warriors turned, along with the making of paper-covered umbrellas. Gradually such work was taken up by ordinary people and various shapes and sizes of lanterns were created. These lanterns remained a primary source of artificial light throughout the period. The practical need for lanterns almost disappeared toward the end of the nineteenth century, however, with the introduction of gas and electricity for street lighting. Nevertheless, this did not signal the end of the lantern. Elegant lanterns, to be hung rather than carried and decorated with painted flowers and grasses, began to be made. These decorative lanterns are used at the summer Festival of the Dead (Obon) and still form part of the everyday scene in Japan.

There are several explanations as to exactly how and when Gifu *chōchin* were first made. The generally accepted version attributes their creation to the lantern maker Morioka Jūzō and his son, who were supplying lanterns to the Nagoya clan during the second half of the eighteenth century. During the first twenty years of the next century, the skills of this craft became firmly established. It was not until the end of the nineteenth century, however, that distribution was expanded, especially after the opening of the Tōkaidō rail line running between Tokyo and Osaka. At this stage, the lanterns were still being made in workshops operated by the wholesalers, albeit with a good deal of organized division of labor. From 1919, however, craftsmen themselves started to specialize. Some prepared the bamboo, others applied the paper, and still others painted on the decorations.

Of all the lanterns being made today, the egg-shaped ones decorated with the "seven grasses of autumn" are the most distinctive. Other types are either round, cylindrical, or *natsumegata*. This is a shape similar to the egg-shaped lanterns but with a more swelling form; its name is derived from the container used for powdered tea in the tea ceremony. Today, the uses of these lanterns are mostly decorative, but some are still used in the traditional way, for example to light the way for guests through a tea garden.

To make a lantern, strips of the bamboo are first wound round a wooden form. This is composed of eight or ten crescent-shaped segments that are put together to form a set. Following the shape of the form, threads are attached to each ring of bamboo and small pieces of paper are applied for strengthening. The whole form is then covered with either a plain paper or a woodblock-printed paper. After the whole thing is dry, the wooden form is collapsed and removed. Any hand-painting or touching up is then done, and rings of tin or pine are attached to the top and bottom. In appropriate cases, painted or unadorned wooden handles and stands are then fixed and the tassels are attached to finish.

The main areas in which lanterns are currently being produced are Gifu, Yame (Fukuoka Prefecture), and Nagoya. Lanterns of various kinds are also being made throughout the country. The decorative charm of Gifu *chōchin* stands out, however, adding a soft, ephemeral quality to Japan's hot and humid summer evenings that somehow makes the heat that much easier to bear.

IZUMO STONE LANTERNS • *Izumo Ishidōrō*

Shinji-chō, Shimane Prefecture

The notion that the light of a candle is symbolic of life must be at least as old as civilization itself. Placing that light in a form sculptured in stone not only serves to keep the flame alive and safe from the wind but also emphasizes its presence. The stone lanterns found in gardens in Japan are a fine example of this ancient practice.

They did not start life as such familiar garden accessories, however. The lanterns—said to have been introduced into Japan in the eighth century together with Buddhism, which arrived from China through the Korean Peninsula—were originally set up in front of the main buildings of worship at Buddhist temples. Their religious function was to provide a living light at a Buddhist memorial service, although more simply they were seen as items dedicated to the temple as part of the act of worship, and on a still more practical level, they provided illumination. During the Heian period, such lanterns began to appear in the grounds of Shintō shrines as well. With the rise of interest in the tea ceremony during the Momoyama period, they became a feature of tea gardens. This idea was expanded on in the following Edo period, when stone lanterns of various shapes and sizes were used as features in almost any kind of garden.

Lanterns from Izumo, a town in the southwest of Honshū facing the Japan Sea, made their appearance in the eighth century. Their golden age did not come until the Edo period, when the leader of the locally supreme Matsue clan prohibited the removal from his province of the type of sandstone tuff from which the lanterns and some other articles were made. Having therefore created a kind of "closed shop," he made efforts to protect and develop the processing of the stone.

Izumo sandstone is soft and easily worked. It is particularly appealing to the Japanese eye because of its clear, blue-gray coloring and the fact that in Japan's humid climate, moss will grow on it quite happily; this adds an aged look to the stone, bringing it closer to in its natural state. Although soft, the stone usually used stands up well to heat and cold, making it both functionally and aesthetically the most suitable stone for the job.

Initially, the processing and sculpting of the stone is done with a number of different tools, including a hand ax, a small pick, and a tool known as a *sanbonba.* Used simply to hack away at the stone, the *sanbonba* is a flat, bladelike hand pick, with three 5-millimeter-square steel teeth welded to the end of the blade. Finer work is done with chisels.

Individual types of lanterns illustrate particular aspects of Japanese aesthetic thinking. The *rokkaku yukimi* type is short, squat, and hexagonal, but can be made with a wide, spreading roof, used principally in gardens as an accent. Its form is perhaps shown to best advantage after a fall of snow, as suggested by the name *yukimi,* or "snow viewing." The *nuresagi* type gets its name from the image of a heron standing characteristically on one leg with snow piling up on its back. The bell-shaped top of the lantern is vaguely reminiscent of the shape of the bird in this pose, suggestion coming close to reality when the lantern actually has snow on it. Playing on the Japanese fascination with the rustic, the *kusaya* type of lantern is made to resemble a traditional farmhouse in model form, something that can look almost kitsch unless well crafted. The Oribe type takes its name from Furuta Oribe, a master of tea in the Momoyama period, and looks like the lantern that was placed over his grave. This is doubtless a measure of the lantern's subsequent strong connections with "the way of tea." Much more indicative of the craft's origins, however, is the pagoda type of lantern, mimicking one of the essential buildings in a temple complex, but again scaled down so as to look like a pagoda in the miniature landscape of a Japanese garden.

Not only lanterns but also such artifacts as stone basins for rinsing the hands, found specifically in tea gardens, or the guardian dog (*komainu*) statues used at Shintō shrines, as well as various other garden ornaments, all seem to be a quintessence of Japanese aesthetic thinking, and therefore deserve a closer inspection than they are usually afforded.

GIFU UMBRELLAS • *Gifu Wagasa*

Gifu, Gifu Prefecture

Traditionally, an umbrella, or *kasa*, was not the only thing to keep off the rain and snow or the rays of the sun. One early type of *kasa* was worn on the head! The word *kasa* is used for both this hat and an umbrella, although the characters are, of course, different.

The origins of Gifu *wagasa* date back to the beginning of the Edo period. In 1639, Matsudaira Tanbanokami Mitsushige, a member of a family directly related to the ruling Tokugawas, moved from Akashi in Banshū province, west of Osaka, to the castle town of Gifu. He took with him some of the *kasa* makers who had been working in Akashi, and it was these artisans who laid the foundations of umbrella-making in the town. In 1756, when the local Kanō clan found itself in some financial difficulties, the lower-ranking samurai were encouraged to start making umbrellas at home in order to alleviate the situation. Umbrella-making flourished, but rather than actually covering the umbrellas, it seems that these samurai were engaged in making the struts and hubs, something that was more difficult for ordinary people to do themselves. Such work became a living for these warriors, who were to all intents and purposes out of a job in these peaceful times. Yet money was not their sole incentive. They were also practicing the craft as a discipline to which they could completely devote themselves, rather as if training themselves for battle. Consequently, they become quite skilled and the quality of the craft steadily improved.

The basic materials for making *wagasa* are the same as they have always been: *madake*, a type of bamboo, and Mino *washi*, a type of handmade paper. After the skin is taken off the bamboo and its nodes are removed, it is initially split, then split again to the required width. The main struts are warmed over a fire and straightened. The secondary struts are then dyed and holes are made in them. When the nodes are removed from the shaft, the hub and end can be fixed. All the struts are then attached to the hub. Paper is fixed around the end and four pieces of fan-shaped paper are laid on the open struts and glued. An undercoat of persimmon juice is now applied, after which the paper is painted with a mixture of paulownia seed oil, clear kerosene, and linseed oil in the proportions of 2:2:6. The umbrella is then put outside in the sun to dry. Thereafter, the secondary struts are joined together with thread, the umbrella is folded, and the outside is coated with natural lacquer. A grip is made by winding rattan around the bamboo shaft, and lastly, the metal fittings are attached.

One of the most characteristic of the Gifu *wagasa* is known as a *janome gasa*. This is a red umbrella with a white "snake-eye ring" (*janome*) on it. These are often very large and act as sunshades outside traditional refreshment places in parks. Other more conventionally sized ones can be found at traditional inns for the use of the guests, and still others are used in classical dancing. Whatever their use, the *wagasa* will always be one of the most compelling manifestations of traditional Japan.

MARUGAME ROUND FANS • *Marugame Uchiwa*

Marugame,* Kagawa Prefecture

Marugame has always been famous for its *uchiwa*. In the past these rigid, nonfolding fans were made only of paper and bamboo. But today the majority are made of plastic and are of a quality that leaves much to be desired. Nonetheless, two traditional types of *uchiwa* are still being made. One is the round-handled *maru-e*, and the other, more commonly seen type is the flat-handled *hira-e*.

The port of Marugame was one of the places at which pilgrims to the famous Kotohiragū Shrine disembarked. Located more or less in the center of Shikoku, Kotohiragū has been one of Japan's most popular shrines for many centuries and is dedicated to a god who is believed to give protection to seafarers and voyagers. In 1633, the local Ikoma clan invited some skilled artisans from Yamato, on the Kii Peninsula, to come to the area to make *maru-e* fans using one of the robust varieties of "male" bamboo found locally. The *uchiwa* they made were decorated with the character for money (*kane*), which is also the first character for the name of Kotohiragu. These fans, sold to the pilgrims, are thought to mark the beginnings of the Marugame *uchiwa*.

Most of the *uchiwa* being produced in Marugame toward the end of the nineteenth century were little more than skeletons that were shipped elsewhere for finishing. However, from about 1897, the fans were completed before being distributed. After the Russo-Japanese War of 1904–1905, *uchiwa* began to be extensively used for advertising and other promotional purposes, and this stimulated the mass production of flat-handled *hira-e* fans. The invention in 1913 of machines to shape and cut holes in the fans increased productivity and led to a further lowering of prices. The *hira-e uchiwa* thus became the main product.

In traditional fans, the ribs are made of a bamboo called *madake* and the paper is good-quality *washi*. To make a flat-handled fan, the bamboo is first cut to length, then the part that will form the ribs is thinned. A hole is made in the handle and a piece of fine bamboo is passed through it and bowed. Both the handle and the ribs are then fixed with thread. The ends of the ribs are cut to follow the shape described by the bowed loop of bamboo, and the paper is glued on. After the edge has been finished off and strengthened with a narrow strip of paper, a rubber roller is used to press the paper down tightly so that the ribs are expressed.

Given the high level of craft needed to produce them, it is inevitable that the traditionally made *uchiwa* are more appealing than the plastic alternatives. Marugame fans have a practical and unaffected elegance that befits such an article and ensures them a lasting place in the Japanese summer.

KYOTO FOLDING FANS • *Kyō Sensu*

Kyoto

The type of folding fan (*sensu*) that is available today was first made in the Gojō district of Kyoto during the 1190s. Originally made with leaves of cypress, it was used only by courtiers, people of noble birth, and the upper echelons of society. It was not until the twelfth century that ordinary people began to use them. After the Muromachi period, fans developed in step with the fashions of ikebana, dance, the tea ceremony, and *kōdō* (the pastime of incense smelling).

Sensu are made with a variety of end uses in mind. Some are for use in the summer (*natsu ōgi*), others are made especially for use at the tea ceremony (*cha ōgi*), and still others are associated with incense smelling (*kō ōgi*). One type is used for traditional dance and Noh. Further variations are miniature folding fans (*mame ōgi*) and *sensu* used for ceremonial purposes (*shūgi ōgi*).

The five-step process of constructing a fan begins with the making of the bamboo ribs. Bamboo is cut, split, then trimmed to the desired length. A hole is drilled through the ribs for the rivet, then the ribs as a block are planed and shaped. The bamboo is then dried in sunlight, polished and, if called for, lacquered on the bottom. The portion of the ribs to be slid between the two layers of paper is next shaved until it is wafer thin.

The paper mounting is of several layers, with inner pockets to accept the ribs. Decoration is done beforehand: gold or silver leaf or powders can be used to express a design. In the technique called *sunago furi*, glue is first painted on the paper and then the powders are sprinkled on. Another decorative technique is known as *bunkin*. For this the covering paper is folded, the paper is slightly stepped, and glue is rubbed into either the top or the bottom. Gold leaf or powder is then applied later. Designs or pictorial motifs are sometimes painted on by hand, sometimes printed. In the final step, the paper is folded and the ribs are inserted into the pockets with a good deal of speed and dexterity, then the paper is secured to the ribs. The fan is now finished.

Even though the advent of air conditioning has displaced the need for fans in some cases, fans of all kinds are still in constant use. Many stores give away rigid fans to their customers during the hot, humid summer months, and folding fans are still part of the summer scene even in the main urban conurbations. It seems, therefore, that the useful life of the fan is far from over.

EDO "ART" DOLLS • *Edo Kimekomi Ningyō*

Tokyo and Iwatsuki, Saitama Prefecture

With their colorful past, fine details, and relatively complex process of fabrication for something of this size, it seems a gross understatement to call these exquisite ornaments "dolls." But this is the only appropriate word in English to describe anything from the simplest of rag toys right up to something as sophisticated and refined as the Edo *kimekomi ningyō*. In any case, the Japanese *ningyō* (or *hitogata*, as the same characters are sometimes pronounced), which literally means "people model," is equally broad in its usage and is similarly used as a tag for this type of principally ornamental doll.

The Edo "art" doll probably started life when a craftsman in Kyoto made a figure from wood scraps. Having therefore most likely originated in a similar way to the *kokeshi*, it took the name of the location where it was first made, namely near the Kamogawa River, hence the initial name Kamo *ningyō*, or even Kamogawa *ningyō*. Because the clothes were tucked into grooves in the body, it later became known as a *kimekomi ningyō*, *kime* referring to the grooves in the wood made to accept the ends of the fabric of the clothes, and *komi* meaning in this instance "to tuck in."

Such was the movement of people during the Edo period that these ornaments soon found their way to Edo (present-day Tokyo) and it was not long before the technique, too, was conveyed to this thriving center of civilization. The greatly increased demand for the dolls in Edo stimulated further developments in the fabrication process.

It was in Edo that the *tōso* technique came into use, although its exact origins and date of development are not clear. Simply speaking, this involves combining fine paulownia sawdust with fresh funorin and pressing the mixture into a mold, thereby making it possible to produce larger numbers of dolls more quickly. The production process is somehow typical of the Edo craftsman's skill and flare for inventive development. First, clay is used to make the heads, bodies, hands, and feet, from which a pair of molds are made using sulfur or aluminum. The resulting molds are then filled with the sawdust and funorin mix. When dry, the parts are removed from the molds, pieced together, and have their join lines tidied up. The heads, for which unglazed pottery is sometimes used, are painted with a ground, and then the eyes, nose, and mouth are shaped with a knife. A final topcoat of color is applied, the eyes and eyebrows are painted in, and color is added to the cheeks and lips. Finally, the head is completed by implanting silk hair. As in the original method using a solid wooden body, so with the wood-composite body grooves are cut to accept the ends of the dresses, some adhesive having first been set into the grooves. The dresses are made from woven silk or cotton cloth that is cut from a pattern. As can be imagined, this way of dressing the doll gives the illusion that the clothes are being worn and adds significantly to the overall charm of these figures, which also display fine facial features and delicate detail.

The types of dolls made are mainly Kabuki and Noh theater figures, those displayed at the Doll Festival (Hina Matsuri) for girls held in March, figures called *gogatsu ningyō* associated with the Children's Day (Kodomo no Hi) for boys in May, and portrayals of characters from traditional stories or simply people of the moment. The continued existence of the Edo *kimekomi ningyō* must have something to do with the fact that at the end of the nineteenth century, two competing factions became established, handing down the artistry and techniques associated with each style of doll-making from one generation to the next. What started in an odd moment has evolved into a highly developed object, which in itself is a story particularly representative of the Japanese way of doing things.

EDO BATTLEDORES • *Edo Oshie Hagoita*

Tokyo

Battledore and shuttlecock, or *hanetsuki*, is one of the more charming traditional games of Japan. The object is simply to hit the shuttlecock back and forth, and the loser has their face smudged with charcoal or white powder as a penalty. This can become very amusing when the game is played by young girls dressed in their finest kimono at the New Year. The game seems to have originated as a charm to protect children from evil and the mosquito, a prevalent pest through the hot, humid summer months and into the autumn. The *hane*, or shuttlecock, is said by some to represent the dragonfly, which eats these insects.

The history of the *hagoita* itself can be traced back through records to the fifth day of the first month in 1432, when the imperial family, court officials, and ladies employed at the court were divided into teams of men and women to play a game called *kogi no ko shōbu*. This was the old form of *hanetsuki*. At the time, there were several types of battledore: some to which illustrations were applied directly; some with raised images using paper, fabric, and a white paste-like pigment; and others with gold and silver leaf.

The Edo *oshie hagoita* seems to have appeared sometime at the beginning of the nineteenth century, a period in Japan when the merchant classes were in ascendance and *ukiyoe* woodblock prints were widely admired. Battledores adorned by the *ukiyoe* artists with images of the Kabuki actors in their stage roles were particularly sought after. Come the end of the year, people would vie with each other to obtain these depictions of popular performers, which were sold at the Battledore Market (*hagoita ichi*) held in the grounds of Tokyo's Sensōji temple. These decorative battledores, which even today depict famous Kabuki characters and geisha as well as contemporary popular singers and personalities, are made to be admired rather than played with, although some still are made for that purpose.

The paddle-shaped battledore is made of the very light, close-grained paulownia wood (*kiri*). To make the padded images, silk and cotton are usually employed, pieces of fabric being cut to match thick paper patterns. The fabric is attached to these and then cotton wadding is inserted between the two before a rice paste is applied, using an iron to form the required three-dimensional effect. The eyes, eyebrows, nose, and mouth are then painted on and the silk hair implanted. All the pieces are then organized and glued to another piece of paper before being attached to the paddle with brass tacks.

The Battledore Market (Hagoita Ichi) is still a popular event in Tokyo's calendar, and those who have become proud owners of one of these brightly decorated *hagoita* are forced to carry them aloft for fear that they might get damaged among the crowds that throng the limited temple grounds and surrounding streets. This craft, so prized in the past, still holds a special place in the hearts of the people.

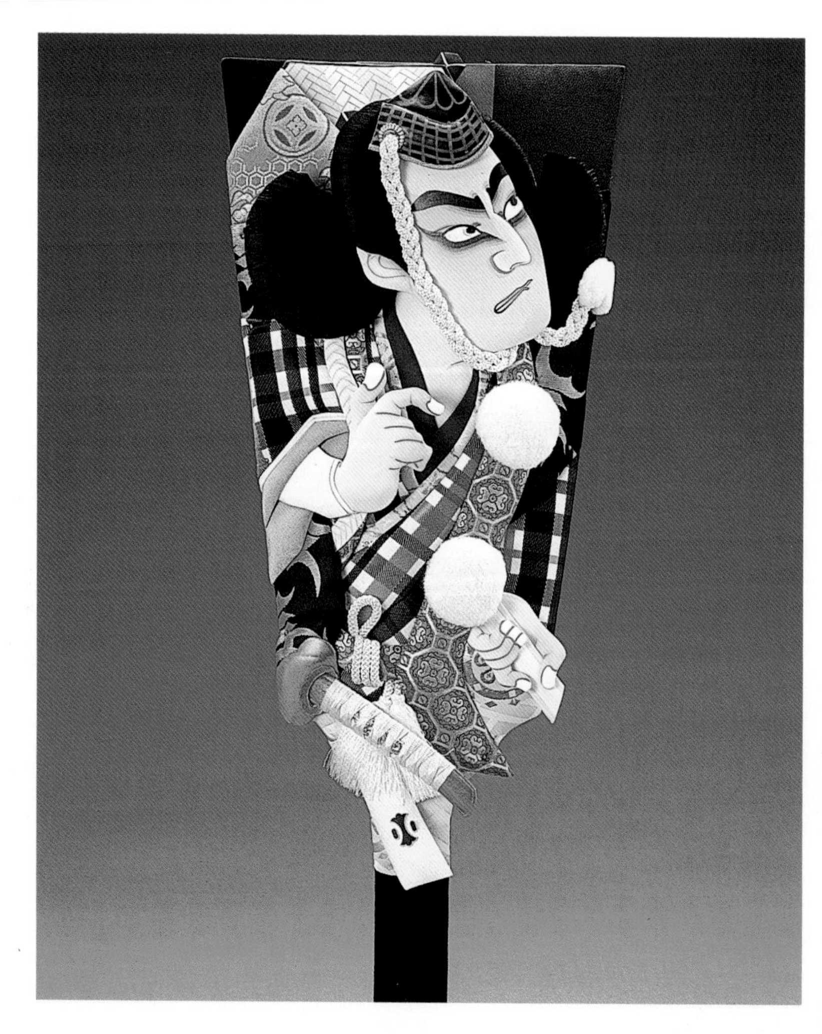

MIHARU DOLLS • *Miharu Hariko*

Miharu-machi and Kōriyama, Fukushima Prefecture

Winter is a very slack season for farmers in most parts of Japan. It is not uncommon, even today, for a farmer to seek other kinds of employment in the winter months to help pay his bills.

It was in this spirit of fostering work besides agriculture that the leader of the Miharu clan, Akita Kiyosue, invited doll makers from Edo (now Tokyo) to this northern province in the early part of the eighteenth century. Kiyosue's aim was to make better use of the available labor during the seasonal lull, and the result was the development of the Miharu *hariko* papier-mâché dolls. Before this, the minor fiefdom of Takashiba, a part of present-day Kōriyama, had been famous for its Miharu *ningyō*, and this has created some confusion with the dolls later produced at Miharu itself. What are now called Miharu *hariko* started out by being known as Takashiba *ningyō* or Miharu *ningyō*. Eventually a name combining the method of fabrication and the name of the locality stuck, establishing the reputation of the Miharu *hariko*. With the name set and the support of the clan, techniques were further refined, and by the beginning of the nineteenth century, Miharu *hariko* were known nationwide. During this golden age for the craft, upward of five hundred different types of dolls were being made. Today there are less of the highly ornate types of figures and animals than there used to be, but a good number are still available. Among these, the *tama usagi* (a rabbit with a cute expression), the portrayal of the god Tenjin riding on a bullock, and representations of Kabuki and other popular characters all manifest the traditions of this uncomplicated folkcraft.

As with most other types of papier-mâché craft, several layers of small pieces of paper, in this case *washi*, are fastened with rice paste to the outside of a wooden form or to the inside of a wooden mold. A further coat of the paste is applied to bind all the pieces of paper together, and sometimes a top-quality paper is used as a final covering. The form is removed by slitting the paper, then more paper is pasted on to cover the slit. Next, a white undercoat made from powdered shells is applied, and finally the dolls are painted. The finished products retain an element of primitive expression to rank them among the world's best folk art.

MIYAGI KOKESHI DOLLS
Miyagi Dentō Kokeshi

Naruko-chō, Zaō-machi, and Shiroishi, Miyagi Prefecture

The cylindrical *kokeshi* doll is well known outside Japan, principally because it is small and light, reasonably priced, and so distinctively Japanese that it makes a popular present or souvenir.

The doll's distinctive cylindrical body, which has neither hands nor feet, has changed very little since it was first created as a toy, probably from leftover scraps of wood, by craftsmen making other wooden articles to sell in one of the many spa towns of Miyagi Prefecture. The making of the *kokeshi* spread to other places, first simply as a toy, and then as a souvenir made in the other northeastern spa towns opened to the general public in the early part of the nineteenth century. In the middle of the twentieth century, new and imaginative types began to appear, and the dolls are now made all over the country to be collected or simply used as ornaments rather than played with.

At the outset, the pieces of this very simple doll were made on a two-person lathe, but in 1885, as demand from souvenir buyers increased, a one-person lathe was introduced. This gave rise to the modern form of *kokeshi*, and as a result of the interest shown in them by collectors and researchers alike, their status as a craft object solidified and their position as a decorative accessory was established.

There are five basic types of traditional Miyagi *kokeshi*, four of which originate from the spa towns of Miyagi. The Naruko Spa gives its name to the *naruko* type, which squeaks when the head is rotated. The *tōgatta* type takes its name from the Tōgatta Spa, while the *sakunami* type is also named after a spa town. Similarly, the *hijiori* type comes from the Hijiori Spa in Yamagata Prefecture. The odd doll out, as far as naming is concerned, is the *yajirō* type from the Kamasaki Spa. Although the shapes of all five types vary slightly, what they share is a beautifully simplified form and elegant outline. The shapes of the heads are usually either flattened spheres, or spheres narrowing toward the neck. The shoulders are sometimes rounded, sometimes stepped, but usually expressed in some way. There are also several styles in which the hair is painted. On the bodies, flowers such as the chrysanthemum, peony, and iris are incorporated in various designs, often together with rings that are painted on when the body is still on the lathe. In their traditional forms, however, almost all of the shapes and designs follow an accepted pattern.

The dolls are fashioned from dogwood, maple, zelkova, cherry, or Japanese pagoda, among others. Some dolls are fabricated in one piece, the heads being turned together with the bodies. Other designs call for the head to be turned separately, then attached to the main body. In this case, two methods may be used. One calls for a part of the head to be turned as a plug on a horizontal lathe, then forced into a hole only just big enough to accept it. This tight fit produces the squeak when the head is twisted. The other method utilizes a peg either projecting from the body or inserted into a hole in the body, onto which the head is fitted. Once the form is finished, the faces, hair, and designs are painted on with either *sumi*—black ink—or stain. After decoration, the doll receives a wax polish.

The story of the *kokeshi* is far from finished. Contemporary versions and innovations go on being crafted, building on the tradition of what might originally have been made by someone to keep a child amused when the snow lay deep outside.

LIST OF REGIONAL CRAFT MUSEUMS, SHOWROOMS, AND CRAFT CENTERS

Many of the places below exhibit and sell items of local manufacture. However, the availability of salable goods and the content of exhibitions vary from place to place, and both museums and showrooms are sometimes closed for short periods of time or accessible only with an appointment. It is recommended that you call ahead in Japanese before making a special trip. Any correspondence should also be in Japanese.

Some crafts, particularly those centered in larger cities, do not have museums or showrooms devoted to the craft. For Kyoto crafts, the Kyoto Museum of Traditional Industry listed under Kyoto Ware displays a number of traditional Kyoto items. In some instances, commercial enterprises have been listed as alternatives. Additional information may be obtained by contacting the appropriate agency noted in the List of Craft Associations *that follows.*

Official English names are marked with an asterisk. All other names are equivalents provided for the reader's reference only.

CERAMICS 陶磁器 *Tōjiki*

KUTANI WARE • *Kutani Yaki*

MUSEUM FOR TRADITIONAL PRODUCTS AND CRAFTS*
石川県立伝統産業工芸館
〒920 1-1 Kenroku-machi, Kanazawa, Ishikawa Prefecture
〒920 石川県金沢市兼六町1-1
☎ 0762-62-2020

TERAI-MACHI KUTANI POTTERY CENTER
寺井町九谷焼陶芸館
〒923-11 9 Aza Izumidai-minami, Terai-machi, Nomi-gun, Ishikawa Prefecture
〒923-11 石川県能美郡寺井町字泉台南9
☎ 0761-58-6300

MASHIKO WARE • *Mashiko Yaki*

POTTERY MESSE MASHIKO
陶芸メッセ益子
〒321-42 3021 Ōaza Mashiko, Mashiko-machi, Haga-gun, Tochigi Prefecture
〒321-42 栃木県芳賀郡益子町大字益子3021
☎ 0285-72-7555

THE MASHIKO REFERENCE COLLECTION MUSEUM*
益子参考館
〒321-42 3388 Ōaza Mashiko, Mashiko-machi, Haga-gun, Tochigi Prefecture
〒321-42 栃木県芳賀郡益子町大字益子3388
☎ 0285-72-5300

MINO WARE • *Mino Yaki*

TOKI MINO WARE TRADITIONAL INDUSTRY HALL*
土岐市美濃焼伝統産業会館
〒509-51 1429-8 Aza Kitayama, Izumi-chō Kujiri, Toki, Gifu Prefecture
〒509-51 岐阜県土岐市泉町久尻字北山1429-8
☎ 0572-55-5527

TOKI MUNICIPAL INSTITUTE OF CERAMICS*
セラテクノ土岐
〒509-51 287-3-1 Hida-chō Hida, Toki, Gifu Prefecture
〒509-51 岐阜県土岐市肥田町肥田287-3-1
☎ 0572-59-8312

HAGI WARE • *Hagi Yaki*

YAMAGUCHI REGIONAL CENTER
山口ふるさと伝承総合センター
〒753 12 Ōaza Shimotate-kōji, Yamaguchi, Yamaguchi Prefecture
〒753 山口県山口市大字下竪小路12
☎ 0839-28-3333

HAGI WARE GALLERY
萩焼資料館
〒758 Horiuchi Jōshi, Hagi, Yamaguchi Prefecture
〒758 山口県萩市堀内城跡
☎ 0838-25-8981

KYOTO WARE, KIYOMIZU WARE • *Kyō Yaki, Kiyomizu Yaki*

KYOTO CITY INDUSTRIAL EXHIBITION HALL*
京都伝統産業ふれあい館 京都市勧業館「みやこめっせ」内
〒606 9-1 Seishōji-chō, Okazaki, Sakyō-ku, Kyoto
〒606 京都府京都市左京区岡崎成勝寺町9-1
☎ 075-222-3337 (temporary; 京都市伝統産業課)

THE MUSEUM OF KYOTO*
京都文化博物館

〒604 Sanjō-takakura, Nakagyō-ku, Kyoto
〒604 京都府京都市中京区三条高倉
☎ 075-222-0888

KYOTO PORCELAIN CENTER
京都陶磁器会館
〒605 Higashi-Ōji Higashi-iru, Gojō-zaka, Higashiyama-ku, Kyoto
〒605 京都府京都市東山区五条坂東大路東入
☎ 075-541-1102

SHIGARAKI WARE • *Shigaraki Yaki*

THE TRADITIONAL CRAFT CENTER*
信楽伝統産業会館
〒529-18 1142 Ōaza Nagano, Shigaraki-chō, Kōka-gun, Shiga Prefecture
〒529-18 滋賀県甲賀郡信楽町大字長野1142
☎ 0748-82-2345

THE SHIGARAKI CERAMIC CULTURAL PARK*
滋賀陶芸の森
〒529-18 2188-7 Chokushi, Shigaraki-chō, Kōka-gun, Shiga Prefecture
〒529-18 滋賀県甲賀郡信楽町勅旨2188-7
☎ 0748-83-0909

BIZEN WARE • *Bizen Yaki*

BIZEN WARE TRADITIONAL CRAFT CENTER
備前焼伝統産業会館
〒705 1657-2 Inbe, Bizen, Okayama Prefecture
〒705 岡山県備前市伊部1657-2
☎ 0869-64-1001

BIZEN WARE POTTERY MUSEUM
備前陶芸美術館
〒705 1659-6 Inbe, Bizen, Okayama Prefecture
〒705 岡山県備前市伊部1659-6
☎ 0869-64-1400

ARITA WARE • *Arita Yaki*

IMARI ARITA TRADITIONAL CRAFT CENTER*
伊万里・有田焼伝統産業会館
〒848 222 Ōkawachi-yama Hei, Ōkawachi-chō, Imari, Saga Prefecture
〒848 佐賀県伊万里市大川内町大川内山丙222
☎ 0955-22-6333

ARITA PORCELAIN PARK*
有田ポーセリンパーク
〒844 370-2 Chūbu-Otsu, Arita-machi, Nishi-Matsuura-gun, Saga Prefecture
〒844 佐賀県西松浦郡有田町中部乙370-2
☎ 0955-42-6100

KYŪSHŪ CERAMIC MUSEUM*
九州陶磁文化会館
〒844 3100-1 Chūbu-Hei, Arita-machi, Nishimatsuura-gun, Saga Prefecture
〒844 佐賀県西松浦郡有田町中部丙3100-1
☎ 0955-43-3681

SAKAIDA KAKIEMON GALLERY
酒井田柿右衛門ギャラリー
〒844 352 Seibu-Tei, Arita-machi, Nishimatsuura-gun, Saga Prefecture
〒844 佐賀県西松浦郡有田町西部丁352
☎ 0955-43-2267

IMAIZUMI IMAEMON GALLERY
今泉今右衛門ギャラリー
〒844 1590 Akae-machi, Arita-machi, Nishimatsuura-gun, Saga Prefecture
〒844 佐賀県西松浦郡有田町赤絵町1590
☎ 0955-42-3101

TOKONAME WARE • *Tokoname Yaki*

TOKONAME POTTERY CENTER
常滑市陶磁器会館
〒479 3-8 Sakae-machi, Tokoname, Aichi Prefecture
〒479 愛知県常滑市栄町3-8
☎ 0569-35-2033

TOKONAME MUNICIPAL POTTERY INSTITUTE
常滑市立陶芸研究所展示室
〒479 7-22 Okujō, Tokoname, Aichi Prefecture
〒479 愛知県常滑市奥条7-22
☎ 0569-35-3970

TOBE WARE • *Tobe Yaki*

TOBE POTTERY TRADITIONAL INDUSTRY HALL*
砥部焼伝統産業会館
〒791-21 335 Ōminami, Tobe-chō, Iyo-gun, Ehime Prefecture
〒791-21 愛媛県伊予郡砥部町大南335
☎ 0899-62-6600

TOBE WARE POTTERY CENTER
砥部焼陶芸館
〒791-21 83 Miyauchi, Tobe-chō, Iyo-gun, Ehime Prefecture
〒791-21 愛媛県伊予郡砥部町宮内83
☎ 0899-62-3900

TSUBOYA WARE • *Tsuboya Yaki*

NAHA TRADITIONAL CRAFT CENTER
那覇市伝統工芸館
〒901-01 1-1 Aza Tōma, Naha, Okinawa Prefecture
〒901-01 沖縄県那覇市字当間1-1
☎ 098-858-6655

TSUBOYA POTTERY CENTER
壺屋陶器会館
〒902 1-21-14 Tsuboya, Naha, Okinawa Prefecture
〒902 沖縄県那覇市壺屋1-21-14
☎ 0988-66-3284

KIKUMA TILES • *Kikuma Gawara*

KIKUMA TILE CENTER
菊間瓦会館
〒799-23 3037 Hama, Kikuma-machi, Ochi-gun, Ehime Prefecture
〒799-23 愛媛県越智郡菊間町浜3037
☎ 0898-54-3450

TEXTILES 染織 *Senshoku*

KIRYŪ FABRICS • *Kiryū Ori*

KIRYŪ LOCAL INDUSTRY PROMOTION CENTER*
桐生地域地場産業振興センター
〒376 2-5 Orihime-chō, Kiryū, Gunma Prefecture

〒376　群馬県桐生市織姫町2–5
☎ 0277–46–1011

MORIHIDE TEXTILE CENTER · YUKARI
森秀織物参考館・紫（ゆかり）
〒376　4–2-24 Higashi, Kiryū, Gunma Prefecture
〒376　群馬県桐生市東4–2–24
☎ 0277–45–3111

OITAMA PONGEE • *Oitama Tsumugi*

YONEZAWA TEXTILE ARCHIVES
米沢織物歴史資料館
〒992　1–1-5 Montō-machi, Yonezawa, Yamagata Prefecture
〒992　山形県米沢市門東町1–1–5
☎ 0238–23–3006

YAMAGISHI TEXTILE
山岸織物
〒992　20457–12 Akakuzure, Yonezawa, Yamagata Prefecture
〒992　山形県米沢市赤崩20457–12
☎ 0238–38–2878

YŪKI PONGEE • *Yūki Tsumugi*

YŪKI PONGEE CENTER
結城紬染織資料館
〒307　12 Ōaza Yūki, Yūki, Ibaragi Prefecture
〒307　茨城県結城市大字結城12
☎ 0296–33–3111

OJIYA RAMIE CREPE • *Ojiya Chijimi*

OJIYA TRADITIONAL CRAFT CENTER
小千谷市伝統産業会館
〒947　1–8-25 Jōnai, Ojiya, Niigata Prefecture
〒947　新潟県小千谷市城内1–8-25
☎ 0258–83–2329

SHINSHŪ PONGEE • *Shinshū Tsumugi*

AIZUYA
会津屋
〒390　7–6 Habaue, Matsumoto, Nagano Prefecture
〒390　長野県松本市巾上7–6
☎ 0263–32–3728

NAMIYA TEXTILE
那美屋織物
〒395　2 Kodenma-chō, Iida, Nagano Prefecture
〒395　長野県飯田市小伝馬町2
☎ 0265–24–7380

NISHIJIN FABRICS • *Nishijin Ori*

NISHIJIN TEXTILE CENTER*
西陣織会館
〒602　Imadegawa Minami-iru, Horikawa-dōri, Kamigyō-ku, Kyoto
〒602　京都府京都市上京区堀川通今出川南入ル
☎ 075–451–9231

YUMIHAMA IKAT • *Yumihama Gasuri*

TOTTORI PREFECTURE TOURIST CENTER
鳥取県物産観光センター
〒680　160 Suehiro-onsen-chō, Tottori, Tottori Prefecture
〒680　鳥取県鳥取市末広温泉町160
☎ 0857–29–0021

ASIAN ARTIFACTS MUSEUM
アジア博物館
〒683–01　57 Ōshinozu-chō, Yonago, Tottori Prefecture
〒683–01　鳥取県米子市大篠津町57
☎ 0859–25–1251

HAKATA WEAVE • *Hakata Ori*

HAKATA TEXTILE CENTER
博多織伝産展示館
〒812　1–14–12 Hakataeki-minami, Hakata-ku, Fukuoka, Fukuoka Prefecture
〒812　福岡県福岡市博多区博多駅南1–14–12
☎ 092–472–0761

TRUE ŌSHIMA PONGEE • *Honba Ōshima Tsumugi*

TRUE ŌSHIMA PONGEE TRADITIONAL CRAFT CENTER
本場大島紬伝統産業会館（奄美群島大島紬総合会館）
〒894　15 Minato-machi, Naze, Kagoshima Prefecture
〒894　鹿児島県名瀬市港町15
☎ 0997–52–3411

ŌSHIMA TSUMUGI NO SATO*
本場大島紬の里
〒891–01　1–8 Nan'ei, Kagoshima, Kagoshima Prefecture
〒891–01　鹿児島県鹿児島市南栄1–8
☎ 0992–68–0331

MIYAKO RAMIE • *Miyako Jōfu*

MIYAKO TRADITIONAL CRAFT CENTER
宮古伝統的工芸品研究センター
〒906　3 Aza Nishizato, Hirara, Okinawa Prefecture
〒906　沖縄県平良市字西里3
☎ 09807–2-8022

SHURI FABRICS • *Shuri Ori*

SHURI FABRIC CENTER
首里織工芸館
〒903　2–64 Shuritōbaru-chō, Naha, Okinawa Prefecture
〒903　沖縄県那覇市首里桃原町2–64
☎ 098–887–2746

KIJOKA ABACA • *Kijoka no Bashōfu*

ŌGIMI-SON ABACA CENTER
大宜味村立芭蕉布会館
〒905–13　454 Aza Kijoka, Ōgimi-son, Kunigami-gun, Okinawa Prefecture
〒905–13　沖縄県国頭郡大宜味村字喜如嘉454
☎ 0980–44–3033

YONTANZA MINSAA • *Yomitanzan Minsaa*

YOMITAN HISTORICAL AND FOLK CRAFT MUSEUM
読谷村立歴史民族資料館
〒904–03　708–6 Aza Zakimi, Yomitan-son, Nakagami-gun, Okinawa Prefecture
〒904–03　沖縄県中頭郡読谷村字座喜味708–6
☎ 098–958–3141

YOMITAN-SON TRADITIONAL CRAFT CENTER
読谷村伝統工芸総合センター
〒904–03　2974–2 Aza Zakimi, Yomitan-son, Nakagami-gun, Okinawa Prefecture
〒904–03　沖縄県中頭郡読谷村字座喜味2974–2
☎ 098–958–4674

ISESAKI IKAT • *Isesaki Gasuri*

ISESAKI TEXTILE CENTER
伊勢崎織物総合展示場
〒372 31–9 Kuruwa-chō, Isesaki, Gunma Prefecture
〒372 群馬県伊勢崎市曲輪町31–9
☎ 0270–25–2700

TSUGARU KOGIN STITCHING • *Tsugaru Kogin*

HIROSAKI KOGIN STITCHING STUDIO
弘前こぎん研究所
〒036 61 Zaifu-chō, Hirosaki, Aomori Prefecture
〒036 青森県弘前市在府町61
☎ 0172–32–0595

TOKYO STENCIL DYEING • *Tokyo Somekomon*

FUTABAEN
二葉苑
〒161 2-3-2 Kami Ochiai, Shinjuku-ku, Tokyo
〒161 東京都新宿区上落合2–3–2
☎ 03–3368–8133

KAGA YŪZEN DYEING • *Kaga Yūzen*

KAGA YŪZEN TRADITIONAL CRAFT CENTER
加賀友禅伝統産業会館
〒920 8–8 Koshō-machi, Kanazawa, Ishikawa Prefecture
〒920 石川県金沢市小将町8–8
☎ 0762–24–5511

KYOTO YŪZEN DYEING • *Kyō Yūzen*
KYOTO STENCIL DYEING • *Kyō Komon*

KYOTO YŪZEN CULTURAL HALL*
京都友禅文化会館
〒615 6 Mameda-chō, Nishi-Kyōgoku, Ukyō-ku, Kyoto
〒615 京都府京都市右京区西京極豆田町6
☎ 075–311–0025

TOKYO YUKATA STENCIL DYEING • *Tokyo Honzome Yukata*

OKAMURA DYEING FACTORY
株式会社 岡村染工場
〒120 29–6 Senjunakai-chō, Adachi-ku, Tokyo
〒120 東京都足立区千住中居町29–6
☎ 03–3881–0111

KYOTO EMBROIDERY • *Kyō Nui*

NISHI EMBROIDERY
有限会社 西刺繍
〒600 Kōtake-chō, Matsubara-agaru, Kawaramachi-dōri, Shimogyō-ku, Kyoto
〒600 京都府京都市下京区河原町通松原上ル幸竹町
☎ 075–361–5494

NAGAKUSA EMBROIDERY
有限会社 長艸刺繍
〒600 562–1 Suhama-chō, Uramon-sagaru, Kamichōja-machi-dōri, Kamigyō-ku, Kyoto
〒600 京都府京都市上京区上長者町通裏門下ル須浜町562–1
☎ 075–451–3391

IGA BRAIDED CORDS • *Iga Kumihimo*

IGA KUMIHIMO CENTER
伊賀くみひもセンター
〒518 1929–10 Shijūku-chō, Ueno, Mie Prefecture
〒518 三重県上野市四十九町1929–10
☎ 0595–23–8038

LACQUER WARE 漆器 *Shikki*

WAJIMA LACQUER • *Wajima Nuri*

WAJIMA LACQUER WARE CENTER
輪島漆器会館
〒928 55, 24 bu Kawai-machi, Wajima, Ishikawa Prefecture
〒928 石川県輪島市河井町24部55
☎ 0768–22–2155

ISHIKAWA WAJIMA URUSHI ART MUSEUM*
石川県輪島漆芸美術館
〒928 11 Shijūgari, Mitomori-chō, Wajima, Ishikawa Prefecture
〒928 石川県輪島市水守町四十苅11
TEL: 0768–22–9788

ŌMUKAI KŌSHŪDŌ GALLERY
株式会社 大向高洲堂
〒928 6–45 Futatsuya-machi, Wajima, Ishikawa Prefecture
〒928 石川県輪島市二つ屋町6–45
TEL: 0768–22–1313

AIZU LACQUER • *Aizu Nuri*

FUKUSHIMA PREFECTURE TRADITIONAL CRAFT CENTER
福島県伝統産業会館
〒965 1–7-3 Ōmachi, Aizuwakamatsu, Fukushima Prefecture
〒965 福島県会津若松市大町1–7-3
☎ 0242–24–5757

SHIROKIYA LACQUER WARE GALLERY
株式会社 白木屋漆器店
〒965 1–2-10 Ōmachi, Aizuwakamatsu, Fukushima Prefecture
〒965 福島県会津若松市大町1–2-10
☎ 0242–22–0203

KISO LACQUER • *Kiso Shikki*

KISO CRAFT CENTER
木曽くらしの工芸館
〒399–63 2272–7 Aza Nagase, Hirasawa, Narakawa-mura, Kiso-gun, Nagano Prefecture
〒399–63 長野県木曽郡楢川村平沢字長瀬2272–7
☎ 0264–34–3888

GALLERY CHIKIRIYA
有限会社 ちきりや手塚万右衛門商店
〒399–63 1736–1 Hirasawa, Narakawa-mura, Kiso-gun, Nagano Prefecture
〒399–63 長野県木曽郡楢川村平沢1736–1
☎ 0264–34–2002

HIDA SHUNKEI LACQUER • *Hida Shunkei*

HIDA TAKAYAMA SHUNKEI CENTER
飛騨高山春慶会館
〒506 1 Kanda-machi, Takayama, Gifu Prefecture
〒506 岐阜県高山市神田町1丁目
☎ 0577–32–3373

HIDA LOCAL INDUSTRY PROMOTION CENTER
飛騨地域地場産業振興センター

〒506　5–1 Tenma-chō, Takayama, Gifu Prefecture
〒506　岐阜県高山市天満町5–1
☎ 0577–35–0370

KISHŪ LACQUER • *Kishū Shikki*

KISHŪ LACQUER WARE TRADITIONAL CRAFT CENTER
紀州漆器伝統産業会館
〒642　222 Funao, Kainan, Wakayama Prefecture
〒642　和歌山県海南市船尾222
☎ 0734–82–0322

MURAKAMI CARVED LACQUER • *Murakami Kibori Tsuishu*

MURAKAMI TSUISHU CRAFT CENTER
村上堆朱工芸館
〒958　1–8–24 Midori-machi, Murakami, Niigata Prefecture
〒958　新潟県村上市緑町1–8–24
☎ 0254–53–2478

KAMAKURA LACQUER • *Kamakura Bori*

KAMAKURA BORI CENTER
鎌倉彫会館
〒248　2–15–13 Komachi, Kamakura, Kanagawa Prefecture
〒248　神奈川県鎌倉市小町2–15–13
☎ 0467–25–1500

KAMAKURA BORI MUSEUM
鎌倉彫資料館
〒248　2–21–20 Komachi, Kamakura, Kanagawa Prefecture
〒248　神奈川県鎌倉市小町2–21–20
☎ 0467–25–1502

YAMANAKA LACQUER • *Yamanaka Shikki*

YAMANAKA LACQUER WARE TRADITIONAL CRAFT CENTER
山中漆器伝統産業会館
〒922–01　268–2, Tsukatani-machi-I, Yamanaka-machi, Enuma-gun, Ishikawa Prefecture
〒922–01　石川県江沼郡山中町塚谷町イ268–2
☎ 07617–8–0305

MUSEUM FOR TRADITIONAL PRODUCTS AND CRAFTS*
石川県立伝統産業工芸館
〒920　1–1 Kenroku-machi, Kanazawa, Ishikawa Prefecture
〒920　石川県金沢市兼六町1–1
☎ 0762–62–2020

KAGAWA LACQUER • *Kagawa Shikki*

KAGAWA PREFECTURE PROMOTION CENTER
香川県商工奨励館
〒760　1–20–16 Ritsurin-chō, Takamatsu, Kagawa Prefecture
〒760　香川県高松市栗林町1–20–16　栗林公園内
☎ 0878–33–7411

KAGAWA PREFECTURE INDUSTRY CENTER
香川県産業会館
〒760　2–2-2 Fukuoka-chō, Takamatsu, Kagawa Prefecture
〒760　香川県高松市福岡町2–2-2
☎ 0878–51–5669

INTERIOR MORISHIGE MARUGAME GALLERY
インテリア·モリシゲ丸亀町店
〒760　2–4 Marugame-chō, Takamatsu, Kagawa Prefecture
〒760　香川県高松市丸亀町2–4
☎ 0878–51–3617

TSUGARU LACQUER • *Tsugaru Nuri*

TSUGARU LACQUER WARE STUDIOS COOP*
津軽塗団地協同組合会館
〒036　2–3-10 Kanda, Hirosaki, Aomori Prefecture
〒036　青森県弘前市神田2–3–10
☎ 0172–33–1188

TANAKAYA
田中屋
〒036　Ichibanchi-kado, Hirosaki, Aomori Prefecture
〒036　青森県弘前市一番地角
☎ 0172–33–6666

BAMBOO CRAFT　竹工品　*Chikkōhin*

SURUGA BASKETRY • *Suruga Take Sensuji Zaiku*

SHIZUOKA INDUSTRY CENTER
静岡市産業展示館
〒422　2992 Nakajima, Shizuoka, Shizuoka Prefecture
〒422　静岡県静岡市中島2992
☎ 054–287–5550

CHIKUDAI STUDIO
ちくだい工房
〒422　139 Takyo, Shizuoka, Shizuoka Prefecture
〒422　静岡県静岡市立帆139
☎ 054–278–3270

NENRINBŌ
燃林房
〒420　130–26 Yanagi-chō, Shizuoka, Shizuoka Prefecture
〒420　静岡県静岡市柳町130–26
☎ 0542–71–0631

BEPPU BASKETRY • *Beppu Take zaiku*

BEPPU TRADITIONAL BAMBOO CRAFT CENTER
別府市竹細工伝統産業会館
〒874　8–3 Higashi Sōen-chō, Beppu, Ōita Prefecture
〒874　大分県別府市東荘園町8–3
☎ 0977–23–1072

IWAO CHIKURAN
岩尾竹籃
〒874　1–5 Hikari-chō, Beppu, Ōita Prefecture
〒874　大分県別府市光町1–5
☎ 0977–22–4074

ŌITA INDUSTRIAL RESEARCH INSTITUTE, BEPPU INDUSTRIAL ARTS RESEARCH DIVISION*
大分県産業科学技術センター　別府産業工芸試験所
〒874　3–3 Higashi Sōen-chō, Beppu, Ōita Prefecture
〒874　大分県別府市東荘園町3–3
☎ 0977–22–0208

TAKAYAMA TEA WHISKS • *Takayama Chasen*

SUICHIKUEN
翠竹園
〒630–01　5621 Takayama-chō, Ikoma, Nara Prefecture
〒630–01　奈良県生駒市高山町5621
☎ 07437–8–0067

KUBO RYŌSAI
久保良斎
〒630–01　6659–3 Takayama-chō, Ikoma, Nara Prefecture
〒630–01　奈良県生駒市高山町6659–3
☎ 07437–8–0059

MIYAKONOJŌ BOWS • *Miyakonojō Daikyū*

MIYAKONOJŌ LOCAL CRAFT PROMOTION CENTER
都城圏域地場産業振興センター
〒885　5225-1 Tohoku-chō, Miyokonojō, Miyazaki Prefecture
〒885　宮崎県都城市都北町5225-1
☎ 0986-38-4561

JAPANESE PAPER 和紙 *Washi*

ECHIZEN PAPER • *Echizen Washi*

PAPYRUS CENTER
パピルス館
〒915-02　8-44 Shinzaike, Imadate-chō, Imadate-gun, Fukui Prefecture
〒915-02　福井県今立郡今立町新在家8-44
☎ 0778-42-1363

WASHI PAPER CENTER
和紙の里会館
〒915-02　11-12 Shinzaike, Imadate-chō, Imadate-gun, Fukui Prefecture
〒915-02　福井県今立郡今立町新在家11-12
☎ 0778-42-0016

AWA PAPER • *Awa Washi*

AWAGAMI FACTORY*
阿波和紙伝統産業会館
〒779-34　141 Aza Kawahigashi, Yamakawa-chō, Oe-gun, Tokushima Prefecture
〒779-34　徳島県麻植郡山川町字川東141
☎ 0883-42-6120

TOKUSHIMA CRAFT CENTER
徳島工芸村
〒770　1-1 Higashi-Hamabōji, Yamashiro-chō, Tokushima, Tokushima Prefecture
〒770　徳島県徳島市山城町東浜傍示1-1
☎ 0886-24-5000

TOSA PAPER • *Tosa Washi*

TOSA WASHI VILLAGE COUR AUX DONS*
土佐和紙伝統産業会館（紙の博物館）
〒781-21　110-1 Saiwai-chō, Ino-chō, Agawa-gun, Kōchi Prefecture
〒781-21　高知県吾川郡伊野町幸町110-1
☎ 0888-93-0886

TOSA PAPER CRAFT CENTER
土佐和紙工芸村
〒781-21　1226 Kashiki, Ino-chō, Agawa-gun, Kōchi Prefecture
〒781-21　高知県吾川郡伊野町鹿敷1226
☎ 0888-92-1001

UCHIYAMA PAPER • *Uchiyama Gami*

IIYAMA FOLK ART MUSEUM
飯山市民芸館
〒389-22　Sekizawa, Ōaza Mizuho, Iiyama, Nagano Prefecture
〒389-22　長野県飯山市大字瑞穂関沢
☎ 0269-65-2501

IIYAMA TRADITIONAL CRAFT CENTER
飯山市伝統産業会館
〒389-22　1436-1 Ōaza Iiyama, Iiyama, Nagano Prefecture
〒389-22　長野県飯山市大字飯山1436-1
☎ 0269-62-4019

INSHŪ PAPER • *Inshū Washi*

YAMANE PAPER GALLERY
山根和紙資料館
〒689-05　128-5 Yamane, Aoya-chō, Ketaka-gun, Tottori Prefecture
〒689-05　鳥取県気高郡青谷町山根128-5
☎ 0857-86-0011

SAJI PAPER GALLERY (KAMING SAJI*)
佐治和紙民芸館
〒689-13　Kamo, Saji-son, Yazu-gun, Tottori Prefecture
〒689-13　鳥取県八頭郡佐治村加茂
☎ 0858-89-1816

WOODCRAFT 木工品 *Mokkōhin*

IWAYADŌ CHESTS • *Iwayadō Tansu*

IWAYADŌ TANSU CHEST GALLERY
岩谷堂箪笥展示館
〒023-11　68-1 Aza Ebishima, Odaki, Esashi, Iwate Prefecture
〒023-11　岩手県江刺市愛宕字海老島68-1
☎ 0197-35-0275

FUJISATO WOODCRAFT GALLERY
匠の森
〒023-17　185 Kanizawa, Tawara, Esashi, Iwate Prefecture
〒023-17　岩手県江刺市田原字蟹沢185
☎ 0197-35-7711

KAMO PAULOWNIA CHESTS • *Kamo Kiri Tansu*

KAMO KIRITANSU, INCORPORATED
加茂桐たんす株式会社
〒959-13　398-1 Ōaza Urasuda, Kamo, Niigata Prefecture
〒959-13　新潟県加茂市大字後須田398-1
☎ 0256-52-6740

KYOTO WOODWORK • *Kyō Sashimono*

ENAMI, INCORPORATED
有限会社　江南
〒600　89 Ebisuno-chō, Aino-machi Nishi-iru, Rokujō-dōri, Shimogyō-ku, Kyoto
〒600　京都府京都市下京区六條通間之町西入夷之町89
☎ 075-361-2816

ŌDATE BENTWOOD WORK • *Ōdate Magewappa*

DENSHŌ, INCORPORATED
株式会社　伝承
〒010　88-3 Aza Nomura, Soto-Asahikawa, Akita, Akita Prefecture
〒010　秋田県秋田市外旭川字野村88-3
☎ 0188-80-1101

AKITA CEDAR BOWLS AND BARRELS • *Akita Sugi Oke Taru*

AKITA PREFECTURE INDUSTRY CENTER
秋田県産業会館
〒010　Atorion B1F, 2-3-8 Naka-dōri, Akita, Akita Prefecture
〒010　秋田県秋田市中通2-3-8　アトリオン地下1階
☎ 0188-36-7830

TARUTOMI KAMATA
樽富かまた
〒016　4–3 Suehiro-chō, Noshiro, Akita Prefecture
〒016　秋田県能代市末広町4–3
☎ 0185–52–2539

HAKONE MARQUETRY • *Hakone Yosegi Zaiku*

HATAJUKU YOSEGI CRAFT CENTER
畑宿寄木会館
〒250–03　103 Hatajuku, Hakone-machi, Ashigarashimo-gun, Kanagawa Prefecture
〒250–03　神奈川県足柄下郡箱根町畑宿103
☎ 0460–5–8170

HONMA WOODCRAFT GALLERY
本間美術館
〒250–03　84 Yumoto, Hakone-machi, Ashigarashimo-gun, Kanagawa Prefecture
〒250–03　神奈川県足柄下郡箱根町湯本84
☎ 0460–5–5646

NAGISO TURNERY • *Nagiso Rokuro Zaiku*

NAGISO TURNERY ASSOCIATION
南木曾ろくろ工芸協同組合
〒399–53　4689 Azuma, Nagiso-chō, Kiso-gun, Nagano Prefecture
〒399–53　長野県木曽郡南木曾町吾妻4689
☎ 0264–58–2434

TAKAYAMA WOODCARVING • *Takayama Ichii-Ittōbori*

WANI HISAYUKI
和丹久幸
〒506　1–296–6 Morishita-chō, Takayama, Gifu Prefecture
〒506　岐阜県高山市森下町1–296–6
☎ 0577–33–0744

SUZUKI WOODCARVING
鈴木彫刻
〒506　1-chōme Hatta-chō, Takayama, Gifu Prefecture
〒506　岐阜県高山市初田町1丁目
☎ 0577–32–1367

INAMI WOODCARVING • *Inami Chōkoku*

INAMI WOODCARVING TRADITIONAL CRAFT CENTER
井波彫刻伝統産業会館
〒932–02　700–111 Inami, Inami-machi, Higashi-Tonami-gun, Toyama Prefecture
〒932–02　富山県東礪波郡井波町井波700–111
☎ 0763–82–5158

INAMI WOODCARVING CENTER
井波彫刻総合会館
〒932–02　733 Kitakawa, Inami-machi, Higashi-Tonami-gun, Toyama Prefecture
〒932–02　富山県東礪波郡井波町北川733
☎ 0763–82–5158

KISO OROKU COMBS • *Kiso Oroku Gushi*

OROKU COMBS CENTER ŌTSUTAYA
お六櫛センター大つたや
〒399–62　1224 Yabuhara, Kiso-mura, Kiso-gun, Nagano Prefecture
〒399–62　長野県木曽郡木祖村薮原1224
☎ 0264–36–2205

AKITA CHERRY-BARK WORK • *Akita Kabazaiku*

KAKUNODATE CHERRY-BARK CRAFT CENTER
角館町樺細工伝承館（ふるさとセンター）
〒014–03　10–1 Shimo-chō Omote-machi, Kakunodate-machi, Senboku-gun, Akita Prefecture
〒014–03　秋田県仙北郡角館町表町下町10–1
☎ 0187–54–1700

FUJIKI DENSHIRŌ, INCORPORATED
株式会社　藤木伝四郎商店
〒014–03　45 Shimoshin-chō, Kakunodate-machi, Senboku-gun, Akita Prefecture
〒014–03　秋田県仙北郡角館町下新町45
☎ 0187–54–1151

METALWORK　金工品　*Kinkōhin*

NANBU CAST IRONWORK • *Nanbu Tekki*

MORIOKA REGIONAL INDUSTRY PROMOTION CENTER
盛岡地域地場産業振興センター（盛岡手づくり工芸村）
〒020　64–102 Aza Oirino, Tsunagi, Morioka, Iwate Prefecture
〒020　岩手県盛岡市繋字尾入野64–102
☎ 0196–89–2336

MIZUSAWA CAST IRON FOUNDRY COOPERATIVE ASSOCIATION*
水沢鋳物工業協同組合
〒023　18 Aza Namiyanagi, Hada-chō, Mizusawa, Iwate Prefecture
〒023　岩手県水沢市羽田町字並柳18
☎ 0197–24–1551

IWACHŪ CASTING WORKS*
岩鋳鉄器館
〒020　2–23–9 Minami Senboku, Morioka, Iwate Prefecture
〒020　岩手県盛岡市南仙北2–23–9
☎ 0196–35–2501

TSUBAME BEATEN COPPERWARE • *Tsubame Tsuiki Dōki*

TSUBAME INDUSTRY ARCHIVES
燕市産業史料館
〒959–12　4330–1 Ōaza Ōmagari, Tsubame, Niigata Prefecture
〒959–12　新潟県燕市大字大曲4330–1
☎ 0256–63–7666

GYOKUSENDŌ GALLERY
株式会社　玉川堂
〒959–12　2 Chūō-dōri, Tsubame, Niigata Prefecture
〒959–12　新潟県燕市中央通2
☎ 0256–62–2015

TAKAOKA CASTING • *Takaoka Dōki*

TAKAOKA'S REGIONAL INDUSTRIAL CENTER*
高岡地域地場産業センター
〒933　1–1 Kaihatsu-honmachi, Takaoka, Toyama Prefecture
〒933　富山県高岡市開発本町1–1
☎ 0766–25–8283

TAKAOKA CASTING GALLERY
高岡銅器展示館
〒933　53–1 Toide Sakae-machi, Takaoka, Toyama Prefecture
〒933　富山県高岡市戸出栄町53–1
☎ 0766–63–5556

OSAKA NANIWA PEWTER WARE • ***Osaka Naniwa Suzuki***

SUZUHAN PEWTER GALLERY
錫半ショールーム
〒541 3-6-9 Minami Kyūhōji-machi, Chūō-ku, Osaka
〒541 大阪市中央区南久宝寺町3-6-9
☎ 06-251-8031

TOKYO SILVERSMITHERY • ***Tokyo Ginki***

ŌBUCHI SILVERWARE GALLERY
株式会社 大淵銀器ショールーム
〒110 3-1-13 Higashi Ueno, Taitō-ku, Tokyo
〒110 東京都台東区東上野3-1-13
☎ 03-3847-7711

SAKAI FORGED BLADES • ***Sakai Uchi Hamono***

SAKAI METAL WORK CENTER
堺金物会館
〒590 1-1-24 Shukuya-chō Nishi, Sakai, Osaka
〒590 大阪府堺市宿屋町西1-1-24
☎ 0722-27-1001

SAKAI FORGED BLADES GALLERY
堺刀司資料館
〒590 1-11 Nishi 2-chō, Sakurano-chō, Sakai, Osaka
〒590 大阪府堺市桜之町西2丁1-11
☎ 0722-38-0888

MIZUNO FORGED BLADES STUDIO
水野鍛錬所
〒590 1-27 Nishi 1-chō, Sakurano-chō, Sakai, Osaka
〒590 大阪府堺市桜之町西1丁1-27
☎ 0722-29-3253

KYOTO METAL INLAY • ***Kyō Zōgan***

KAWAHITO HANDS
川人象嵌・象嵌の館川人ハンズ
〒603 76 Minami-machi, Tōjiin, Kita-ku, Kyoto
〒603 京都府京都市北区等持院南町76
☎ 075-461-2773

OTHER CRAFTS 諸工芸品 *Shokōgeihin*

BANSHŪ ABACUS • ***Banshū Soroban***

ONO TRADITIONAL CRAFT CENTER
小野市伝統産業会館
〒675-13 806-1 Ōji-chō, Ono, Hyōgo Prefecture
〒675-13 兵庫県小野市王子町806-1
☎ 07946-2-3121

MIYAMOTO KAZUHIRO
宮本一広
〒675-13 1113 Tenjin-chō, Ōno, Hyōgo Prefecture
〒675-13 兵庫県小野市天神町1113
☎ 0794-62-6419

FUKUYAMA PLANE HARP • ***Fukuyama Koto***

OGAWA MUSICAL INSTRUMENT, INCORPORATED
小川楽器製造株式会社
〒720 3-2-8 Miyoshi-machi, Fukuyama, Hiroshima Prefecture
〒720 広島県福山市三吉町3-2-8
☎ 0849-24-1150

KŌSHŪ LACQUERED DEERHIDE • ***Kōshū Inden***

UEHARA YŪSHICHI INDEN GALLERY
株式会社 印傳屋上原勇七ショールーム
〒400 201 Aria, Kawada-chō, Kōfu, Yamanashi Prefecture
〒400 山梨県甲府市川田町アリア201
☎ 0552-20-1661

YAMANASHI TRADITIONAL CRAFT CENTER
山梨伝統産業会館
〒406 1569 Yokkaichiba, Isawa-chō, Higashi-Yatsushiro-gun, Yamanashi Prefecture
〒406 山梨県東八代郡石和町四日市場1569
☎ 0552-63-6741

NARA SUMI INK • ***Nara Zumi***

KOBAIEN
古梅園
〒630 7 Tsubai-chō, Nara, Nara Prefecture
〒630 奈良県奈良市椿井町7
☎ 0742-23-2965

BOKU-UN-DŌ GALLERY
株式会社 墨運堂内 墨の資料館
〒630 1-5-35 Rokujō, Nara, Nara Prefecture
〒630 奈良県奈良市六条1-5-35
☎ 0742-41-7155

NARA BRUSHES • ***Nara Fude***

AKASHIYA INCORPORATED
株式会社 あかしや
〒630 4-1 Shijō Ōji, Nara, Nara Prefecture
〒630 奈良県奈良市四条大路4-1
☎ 0742-33-6181

OGATSU INKSTONES • ***Ogatsu Suzuri***

OGATSU INKSTONE TRADITIONAL CRAFT CENTER (STONE PLAZA*)
雄勝硯伝統産業会館
〒986-13 53-1 Aza Tera, Ōaza Ogatsu, Ogatsu-chō, Monou-gun, Miyagi Prefecture
〒986-13 宮城県桃生郡雄勝町大字雄勝字寺53-1
☎ 0225-57-3211

GYŌTOKU ŌKARAHAFŪ PORTABLE SHRINE • ***Gyōtoku Ōkarahafū Mikoshi***

NAKADAI PORTABLE SHRINE
中台神輿製作所
〒272-01 21-3 Honjio, Ichikawa, Chiba Prefecture
〒272-01 千葉県市川市本塩21-3
☎ 0473-57-2061

KYOTO HOUSEHOLD BUDDHIST ALTARS, FITTINGS • ***Kyō Butsudan, Kyō Butsugu***

MATSUMOTO-YA
松本屋
〒600 Takatsuji-kado, Teramachi-dōri, Shimogyō-ku, Kyoto
〒600 京都府京都市下京区寺町通高辻角
☎ 075-343-1200

KYOTO CANDLES • ***Kyō Rōsoku***

TANJI RENSHŌDŌ
丹治蓮生堂
〒600 114 Nakai-chō, Karasuma Nishi-iru, Shichijō-dōri,

Shimogyō-ku, Kyoto
〒600 京都府京都市下京区七条通烏丸西入中居町114
☎ 075-361-0937

GIFU LANTERNS • *Gifu Chōchin*
GIFU UMBRELLAS • *Gifu Wagasa*

GIFU INDUSTRY CENTER
岐阜県物産展示室
〒500 2-11-1 Rokujō-minami, Gifu, Gifu Prefecture
〒500 岐阜県岐阜市六条南2-11-1
☎ 0582-72-3399

IZUMO STONE LANTERNS • *Izumo Ishidōrō*

SHIMANE PREFECTURE TOURIST CENTER
島根県物産観光館
〒690 191 Tono-machi, Matsue, Shimane Prefecture
〒690 島根県松江市殿町191 島根ふるさと館内
☎ 0852-22-5758

MARUGAME ROUND FANS • *Marugame Uchiwa*

KAGAWA INDUSTRY CENTER
香川県産業会館
〒760 2-2-2 Fukuoka-chō, Takamatsu, Kagawa Prefecture
〒760 香川県高松市福岡町2-2-2
☎ 0878-51-5669

UCHIWA NO MINATO MUSEUM
うちわの港ミュージアム
〒763 307-15 Minato-machi, Marugame, Kagawa Prefecture
〒763 香川県丸亀市港町307-15
☎ 0877-24-7055

KYOTO FOLDING FANS • *Kyō Sensu*

MIYAWAKI BAISEN-AN
宮脇賣扇庵
〒604 Tominokōji Nishi-iru, Rokkaku-dōri, Nakagyō-ku, Kyoto
〒604 京都府京都市中京区六角通富小路西入ル
☎ 075-221-0181

EDO "ART" DOLLS • *Edo Kimekomi Ningyō*

MATARO DOLLS CENTER
真多呂人形会館
〒110 5-15-13 Ueno, Taitō-ku, Tokyo
〒110 東京都台東区上野5-15-13
☎ 03-3833-9661

KAKINUMA DOLL, INCORPORATED
株式会社 柿沼人形
〒343 2-174-4 Hichiza-chō, Koshigaya, Saitama Prefecture
〒343 埼玉県越谷市七左町2-174-4
☎ 0489-64-7877

EDO BATTLEDORES • *Edo Oshie Hagoita*

HAGOITA BATTLEDORE GALLERY
羽子板資料館
〒131 5-43-25 Mukōjima, Sumida-ku, Tokyo
〒131 東京都墨田区向島5-43-25
☎ 03-3623-1305

MIHARU DOLLS • *Miharu Hariko*

FUKUSHIMA PREFECTURE REGIONAL PRODUCTS CENTER
福島県物産館
〒960 4-15 Ōmachi, Fukushima, Fukushima Prefecture
〒960 福島県福島市大町4-15
☎ 0245-22-3948

TAKASHIBA DEKO YASHIKI
高柴デコ屋敷
〒977-01 163 Aza Tateno, Takashiba, Nishida-machi, Kōriyama, Fukushima Prefecture
〒977-01 福島県郡山市西田町高柴字館野163
☎ 0249-71-3176

MIHARU DOLL MUSEUM
三春郷土人形館
〒960 30 Aza Ōmachi, Miharu-machi, Fukushima Prefecture
〒960 福島県福島市三春町字大町30
☎ 0247-62-7053

MIYAGI KOKESHI DOLLS • *Miyagi Dentō Kokeshi*

ZAŌ-MACHI TRADITIONAL CRAFT CENTER
蔵王町伝統産業会館（みやぎ蔵王こけし館）
〒989-09 36-135 Nishi-Urayama, Aza Shinchi, Tōgatta-onsen, Zaō-machi, Katta-gun, Miyagi Prefecture
〒989-09 宮城県刈田郡蔵王町遠刈田温泉字新地西裏山36-135
☎ 0224-34-2385

JAPAN KOKESHI MUSEUM
日本こけし館
〒989-68 74-2 Aza Sutomae, Naruko-chō, Tamazukuri-gun, Miyagi Prefecture
〒989-68 宮城県玉造郡鳴子町字尿前74-2
☎ 0229-83-3600

LIST OF LOCAL CRAFT ASSOCIATIONS

Most of the organizations listed below are locally run and possess limited resources. While they are eager to lend assistance, they rarely have English-speaking staff, so your initial contact, whether by telephone or facsimile, should be in Japanese. An asterisk indicates an official English name. In cases where an English name was not available, an approximation of the Japanese name has been supplied. When appropriate or when a representative association could not be located, a company name has been suggested.

CERAMICS 陶磁器 *Tōjiki*

KUTANI WARE • *Kutani Yaki*

KUTANI PORCELAIN ASSOCIATION, ISHIKAWA PREFECTURE
石川県九谷陶磁器商工業協同組合連合会
〒923-11 25 Aza Terai-Yo, Terai-machi, Nomi-gun, Ishikawa Prefecture
〒923-11 石川県能美郡寺井町字寺井よ25
☎ 0761-57-0125
FAX: 0761-57-0320

MASHIKO WARE • *Mashiko Yaki*

MASHIKO POTTERS' ASSOCIATION
益子焼協同組合
〒321-42 4352-2 Ōaza Mashiko, Mashiko-machi, Haga-gun, Tochigi Prefecture
〒321-42 栃木県芳賀郡益子町大字益子4352-2
☎ 0285-72-3107
FAX: 0285-72-3058

MINO WARE • *Mino Yaki*

MINO WARE ASSOCIATION
美濃焼伝統工芸品協同組合
〒509-51 1429-8 Izumi-chō Kujiri, Toki, Gifu Prefecture
〒509-51 岐阜県土岐市泉町久尻1429-8
☎ 0572-55-5527
FAX: 0572-55-7352

HAGI WARE • *Hagi Yaki*

HAGI MUNICIPAL OFFICE
萩市経済部商工課
〒758 510 Emukai, Hagi, Yamaguchi Prefecture
〒758 山口県萩市江向510
☎ 0838-25-3131
FAX: 0838-26-0716

KYOTO WARE, KIYOMIZU WARE • *Kyō Yaki, Kiyomizu Yaki*

KYOTO FEDERATION OF CERAMICS ASSOCIATION*
京都陶磁器協同組合連合会
〒605 570-3 Shiraito-chō, Higashi-Ōji, Higashi-iru, Gojō-dōri, Higashiyama-ku, Kyoto
〒605 京都府京都市東山区五条通東大路東入ル白糸町570-3 京都陶磁器会館内
☎ 075-541-1102
FAX: 075-541-1195

SHIGARAKI WARE • *Shigaraki Yaki*

ASSOCIATION OF SHIGARAKI CERAMIC INDUSTRY*
信楽陶器工業協同組合
〒529-18 1142 Ōaza Nagano, Shigaraki-chō, Kōka-gun, Shiga Prefecture
〒529-18 滋賀県甲賀郡信楽町大字長野1142
☎ 0748-82-0831
FAX: 0748-82-3473

BIZEN WARE • *Bizen Yaki*

BIZEN POTTERY ASSOCIATION, OKAYAMA PREFECTURE
協同組合岡山県備前焼陶友会
〒705 1657-2 Inbe, Bizen, Okayama Prefecture
〒705 岡山県備前市伊部1657-2
☎ 0869-64-1001
FAX: 0869-64-1002

ARITA WARE • *Arita Yaki*

SAGA PREFECTURAL CERAMIC WARE INDUSTRY COOPERATIVE*
佐賀県陶磁器工業協同組合
〒844 1217 Chūbu-Hei, Arita-machi, Nishi-Matsuura-gun, Saga Prefecture
〒844 佐賀県西松浦郡有田町中部丙1217
☎ 0955-42-3164
FAX: 0955-43-2917

TOKONAME WARE • *Tokoname Yaki*

TOKONAME POTTERY ASSOCIATION
とこなめ焼協同組合
〒479 3-8 Sakae-machi, Tokoname, Aichi Prefecture
〒479 愛知県常滑市栄町3-8
☎ 0569-35-4309
FAX: 0569-34-4952

TOBE WARE • *Tobe Yaki*

IYO CERAMICS ASSOCIATION

伊予陶磁器協同組合
〒791–21　604 Ōminami, Tobe-chō, Iyo-gun, Ehime Prefecture
〒791–21　愛媛県伊予郡砥部町大南604
☎ 0899–62–2018
FAX: 0899–62–6246

TSUBOYA WARE • *Tsuboya Yaki*

TSUBOYA POTTERY COOPERATIVE SOCIETY*
壷屋陶器事業協同組合
〒902　1–21–14 Tsuboya, Naha, Okinawa Prefecture
〒902　沖縄県那覇市壷屋1–21–14
☎ 098–866–3284
FAX: 098–864–1472

KIKUMA TILES • *Kikuma Gawara*

KIKUMA TILES ASSOCIATION
菊間町窯業協同組合
〒799–23　316–2 Hama, Kikuma-chō, Ochi-gun, Ehime Prefecture
〒799–23　愛媛県越智郡菊間町浜316–2
☎ 0898–54–5511
FAX: 0898–54–5511

TEXTILES　染織　*Senshoku*

KIRYŪ FABRICS • *Kiryū Ori*

KIRYŪ TEXTILES WEAVERS' COOPERATIVE ASSOCIATION*
桐生織物協同組合
〒376　5–1 Eiraku-chō, Kiryū, Gunma Prefecture
〒376　群馬県桐生市永楽町5–1
☎ 0277–43–7171
FAX: 0277–47–5517

OITAMA PONGEE • *Oitama Tsumugi*

OITAMA PONGEE TRADITIONAL TEXTILE ASSOCIATION
置賜紬伝統織物協同組合連合会
〒992　1-1-5 Montō-machi, Yonezawa, Yamagata Prefecture
〒992　山形県米沢市門東町1–1–5
☎ 0238–23–3525
FAX: 0238–23–7229

YŪKI PONGEE • *Yūki Tsumugi*

YŪKI PONGEE TEXTILE ASSOCIATION, IBARAGI PREFECTURE
茨城県本場結城紬織物協同組合
〒307　607 Ōaza Yūki, Yūki, Ibaragi Prefecture
〒307　茨城県結城市大字結城607
☎ 02963–2-1108
FAX: 02963–2-1108

TRUE KIHACHIJŌ • *Honba Kihachijō*

KIHACHIJŌ TEXTILE ASSOCIATION
黄八丈織物協同組合
〒100–14　2025 Kashitate, Hachijō-machi, Hachijōjima, Tokyo
〒100–14　東京都八丈島八丈町樫立2025
☎ 04996–7–0516

OJIYA RAMIE CREPE • *Ojiya Chijimi*

OJIYA TEXTILE ASSOCIATION
小千谷織物同業協同組合
〒947　1–8-25 Jōnai, Ojiya, Niigata Prefecture
〒947　新潟県小千谷市城内1–8–25
☎ 0258–83–2329
FAX: 0258–83–2328

SHINSHŪ PONGEE • *Shinshū Tsumugi*

NAGANO PREFECTURAL TEXTILE ASSOCIATION
長野県織物工業組合
〒399　1-7-7 Nomizo-Nishi, Matsumoto, Nagano Prefecture
〒399　長野県松本市野溝西1–7–7
☎ 0263–26–0721
FAX: 0263–26–5350（試験場）

NISHIJIN FABRICS • *Nishijin Ori*

NISHIJIN TEXTILE INDUSTRIAL COOPERATIVE*
西陣織工業組合
〒602　414 Tatemonzen-chō, Imadegawa Minami-iru, Horikawa-dōri, Kamigyō-ku, Kyoto
〒602　京都府京都市上京区堀川通今出川南入ル堅門前町414
☎ 075–432–6131
FAX: 075–414–1521

YUMIHAMA IKAT • *Yumihama Gasuri*

YUMIHAMA IKAT ASSOCIATION, TOTTORI PREFECTURE
鳥取県弓浜絣協同組合
〒683　3001–3 Yomi-chō, Yonago, Tottori Prefecture
〒683　鳥取県米子市夜見町3001–3　鳥取県工業試験場生産技術科
☎ 0859–29–0851
FAX: 0859–29–5482

HAKATA WEAVE • *Hakata Ori*

HAKATA TEXTILE INDUSTRIAL ASSOCIATION*
博多織工業組合
〒812　1–14–12 Hakataeki-Minami, Hakata-ku, Fukuoka, Fukuoka Prefecture
〒812　福岡県福岡市博多区博多駅南1–14–12
☎ 092–472–0761
FAX: 092–472–1254

TRUE ŌSHIMA PONGEE • *Honba Ōshima Tsumugi*

FEDERATION OF KAGOSHIMA PREFECTURE ŌSHIMA TSUMUGI COOPERATIVE ASSOCIATIONS*
鹿児島県本場大島紬協同組合連合会
〒894　15–1 Minato-machi, Naze, Kagoshima Prefecture
〒894　鹿児島県名瀬市港町15–1
☎ 0997–53–4968
FAX: 0997–53–8255

MIYAKO RAMIE • *Miyako Jōfu*

MIYAKO TEXTILE ASSOCIATION
宮古織物事業協同組合
〒906　3 Aza Nishizato, Hirara, Okinawa Prefecture
〒906　沖縄県平良市字西里3
☎ 09807–2-8022
FAX: 09807–2-8022

SHURI FABRICS • *Shuri Ori*

NAHA TRADITIONAL TEXTILE ASSOCIATION
那覇伝統織物事業協同組合
〒903　2–64 Shuritōbaru-chō, Naha, Okinawa Prefecture
〒903　沖縄県那覇市首里桃原町2–64
☎ 098–887–2746
FAX: 098–885–5674

KIJOKA ABACA • *Kijoka no Bashōfu*

KIJOKA ABACA ASSOCIATION
喜如嘉芭蕉布事業協同組合
〒905–13 1103 Aza Kijoka, Ōgimi-son, Kunigami-gun, Okinawa Prefecture
〒905–13 沖縄県国頭郡大宜味村字喜如嘉1103
☎ 0980–44–3202
FAX: 0980–44–3202

YONTANZA MINSAA • *Yomitanzan Minsaa*

YOMITANZAN MINSAA ASSOCIATION
読谷山花織事業協同組合
〒904–03 2974–2 Aza Zakimi, Yomitan-son, Nakagami-gun, Okinawa Prefecture
〒904–03 沖縄県中頭郡読谷村字座喜味2974–2
☎ 098–958–4674
FAX: 098–958–4674

ISESAKI IKAT • *Isesaki Gasuri*

ISESAKI TEXTILE ASSOCIATION
伊勢崎織物工業組合
〒372 31–9 Kuruwa-chō, Isesaki, Gunma Prefecture
〒372 群馬県伊勢崎市曲輪町31–9
☎ 0270–25–2700
FAX: 0270–24–6347

TSUGARU KOGIN STITCHING • *Tsugaru Kogin*

HIROSAKI KOGIN NEEDLEPOINT CENTER
(有) 弘前こぎん研究所
〒036 61 Zaifu-chō, Hirosaki, Aomori Prefecture
〒036 青森県弘前市在府町61
☎ 0172–32–0595
FAX: 0172–32–0595

TOKYO STENCIL DYEING • *Tokyo Somekomon*

TOKYO ORDER-MADE DYEING ASSOCIATION*
東京都染色工業協同組合
〒169 3–20–12 Nishi-Waseda, Shinjuku-ku, Tokyo
〒169 東京都新宿区西早稲田3–20–12
☎ 03–3208–1521
FAX: 03–3208–1523

KAGA YŪZEN DYEING • *Kaga Yūzen*

KAGAZOME COOPERATIVE ASSOCIATION*
協同組合加賀染振興協会
〒920 8–8 Koshō-machi, Kanazawa, Ishikawa Prefecture
〒920 石川県金沢市小将町8–8
☎ 0762–24–5511
FAX: 0762–24–5533

KYOTO YŪZEN DYEING • *Kyō Yūzen*
KYOTO STENCIL DYEING • *Kyō Komon*
KYOTO TIE-DYEING • *Kyō Kanoko Shibori*

KYOTO CORPORATE FEDERATION OF DYERS & COLORISTS*
京都染色協同組合連合会
〒604 481 Tōrōyama-chō, Shijō-agaru, Nishinotōin-dōri, Nakagyō-ku, Kyoto
〒604 京都府京都市中京区西洞院通四条上ル蟷螂山町481 京染会館3F
☎ 075–255–4496
FAX: 075–255–4496

TOKYO YUKATA STENCIL DYEING • *Tokyo Honzome Yukata*

TOKYO YUKATA ASSOCIATION
東京ゆかた工業協同組合
〒103 9–16 Nihonbashi-kobuna-chō, Chūō-ku, Tokyo
〒103 東京都中央区日本橋小舟町9–16
☎ 03–3661–3862
FAX: 03–3669–0888

KYOTO EMBROIDERY • *Kyō Nui*

KYOTO EMBROIDERY ASSOCIATION*
京都刺繍協同組合
〒600 378–1 Kōtake-chō, Matsubara-agaru, Kawaramachi-dōri, Shimogyō-ku, Kyoto
〒600 京都府京都市下京区河原町通松原上ル幸竹町378–1 西刺繍内
☎ 075–361–5494
FAX: 075–365–0791

IGA BRAIDED CORDS • *Iga Kumihimo*

MIE PREFECTURAL KUMIHIMO COOPERATIVE SOCIETY*
三重県組紐協同組合
〒518 1929–10 Shijuku-chō, Ueno, Mie Prefecture
〒518 三重県上野市四十九町1929–10 伊賀くみひもセンター内
☎ 0595–23–8038
FAX: 0595–24–1015

LACQUER WARE 漆器 *Shikki*

WAJIMA LACQUER • *Wajima Nuri*

WAJIMA URUSHI WARE COOPERATIVE SOCIETY*
輪島漆器商工業協同組合
〒928 55, 24-bu, Kawai-machi, Wajima, Ishikawa Prefecture
〒928 石川県輪島市河井町24部55
☎ 0768–22–2155
FAX: 0768–22–2894

AIZU LACQUER • *Aizu Nuri*

AIZU LACQUER WARE COOPERATION UNION*
会津漆器協同組合連合会
〒965 1-7-3 Ōmachi, Aizuwakamatsu, Fukushima Prefecture
〒965 福島県会津若松市大町1–7-3
☎ 0242–24–5757
FAX: 0242–24–5726

KISO LACQUER • *Kiso Shikki*

KISO SHIKKI INDUSTRIAL COOPERATIVE ASSOCIATION*
木曽漆器工業協同組合
〒399–63 1729 Ōaza Hirasawa, Narakawa-mura, Kiso-gun, Nagano Prefecture
〒399–63 長野県木曽郡楢川村大字平沢1729
☎ 0264–34–2113
FAX: 0264–34–2312

HIDA SHUNKEI LACQUER • *Hida Shunkei*

HIDA SHUNKEI LACQUER WARE ASSOCIATION
飛騨春慶連合協同組合
〒506 Kamisanno-machi, Takayama, Gifu Prefecture
〒506 岐阜県高山市上三之町 福田屋内
☎ 0577–32–0065
FAX: 0577–35–1727

KISHŪ LACQUER • *Kishū Shikki*

WAKAYAMA-KEN LACQUER WARE COMMERCIAL AND INDUSTIAL COOPERATIVE ASSOCIATION*
和歌山県漆器商工業協同組合
〒642 222 Funao, Kainan, Wakayama Prefecture
〒642 和歌山県海南市船尾222
☎ 0734-82-0322
FAX: 0734-83-2341

MURAKAMI CARVED LACQUER • *Murakami Kibori Tsuishu*

MURAKAMI TSUISHU LACQUER WARE ASSOCIATION
村上堆朱事業協同組合
〒958 3-1-17 Matsubara-chō, Murakami, Niigata Prefecture
〒958 新潟県村上市松原町3-1-17
☎ 0254-53-1745
FAX: 0254-53-3053

KAMAKURA LACQUER • *Kamakura Bori*

TRADITIONAL KAMAKURA BORI ASSOCIATION
伝統鎌倉彫事業協同組合
〒248 3-4-7 Yuigahama, Kamakura, Kanagawa Prefecture
〒248 神奈川県鎌倉市由比ヶ浜3-4-7 神奈川県工芸指導所鎌倉支所内
☎ 0467-23-0154
FAX: 0467-23-0154

YAMANAKA LACQUER • *Yamanaka Shikki*

YAMANAKA LACQUER WARE COOPERATIVE ASSOCIATION*
山中漆器連合協同組合
〒922-01 268-2 Tsukatani-machi-I, Yamanaka-machi, Enuma-gun, Ishikawa Prefecture
〒922-01 石川県江沼郡山中町塚谷町イ268-2
☎ 07617-8-0305
FAX: 07617-8-5205

KAGAWA • *Kagawa Shikki*

KAGAWA LACQUER WARE ASSOCIATION
香川県漆器工業協同組合
〒761-01 1595 Kasuga-chō, Takamatsu, Kagawa Prefecture
〒761-01 香川県高松市春日町1595
☎ 0878-41-9820
FAX: 0878-41-9854

TSUGARU LACQUER • *Tsugaru Nuri*

AOMORI PREFECTURAL LACQUER WARE ASSOCIATION
青森県漆器協同組合連合会
〒036 2-4-9 Ōaza Kanda, Hirosaki, Aomori Prefecture
〒036 青森県弘前市大字神田2-4-9 津軽塗福祉センター内
☎ 0172-35-3629
FAX: 0172-33-1189

BAMBOO CRAFT 竹工品 *Chikkōhin*

SURUGA BASKETRY • *Suruga Take Sensuji Zaiku*

SHIZUOKA BAMBOO CRAFT COOPERATIVE ASSOCIATION*
静岡竹工芸協同組合
〒420 22 Hachiban-chō, Shizuoka, Shizuoka Prefecture
〒420 静岡県静岡市八番町22
☎ 054-252-4924
FAX: 054-273-2679

BEPPU BASKETRY • *Beppu Take Zaiku*

BEPPU BAMBOO CRAFT ASSOCIATION
別府竹製品協同組合
〒874 5-Kumi, Ōhata-machi, Beppu, Ōita Prefecture
〒874 大分県別府市大畑町5組 早野久雄クラフト研究室内
☎ 0977-23-6043
FAX: 0977-23-6093

TAKAYAMA TEA WHISKS • *Takayama Chasen*

TAKAYAMA TEA WHISK ASSOCIATION, NARA PREFECTURE
奈良県高山茶筌生産協同組合
〒630-01 6439-3 Takayama-chō, Ikoma, Nara Prefecture
〒630-01 奈良県生駒市高山町6439-3
☎ 07437-8-0034
FAX: 07437-9-1851

MIYAKONOJŌ BOWS • *Miyakonojō Daikyū*

SOUTH KYŪSHŪ ARCHERY BOW ASSOCIATION
全日本弓道具南九州地区協会
〒885 8-14 Tsumagaoka-chō, Miyakonojō, Miyazaki Prefecture
〒885 宮崎県都城市妻ケ丘町8-14
☎ 0986-22-4604
FAX: 0986-22-4009

JAPANESE PAPER 和紙 *Washi*

ECHIZEN PAPER • *Echizen Washi*

FUKUI-KEN JAPANESE PAPER INDUSTRIAL COOPERATIVE*
福井県和紙工業協同組合
〒915-02 11-11 Ōtaki, Imadate-chō, Imadate-gun, Fukui Prefecture
〒915-02 福井県今立郡今立町大滝11-11
☎ 0778-43-0875
FAX: 0778-43-1142

AWA PAPER • *Awa Washi*
AWAGAMI FACTORY*
阿波手漉和紙商工業協同組合
〒779-34 141 Aza Kawahigashi, Yamakawa-chō, Oe-gun, Tokushima Prefecture
〒779-34 徳島県麻植郡山川町字川東141 阿波和紙伝統産業会館内
☎ 08834-2-6120
FAX: 08834-2-6120

TOSA PAPER • *Tosa Washi*

KŌCHI PREFECTURAL HANDMADE PAPER COOP UNION*
高知県手すき和紙協同組合
〒781-21 287-4 Hakawa, Ino-chō, Agawa-gun, Kōchi Prefecture
〒781-21 高知県吾川郡伊野町波川287-4
☎ 0888-92-4170
FAX: 0888-92-4168

UCHIYAMA PAPER • *Uchiyama Gami*

UCHIYAMA HANDMADE PAPER INDUSTRY ASSOCIATION, NORTH NAGANO
北信内山紙工業協同組合
〒389-23 6385 Ōaza Mizuho, Iiyama, Nagano Prefecture
〒389-23 長野県飯山市大字瑞穂6385
☎ 0269-65-2511
FAX: 0269-65-2601

INSHŪ PAPER • *Inshū Washi*

SAJI INSHŪ HANDMADE PAPER ASSOCIATION
佐治因州和紙協同組合
〒689–13 37–3 Fukuzono, Saji-son, Yazu-gun, Tottori Prefecture
〒689–13 鳥取県八頭郡佐治村福園37–3 佐治村商工会内
☎ 0858–88–0540
FAX: 0858–88–0540

WOODCRAFT 木工品 *Mokkōhin*

IWAYADŌ CHESTS • *Iwayadō Tansu*

IWAYADO TANSU CHESTS ASSOCIATION
岩谷堂箪笥生産協同組合
〒023–11 68–1 Aza Ebishima, Odaki, Esashi, Iwate Prefecture
〒023–11 岩手県江刺市愛宕字海老島68–1
☎ 0197–35–0275
FAX: 0197–35–0972

KAMO PAULOWNIA CHESTS • *Kamo Kiri Tansu*

KAMO PAULOWNIA CHESTS ASSOCIATION
加茂箪笥協同組合
〒959–13 2-2-4 Saiwai-chō, Kamo, Niigata Prefecture
〒959–13 新潟県加茂市幸町2–2–4 加茂市産業センター2F
☎ 0256–52–0445
FAX: 0256–52–0428

KYOTO WOODWORK • *Kyō Sashimono*

KYOTO JOINERY CRAFT ASSOCIATION
京都木工芸協同組合
〒600 89 Ebisuno-chō, Aino-machi Nishi-iru, Rokujō-dōri, Shimogyō-ku, Kyoto
〒600 京都府京都市下京区六条通間之町西入ル夷之町89 （有）江南ビル
☎ 075–361–2816
FAX: 075–351–8657

ŌDATE BENTWOOD WORK • *Ōdate Magewappa*

ŌDATE BENTWOOD WARE ASSOCIATION
大館曲ワッパ協同組合
〒017 1-3-1 Onari-chō, Ōdate, Akita Prefecture
〒017 秋田県大館市御成町1–3–1
☎ 0186–49–5221
FAX: 0186–49–5221

AKITA CEDAR BOWLS AND BARRELS • *Akita Sugi Oke Taru*

AKITA CEDAR BOWLS AND BARRELS ASSOCIATION
秋田杉桶樽協同組合
〒010 1–5 Kyokuhokusakae-machi, Akita, Akita Prefecture
〒010 秋田県秋田市旭北栄町1–5 （社）秋田木材会館
☎ 0188–64–2761
FAX: 0188–64–2756

HAKONE MARQUETRY • *Hakone Yosegi Zaiku*

ODAWARA & HAKONE TRADITIONAL MARQUETRY ASSOCIATION
小田原箱根伝統寄木協同組合
〒250 1–21 Jōnai, Odawara, Kanagawa Prefecture
〒250 神奈川県小田原市城内1–21 商工会館5F （社）箱根物産連合会内
☎ 0465–22–4896
FAX: 0465–23–4531

NAGISO TURNERY • *Nagiso Rokuro Zaiku*

NAGISO TURNED GOODS ASSOCIATION
南木曾ろくろ工芸協同組合
〒399–53 4689 Azuma, Nagiso-machi, Kiso-gun, Nagano Prefecture
〒399–53 長野県木曽郡南木曽町吾妻4689
☎ 0264–58–2434
FAX: 0264–58–2434

TAKAYAMA WOODCARVING • *Takayama Ichii-Ittōbori*

HIDA WOODCARVING ASSOCIATION
飛騨一位一刀彫協同組合
〒506 1–296–6 Morishita-machi, Takayama, Gifu Prefecture
〒506 岐阜県高山市森下町1–296–6
☎ 0577–33–0744
FAX: 0577–33–0744

INAMI WOODCARVING • *Inami Chōkoku*

INAMI WOODCARVING ASSOCIATION
井波彫刻協同組合
〒932–02 700–111 Inami, Inami-machi, Higashi-Tonami-gun, Toyama Prefecture
〒932–02 富山県東砺波郡井波町井波700–111 井波彫刻伝統産業会館内
☎ 0763–82–5158
FAX: 0763–82–5163

KISO OROKU COMBS • *Kiso Oroku Gushi*

YABUHARA OROKU COMBS ASSOCIATION
籔原お六櫛組合
〒399–62 Ōaza Yabuhara, Kiso-mura, Kiso-gun, Nagano Prefecture
〒399–62 長野県木曽郡木祖村大字籔原
☎ 0264–36–2205
FAX: 0264–36–2727

AKITA CHERRY-BARK WORK • *Kabazaiku*

KAKUNODATE CRAFT ASSOCIATION
角館工芸協同組合
〒014–03 18 Aza Tonoyama, Iwase, Kakunodate-machi, Senboku-gun, Akita Prefecture
〒014–03 秋田県仙北郡角館町岩瀬字外ノ山18
☎ 0187–53–2228
FAX: 0187–53–2228

METALWORK 金工品 *Kinkōhin*

NANBU CAST IRONWORK • *Nanbu Tekki*

NAMBU IRON WARE ASSOCIATION, IWATE PREFECTURE
岩手県南部鉄器協同組合連合会
〒020 64–102 Aza Oirino, Tsunagi, Morioka, Iwate Prefecture
〒020 岩手県盛岡市繋字尾入野64–102
☎ 0196–89–2336
FAX: 0196–89–2337

TSUBAME BEATEN COPPER WARE • *Tsubame Tsuiki Dōki*

TSUBAME BUNSUI COPPER WARE ASSOCIATION
燕・分水銅器協同組合
〒959–12 2 Chūō-dōri, Tsubame, Niigata Prefecture
〒959–12 新潟県燕市中央通2

☎ 0256-62-2015
FAX: 0256-64-5945

TAKAOKA CASTING • *Takaoka Dōki*

TAKAOKA COPPER WARE ASSOCIATION
伝統工芸高岡銅器振興協同組合
〒933 1-1 Kaihatsu-honmachi, Takaoka, Toyama Prefecture
〒933 富山県高岡市開発本町1-1 高岡地域地場産業センター内
☎ 0766-24-8565
FAX: 0766-26-0875

OSAKA NANIWA PEWTER WARE • *Osaka Naniwa Suzuki*

PEWTER WARE ASSOCIATION
錫器事業協同組合
〒540 Matsuya Building No. 6, 628, 2-1-31, Nōninbashi, Chūō-ku, Osaka
〒540 大阪市中央区農人橋2-1-31 第6松屋ビル628号
☎ 06-947-2773
FAX: 06-947-2773

TOKYO SILVERSMITHERY • *Tokyo Ginki*

TOKYO ART SILVER WARE ASSOCIATION
東京金銀器工業協同組合
〒110 2-24-4 Higashi-Ueno, Taitō-ku, Tokyo
〒110 東京都台東区東上野2-24-4 東京銀器会館
☎ 03-3831-3317
FAX: 03-3831-3326

SAKAI FORGED BLADES • *Sakai Uchi Hamono*

SAKAI CUTLERY FEDERATION COOPERATIVE*
堺刃物商工業協同組合連合会
〒590 1-1-24 Shukuya-chō Nishi, Sakai, Osaka
〒590 大阪府堺市宿屋町西1-1-24
☎ 0722-27-1001
FAX: 0722-38-8906

KYOTO METAL INLAY • *Kyō Zōgan*

KYOTO METAL INLAY ASSOCIATION
京都府象嵌振興会
〒616 10-3 Setogawa-chō, Saga Tenryūji, Ukyō-ku, Kyoto
〒616 京都府京都市右京区嵯峨天竜寺瀬戸川町10-3
☎ 075-871-2610
FAX: 075-882-0525

OTHER CRAFTS 諸工芸品 *Shokōgeihin*

BANSHŪ ABACUS • *Banshū Soroban*

BANSHŪ ABACUS ASSOCIATION
播州算盤工芸品協同組合
〒675-13 600 Hon-machi, Ono, Hyōgo Prefecture
〒675-13 兵庫県小野市本町600
☎ 07946-2-2108
FAX: 07946-2-2109

FUKUYAMA PLANE HARPS • *Fukuyama Koto*

FUKUYAMA MUSICAL INSTRUMENT ASSOCIATION
福山邦楽器製造業協同組合
〒720 3-2-8 Miyoshi-chō, Fukuyama, Hiroshima Prefecture
〒720 広島県福山市三吉町3-2-8 小川楽器製造（株）内
☎ 0849-24-1150
FAX: 0849-22-0119

KŌSHŪ LACQUERED DEERHIDE • *Kōshū Inden*

KŌSHŪ INDEN ASSOCIATION
甲府印伝商工業協同組合
〒400 3-11-15 Chūō, Kōfu, Yamanashi Prefecture
〒400 山梨県甲府市中央3-11-15
☎ 0552-33-1100
FAX: 0552-32-8428

NARA SUMI INK • *Nara Zumi*

NARA INKSTICK ASSOCIATION
奈良製墨協同組合
〒630 43 Nashihara-chō, Nara, Nara Prefecture
〒630 奈良県奈良市内侍原町43
☎ 0742-23-6589
FAX: 0742-23-6589

NARA BRUSHES • *Nara Fude*

NARA BRUSH ASSOCIATION
奈良毛筆協同組合
〒630 80-1 Nakatsuji-chō, Nara, Nara Prefecture
〒630 奈良県奈良市中辻町80-1
☎ 0742-22-3024
FAX: 0742-24-0356

OGATSU INKSTONES • *Ogatsu Suzuri*

OGATSU INKSTONE ASSOCIATION
雄勝硯生産販売協同組合
〒986-13 53-1 Aza Tera, Ōaza Ogatsu, Ogatsu-chō, Monou-gun, Miyagi Prefecture
〒986-13 宮城県桃生郡雄勝町大字雄勝字寺53-1 雄勝硯伝統産業会館内
☎ 0225-57-2632
FAX: 0255-57-3211

GYŌTOKU ŌKARAHAFU PORTABLE SHRINE • *Gyōtoku Ōkarahafū Mikoshi*

NAKADAI RELIGIOUS ORNAMENTS INCORPORATED
中台神仏具製作所
〒272-01 21-3 Hon-Gyōtoku, Ichikawa, Chiba Prefecture
〒272-01 千葉県市川市本行徳21-3
☎ 0473-57-2061
FAX: 0473-57-0809

KYOTO HOUSEHOLD BUDDHIST ALTARS, FITTINGS • *Kyō Butsudan, Kyō Butsugu*

KYOTO RELIGIOUS ORNAMENTS ASSOCIATION
京都府仏具協同組合
〒600 Marudai Building 4F, Minamigawa, Nishinotōin-Nishi-iru, Shichijō-dōri, Shimogyō-ku, Kyoto
〒600 京都府京都市下京区七条通西洞院西入ル南側 マルダイビル4F
☎ 075-341-2426
FAX: 075-343-2850

KYOTO CANDLES • *Kyō Rōsoku*

NAKAMURA CANDLE INCORPORATED
中村ろうそく（株）
〒604 211 Hashiura-chō, Sanjō-agaru, Higashi-Horikawa-dōri, Nakagyō-ku, Kyoto
〒604 京都府京都市中京区東堀川通三条上ル橋浦町211
☎ 075-221-1621
FAX: 075-251-1705

GIFU LANTERNS • *Gifu Chōchin*

GIFU LANTERN ASSOCIATION
岐阜提灯振興会
〒500 1-chōme, Oguma-chō, Gifu, Gifu Prefecture
〒500 岐阜県岐阜市小熊町1丁目
☎ 0582-63-0111
FAX: 0582-62-0058

IZUMO STONE LANTERNS • *Izumo Ishidōrō*

MATSUE STONE LANTERN ASSOCIATION
松江石灯ろう協同組合
〒690 86 Kuroda-chō, Matsue, Shimane Prefecture
〒690 島根県松江市黒田町86
☎ 0852-24-1815
FAX: 0852-24-1815

GIFU UMBRELLAS • *Gifu Wagasa*

GIFU UMBRELLA ASSOCIATION
岐阜和傘振興会
〒500 7 Yashima-chō, Gifu, Gifu Prefecture
〒500 岐阜県岐阜市八島町7
☎ 0582-71-3958
FAX: 0582-72-0095

MARUGAME ROUND FANS • *Marugame Uchiwa*

KAGAWA FAN ASSOCIATION
香川県うちわ協同組合連合会
〒763 2-3-1 Ōte-chō, Marugame, Kagawa Prefecture
〒763 香川県丸亀市大手町2-3-1 丸亀市役所商工観光課内
☎ 0877-23-2111
FAX: 0877-24-8863

KYOTO FOLDING FANS • *Kyō Sensu*

KYOTO FAN ASSOCIATION
京都扇子団扇商工協同組合
〒600 Hama Bldg. 5F, 583-4, Moto-Siogama-chō, Gojō-sagaru, Kawaramachi-dōri, Shimogyō-ku, Kyoto
〒600 京都府京都市下京区河原町通五条下ル本塩竈町583-4 ハマビル5F
☎ 075-352-3254
FAX: 075-352-3253

EDO "ART" DOLLS • *Edo Kimekomi Ningyō*
EDO BATTLEDORE • *Edo Oshie Hagoita*

TOKYO HINA DOLL ASSOCIATION
東京都雛人形工業協同組合
〒111 2-1-9 Yanagibashi, Taitō-ku, Tokyo
〒111 東京都台東区柳橋2-1-9 東京卸商センター内
☎ 03-3861-3950
FAX: 03-3851-8248

MIHARU DOLLS • *Miharu Hariko*

TAKASHIBA DEKO YASHIKI
高柴デコ屋敷
〒977-01 163 Aza Tateno, Takashiba, Nishida-machi, Kōriyama, Fukushima Prefecture
〒977-01 福島県郡山市西田町高柴字館野163
☎ 0249-71-3176
FAX: 0249-71-3176

MIYAGI KOKESHI DOLLS • *Miyagi Dentō Kokeshi*

NARUKO WOODEN TOY ASSOCIATION
鳴子木地玩具協同組合
〒989-68 122-12 Aza Shin-yashiki, Naruko-chō, Tamazukuri-gun, Miyagi Prefecture
〒989-68 宮城県玉造郡鳴子町字新屋敷122-12
☎ 0229-83-3515
FAX: 0229-83-3512

AN ANNOTATED READING LIST

The following list was assembled with the idea of acknowledging some of the better publications on Japanese craft and providing avenues for further exploration. Some of the books and catalogs below are no longer in print, but may be found in select museums and public libraries in Japan, the United States, Europe, and elsewhere.

GENERAL

Birdsall, Derek, ed. *The Living Treasures of Japan*. London: Wildwood House, 1973.

Features fourteen selected craftsmen, including a papermaker, weavers, dyers, potters, stencil cutters, a woodworker, a lacquer artist, a swordsmith, and a bamboo craftsmen. Text in both English and Japanese based on actual interviews.

Fontein, Jan, ed. *Living National Treasures of Japan*. Tokyo, 1982.

Exhibition catalogue, with color photographs of more than 200 works by "Living National Treasures." Included are ceramics, textiles, dolls, lacquer ware, wood and bamboo crafts, metalwork, swords, and handmade paper. There are summaries of the history of each craft tradition and brief biographies of the artists.

Hauge, Victor and Takako. *Folk Traditions in Japanese Art*. Tokyo: Kodansha International, 1978.

Exhibition catalogue with 231 examples of painting, printmaking, sculpture, ceramics, textiles, lacquer ware, woodwork, metalwork, bamboo ware and basketry, dolls, and toys. Introductory essay contains overviews of the various craft traditions and individual entries for each of the objects.

Hickman, B., ed. *Japanese Crafts: Materials and their Applications*. London: Fine Books Oriental, 1977.

Text emphasizing materials, techniques, and uses of different objects, covering wood, bamboo, and metalwork. More than half of the book is devoted to bows and arrows, swords, and other military accoutrements. Illustrated by a combination of small black-and-white photographs and drawings.

Japan Folk Crafts Museum. *Mingei: Masterpieces of Japanese Folkcraft*. Tokyo: Kodansha International, 1991.

Beautifully illustrated book (158 color plates) featuring traditional textiles, ceramics (includes Hamada, Kawai, and Leach), woodcraft, lacquer ware, metalwork, and some pictorial art. There are essays on the *mingei* movement, the Japan Folk Crafts Museum, and Yanagi Sōetsu as well as descriptive notes to the plates adapted from Yanagi's writings.

Lowe, John. *Japanese Crafts*. London: John Murray, 1983.

Short descriptions, emphasizing general history and technique, of various craft objects such as folding fans, paper umbrellas and lanterns, writing and painting brushes, seals, wooden combs, brooms, kitchen knives, and tatami mats.

Massy, Patricia. *Sketches of Japanese Crafts and the People Who Make Them*. Tokyo: The Japan Times, 1980.

Brief introductions to a host of folkcrafts, including furniture, lacquer ware, pottery, textiles, toys, umbrellas, folding fans, bamboo baskets, and inkstones.

Moes, Robert. *Mingei: Japanese Folk Art*. New York: The Brooklyn Museum, 1985.

Exhibition catalogue covering 115 works including stone and wood sculpture, ceramics, furniture, kitchen utensils, metalwork, basketry, toys, and textiles, with separate sections on Okinawan and Ainu folk art. Good general essays on the history and characteristics of folk art in Japan.

Munsterberg, Hugo. *The Folk Arts of Japan*. Tokyo: Charles E. Tuttle, 1958.

Rudimentary introduction to selected craft traditions including pottery, basketry, lacquer ware, woodcraft, metalwork, toys, textiles, painting and sculpture, and peasant houses. Also discussion of folk art aesthetics and the contemporary folk art movement.

——. *Mingei: Folk Arts of Old Japan*. New York: The Asia Society, 1965.

Exhibition catalogue, with ceramics, lacquer ware, paintings, book illustrations, rubbings, sculpture, toys, textiles, and wood and metal objects (ninety-eight in total). Brief introductory essay on *mingei* traditions and their appreciation in the twentieth century.

Muraoka, Kageo and Okamura Kichiemon. *Folk Arts and Crafts of Japan*. The Heibonsha Survey of Japanese Art, Vol. 26. New York and Tokyo: Weatherhill/Heibonsha, 1973.

Well-illustrated survey (covers ceramics, textiles, woodwork, metalwork, and folk pictures), with essays on the *mingei* movement, the history and appreciation of folk crafts in Japan, and the special characteristics and "beauty" of folk art.

Ogawa, Masataka, et. al. *The Enduring Crafts of Japan: 33 Living National Treasures.* New York and Tokyo: Walker/ Weatherhill, 1968.

Selection of thirty-three potters, weavers, dyers, paper stencil cutters, lacquer craftsmen, metal craftsmen, bamboo craftsmen, and doll makers, accompanied by short biographies and photographs (mostly black-and-white) by Sugimura Tsune of each artist at work.

Rathbun, William, and Michael Knight. *Yo no Bi: The Beauty of Japanese Folk Art.* Seattle: University of Washington Press, 1983.

Catalogue of exhibition of Japanese folk art from the Seattle Art Museum and other Pacific Northwest collections (179 objects). Includes sculpture, woodwork, furniture, metalwork, lacquer ware, bamboo ware and basketry, textiles, paintings, shop signs, and ceramics. Short introductory essay on *mingei* and individual essays on all of the above categories, illustrated by works in the exhibition.

Suzuki, Hisao. *Living Crafts of Okinawa.* New York and Tokyo: Weatherhill, 1973.

Basic introduction to Okinawan craft traditions including carpentry, stonemasonry, textiles, pottery and ceramic roof tiles, lacquer ware, and musical instruments. The strength of this book is the photography (mostly black-and-white) by Sugimura Tsune showing craftspeople at work and their creations.

Yanagi, Sōetsu, adapted by Bernard Leach. *The Unknown Craftsman: A Japanese Insight into Beauty.* New York and Tokyo: Kodansha International, 1972.

Collection of essays by a leader of the Japanese craft movement, focusing on the aesthetics and appreciation of traditional Japanese crafts. Includes seventy-six photographs of craft objects originally in his collection.

CERAMICS

Cort, Louise Allison. *Seto and Mino Ceramics.* Freer Gallery of Art, 1992.

Catalogue of 126 examples of Seto and Mino ceramics in the Freer Gallery collection, divided chronologically and according to kilns and types of ware. Text provides overviews of various traditions, followed by detailed entries on individual pieces. Technology of selected ceramics discussed in detail in an appendix by Pamela B. Vandiver of the Smithsonian's Conservation Analytical Laboratory.

Cort, Louise Allison. *Shigaraki, Potters' Valley.* Tokyo: Kodansha International, 1979.

Major scholarly work covering the evolution of ceramics in Shigaraki from early times up to the present day. Appendices include accounts of Shigaraki made in 1678 and 1872 by potters from other districts, the biography of a Shigaraki woman, a list of kiln sites, and technical information on Shigaraki clays.

Fujioka, Ryōichi. *Shino and Oribe Ceramics.* Japan Arts Library, Vol. 1. Tokyo: Kodansha International, 1977.

Comprehensive survey of Shino and Oribe traditions, abundantly illustrated and accompanied by a short introduction to ceramics and the tea ceremony by the translator, Samuel Morse.

Inumaru, Tadashi and Yoshida Mitsukuni, eds. *Ceramics.* The Traditional Crafts of Japan, Vol. 3. Tokyo: Diamond, 1992

Informative essays on twenty-eight regional ceramic traditions and three production centers of tiles, accompanied by beautiful color photographs, including many close-ups.

Jenyns, Soame. *Japanese Pottery.* London: Faber and Faber, 1971.

Chronological survey of traditional Japanese ceramic wares, with detailed descriptions and small black-and-white photographs. A useful introduction even though knowledge of certain wares has been expanded through recent research and excavations.

Mikami, Tsugio. *The Art of Japanese Ceramics.* Heibonsha Survey of Japanese Art, Vol. 29. Tokyo and New York: Weatherhill and Heibonsha, 1973.

Broad overview of ceramics in Japan, with general discussions of many of the best-known wares and 200 photographs in both black-and-white and color.

Mizuo, Hiroshi. *Folk Kilns I.* Famous Ceramics of Japan, Vol. 3. Tokyo: Kodansha International, 1981.

Introduction to selected folk wares in Honshu, divided by district. Accompanied by ninety-four color photographs with notes.

Moes, Robert. *Japanese Ceramics.* New York: The Brooklyn Museum, 1972.

Illustrated catalogue of exhibition (seventy-five objects) featuring primarily the Brooklyn Museum collection. Short essay on the general characteristics of Japanese ceramics and summaries of the development of different wares in each historical period.

Nakagawa, Sensaku. *Kutani Ware.* Tokyo: Kodansha International, 1979.

Well-illustrated survey of the origins and development of Old Kutani and later (Edo-period) Kutani wares.

Okamura, Kichiemon. *Folk Kilns II.* Famous Ceramics of Japan, Vol. 4. Tokyo: Kodansha International, 1981.

Introduction to some of the folk ceramic traditions in Kyushu, Shikoku, and Okinawa. There are seventy-four color plates with notes.

Rhodes, Daniel. *Tamba Pottery: The Timeless Art of a Japanese Village.* Tokyo: Kodansha International, 1970.

Overview of the development of pottery in Tamba from the Middle Ages up to the present day. Many black-and-white photographs of potters at work as well as of wares.

Sanders, Herbert H., with Tomimoto Kenkichi. *The World of Japanese Ceramics.* Tokyo: Kodansha International, 1967.

Detailed descriptions of the tools and materials used

for making pottery, forming and decorating techniques, and glazing and overglaze enamel decoration, gleaned from personal interviews and kiln visits. Includes appendices with other technical information for potters.

Satō Masahiko. *Kyoto Ceramics*. New York: Weatherhill, 1973.
General overview of the origins and development of ceramics in Kyoto, with special chapters on Ninsei and Kenzan.

Seattle Art Museum. *Ceramic Art of Japan: One Hundred Masterpieces from Japanese Collections*. Seattle: Seattle Art Museum, 1972.
Catalogue of the first major exhibition in America of Japanese ceramics from collections in Japan. Introductory essay with general background on many of the major wares and descriptive notes to the plates.

Simpson, Penny and Sodeoka Kanji. *The Japanese Pottery Handbook*. Tokyo: Kodansha International, 1979.
Contains short but thorough descriptions (and many drawings) of tools and equipment, forming processes, decoration, kilns and firing, etc. English terminology is accompanied by Japanese characters and romanization, making this a handy reference book for those wishing to communicate with Japanese potters.

Yoshiharu, Sawada. *Tokoname*. Famous Ceramics of Japan, Vol. 7. Tokyo: Kodansha International, 1982.
General essay outlining the origins, types, and techniques of Tokoname ware. There are fifty-eight color plates accompanied by brief descriptions.

TEXTILES

Bethe, Monica. "Color: Dyes and Pigments." In *Kosode: 16th-19th Century Textiles from the Nomura Collection*, eds. Naomi Noble, Richard and Margot Paul. New York: Japan Society and Kodansha International, 1984.
Volume provides detailed information on traditional dyes and techniques, and the aesthetics and significance of different colors in Japan.

Brandon, Reiko Mochinaga. *Country Textiles of Japan: The Art of Tsutsugaki*. New York and Tokyo: Weatherhill, 1986.
Published in conjunction with an exhibition at the Honolulu Academy of Arts. In addition to commentaries on each of the forty-eight works, there are essays giving background on the historical development, functions, motifs, and techniques associated with *tsutsugaki*, as well as on cotton and indigo.

Dusenbury, Mary. "Kasuri: A Japanese Textile." *The Textile Museum Journal*, Vol. 17 (1978).
Scholarly treatise on the origins and development of *kasuri* weaving in Japan, including discussion of materials, dyes, and techniques.

Inumaru, Tadashi and Yoshida Mitsukuni, eds. *Textiles I, Textiles II*. The Traditional Crafts of Japan, Vols. 1 and 2. Tokyo: Diamond, 1992.
Lavish color plates accompanied by lengthy entries introducing forty-six textile traditions of Okinawa and other parts of Japan. Introductory essays survey the development of weaving and dyeing.

Nakano, Eisha and Barbara B. Stephan. *Japanese Stencil Dyeing: Paste-Resist Techniques*. New York and Tokyo: Weatherhill, 1982.
Introduction to the *katazome* tradition, with technical instructions on making stencils, preparing the paste and fabric, laying the paste, sizing and dyeing the fabric, etc.

Rathbun, William Jay, ed. *Beyond the Tanabata Bridge: Traditional Japanese Textiles*. Seattle: Seattle Art Museum, 1993.
Exhibition catalogue, with discussions of sixty-one works (illustrated in color) and informative essays on the tradition of folk textiles in Japan, color, bast fibers, *sashiko*, resist-dyeing, *kasuri*, and Okinawan and Ainu textiles.

Tomito, Jun and Noriko. *Japanese Ikat Weaving: The Techniques of Kasuri*. London: Routledge & Kegan Paul, 1982.
Basic introduction to *kasuri*, with technical instructions on weaving and dyeing. Explanations accompanied by drawings and a few black-and-white photographs.

Wada, Yoshiko, et al., *Shibori: The Inventive Art of Japanese Shaped Resist Dyeing*. New York and Tokyo: Kodansha International, 1983.
Historical overview of the *shibori* tradition, followed by detailed descriptions of various techniques (generously illustrated with drawings and photographs). There is also a discussion of the *shibori* craft in contemporary Japan and the West, with examples by twenty-four artists. Appendix with technical information on preparing cloth and dyes, and glossary.

LACQUER WARE

Brommelle, N. S. and Perry Smith, eds. *Urushi: Proceedings of the Urushi Study Group, June 10–27, 1985, Tokyo*. Marina del Rey, California: The Getty Conservation Institute, 1988
Includes scholarly papers dealing with the historical development of lacquer in Japan, conservation, and techniques. Comprehensive glossary of terminology.

Inumaru, Tadashi and Yoshida Mitsukuni, eds. *Lacquerware*. The Traditional Crafts of Japan, Vol. 4. Tokyo: Diamond, 1992
Beautiful color photographs document the lacquer craft traditions of Tsugaru, Hidehira, Jōhōji, Kawatsura, Aizu, Kamakura, Odawara, Murakami, Takaoka, Wajima, Yamanaka, Kanazawa, Echizen, Wakasa, Kiso, Hida Shunkei, Kyoto, Kishū, Ōuchi, Kagawa, and Ryūkyū. In addition to a general essay covering the development of lacquer production in Japan, each regional tradition is discussed individually.

Von Ragué, Beatrix. *A History of Japanese Lacquerwork*. Toronto and Buffalo: University of Toronto Press, 1976.
Chronological survey of lacquer focusing on stylistic development and dating, with photographs of 199 representative works. Includes glossary and list of characters for Japanese names and technical terms. Translation from German.

Watt, James and Barbara Brennan Ford. *East Asian Lacquer: The Florence and Herbert Irving Collection*. New York: The Metropolitan Museum of Art, 1991.
Exhibition catalogue including approximately eighty

examples of Japanese lacquer and twenty examples of lacquer from the Ryūkyū islands. Essays discussing *makie* and *negoro*, Kōdaiji and *namban* lacquer, as well as informative entries on the individual works. Glossary of lacquer terms.

Yonemura, Ann. *Japanese Lacquer*. Washington, D.C.: Freer Gallery of Art, Smithsonian Institution, 1979.

Exhibition catalogue featuring fifty-seven objects from the Freer Gallery of Art collection. Introductory essay outlines the historical development of Japanese lacquer, and individual entries discuss the lacquer techniques and design of each object. Useful glossary of Japanese lacquer terms.

BAMBOO CRAFT

Inuyama, Tadashi and Yoshida Mitsukuni. *Wood and Bamboo*. The Traditional Crafts of Japan, Vol. 5. Tokyo: Diamond, 1992. (*See* Woodcraft)

Kudō, Kazuyoshi. *Japanese Bamboo Baskets*. Form and Function Series. Tokyo: Kodansha International, 1980.

Fine black-and-white photographs of a wide range of baskets, with explanatory captions in English providing information on functions, special features, and provenances.

PAPER

Barrett, Timothy. *Japanese Papermaking: Traditions, Tools and Techniques*. New York: Weatherhill, 1983.

Part 1 describes in detail the traditional papermaking process in Japan, while Part 2 gives step-by-step directions on how to make Japanese-style paper. A valuable source of information on raw materials, tools, equipment, etc.

Hughes, Sukey. *Washi: The World of Japanese Paper*. Tokyo: Kodansha, International, 1982.

Covers the history of papermaking and uses of paper in Japan, with detailed descriptions of materials and processes accompanied by many drawings and photographs. Also included are interviews with contemporary papermakers, a discussion of the aesthetics of *washi*, a catalog of different kinds of papers with short descriptions, and a glossary of terms.

Inuyama, Tadashi and Yoshida Mitsukuni. *Paper and Dolls*. The Traditional Crafts of Japan, Vol. 7. Tokyo: Diamond, 1992

Beautifully illustrated (all-color) book with a general essay outlining the historical development of paper in Japan, and separate essays on ten regional papers, as well as umbrellas, lanterns, fans, stencils, dolls, drums, shamisen, and fishing flies.

WOODCRAFT

Clarke, Rosy. *Japanese Antique Furniture: A Guide to Evaluating and Restoring*. New York and Tokyo: Weatherhill, 1983.

Introduction to traditional Japanese furniture, especially *tansu*, with photographs and discussions of different types and styles. Includes useful information on materials and techniques, and glossary of Japanese terms.

Heineken, Ty and Kiyoko. *Tansu: Traditional Japanese Cabinetry*. New York and Tokyo: Weatherhill, 1981.

Covers the historical development of *tansu* from the Edo to the Taisho periods, regional styles, materials, and construction techniques. Amply illustrated with black-and-white and color photographs as well as drawings.

Inuyama, Tadashi and Yoshida Mitsukuni. *Wood and Bamboo*. The Traditional Crafts of Japan, Vol. 5. Tokyo: Diamond, 1992.

Handsomely illustrated (all-color) book covering twenty-nine regional wood and bamboo craft traditions. Includes *tansu* and other furniture items, household containers and vessels, *go* and *shōgi* boards, abacuses, carved wood transoms, sculpture, masks, bamboo wares and basketry, and tea whisks. Each craft is discussed in detail in individual essays.

Koizumi, Kazuko. *Traditional Japanese Furniture: A Definitive Guide*. Tokyo: Kodansha International, 1986.

A thorough survey of most types of furniture, including *tansu*, screens, shelving, and lanterns. Introductory text on each type of piece, followed by a historical overview tracing the evolution and use of these different types. Extensive color and black-and-white illustrations, a short section on techniques, and an impressive section entitled Illustrated History of Japanese Furniture.

METALWORK

Arts, P.L.W. *Tetsubin: A Japanese Waterkettle*. Groningen: Geldermalsen Publications, 1988.

Scholarly treatise on cast-iron waterkettles, with chapters on the development of *tetsubin* as a tea and household utensil, regional centers of manufacture, guilds, patronage patterns, casting techniques, and metallographic data.

Inuyama, Tadashi and Yoshida Mitsukuni. *Metal and Stone*. The Traditional Crafts of Japan, Vol. 6. Tokyo: Diamond, 1992.

Covers twenty-two regional metal and stone craft traditions, including iron teakettles; silver, bronze, copper, and pewter vessels; forged blades; gold leaf; cloisonne; metal inlay; agate, crystal, and coral carving; and stone lanterns and basins. Each craft is documented by beautiful color photographs and a descriptive, informative text.

OTHER CRAFTS

Inuyama, Tadashi and Yoshida Mitsukuni. *Writing Utensils and Household Buddhist Altars*. The Traditional Crafts of Japan, Vol. 8. Tokyo; Diamond, 1992.

Opening essay with historical overview followed by lavish color plates of inkstones, brushes, ink sticks, household Buddhist altars, and other Buddhist as well as Shinto paraphernalia, produced in various regions. Each of the twenty-six craft traditions is discussed in detail in essays appended at the end.

Usui, Masao. *Japanese Brushes*. Form and Function Series. Tokyo: Kodansha International, 1979.

Black-and-white photographs of a diversity of brushes made by skilled craftsmen in Tokyo, Kyoto, and Nara. Notes to the plates provide information on the type of hair used and the functions of the brushes. The process of making a brush is documented with sequential photographs.

ENGLISH-JAPANESE CRAFTS INDEX

JAPANESE-ENGLISH CRAFTS INDEX